Reliable and Efficient Trajectory Design in UAV-Aided Wireless Networks

Yulin Hu · Xiaopeng Yuan · Yue Cao · Kaitao Meng
Editors

Reliable and Efficient Trajectory Design in UAV-Aided Wireless Networks

Springer

Editors
Yulin Hu
School of Electronic Information
Wuhan University
Wuhan, Hubei, China

Yue Cao
School of Cyber Science and Engineering
Wuhan University
Wuhan, Hubei, China

Xiaopeng Yuan
Chair of Information Theory and Data Analytics
RWTH Aachen University
Aachen, Germany

Kaitao Meng
Department of Electronic and Electrical Engineering
University College London
London, UK

ISBN 978-981-96-5987-6 ISBN 978-981-96-5988-3 (eBook)
https://doi.org/10.1007/978-981-96-5988-3

This work was supported in part by the National Key R&D Program of China under Grant 2023YFE0206600, NSFC Grant 62471341, and the Fundamental Research Funds for the Central Universities under Grant 2042025kf0039.

This Springer imprint is published by the registered company Springer Nature Singapore Pte Ltd.
The registered company address is: 152 Beach Road, #21-01/04 Gateway East, Singapore 189721, Singapore

Contents

Chapter 1
Optimal 1D UAV Trajectory Design for Multiuser Wireless Power Transfer

Yulin Hu, Xinran Wang, Peng Wu, and Xiaopeng Yuan

Abstract In this chapter, we investigate a UAV-enabled wireless power transfer (WPT) network, where an unmanned aerial vehicle (UAV) flies at a constant altitude to wirelessly supply energy to multiple ground nodes arranged in a linear topology. Our goal is to maximize the minimum received energy among all ground nodes by optimizing the UAV one-dimensional (1D) trajectory, subject to the maximum UAV speed constraint. Unlike previous studies that only provided heuristic or locally optimal solutions, this work is the first to propose a globally optimal 1D trajectory solution for the considered problem. Specifically, we first demonstrate that for any given speed-constrained UAV trajectory, an equivalent combination of a maximum-speed trajectory and a speed-free trajectory can be constructed to achieve the same energy reception for all ground nodes. We then reformulate the original speed-constrained trajectory optimization problem into an equivalent speed-free problem, which is subsequently solved using the Lagrange dual method. The resulting optimal UAV trajectory follows a successive hover-and-fly (SHF) structure, where the UAV hovers at a finite set of locations for optimized durations and flies between them at maximum speed. Building on the derived optimal trajectory structure, we also present a low-complexity trajectory design. This approach first reformulates the problem as an equivalent non-convex optimization with only UAV hovering locations and durations as optimization variables. Then, the successive convex approximation (SCA) technique is applied to iteratively refine the trajectory, ensuring convergence to a suboptimal solution with significantly lower computational complexity, regardless of the geographical network size. Numerical results validate the performance of

Y. Hu · X. Wang · P. Wu
School of Electronic Information, Wuhan University, Wuhan, China
e-mail: yulin.hu@whu.edu.cn

X. Wang
e-mail: xinran.wang@whu.edu.cn

P. Wu
e-mail: peng.wu@whu.edu.cn

X. Yuan (✉)
Chair of Information Theory and Data Analytics, RWTH Aachen University, Aachen, Germany
e-mail: yuan@inda.rwth-aachen.de

Y. Hu et al. (eds.), *Reliable and Efficient Trajectory Design in UAV-Aided Wireless Networks*, https://doi.org/10.1007/978-981-96-5988-3_1

the proposed low-complexity trajectory design, demonstrating that it achieves the same energy efficiency as the optimal solution. Moreover, both the optimal and low-complexity designs outperform existing benchmark algorithms across different scenarios, highlighting their effectiveness in UAV-enabled WPT networks.

Keywords Unmanned aerial vehicle · Wireless power transfer · Energy fairness · Trajectory design · Successive hover-and-fly

Abbreviations

1D	One Dimension
2D	Two Dimension
A2G	Aerial-to-Ground
BS	Base Stations
DC	Direct Current
IoT	Internet of Things
LoS	Line-of-Sight
RF	Radio-Frequency
SCA	Successive Convex Approximation
SHF	Successive Hover-and-Fly
UAV	Unmanned Aerial Vehicle
WPT	Wireless Power Transfer

1.1 Introduction

1.1.1 Background

Unmanned aerial vehicles (UAVs) have been widespread applied in areas such as cargo delivery, aerial inspection, and surveillance, owing to their high payload capacity and long endurance. Recently, UAV-enabled wireless applications have garnered significant attention due to their ability to rapidly deploy on-demand wireless networks. With line-of-sight (LoS) aerial-to-ground (A2G) wireless links, UAV-enabled wireless networks can achieve superior performance compared to conventional terrestrial networks [1].

There are primarily two types of UAV-enabled wireless applications: wireless communication and wireless power transfer (WPT). UAVs can function as mobile base stations or relays, enhancing coverage [2–6] and data rate throughput [7–15] for wireless communication. Alternatively, UAVs can also serve as mobile energy transmitters, broadcasting radio frequency (RF) signals to power low-power ground nodes such as sensors and Internet of Things (IoT) devices. Moreover, recent researches have explored UAV-enabled wireless powered communication networks [16–18] and wireless-powered edge computing [19, 20].

1.1.2 Motivation

UAV-enabled wireless power transfer (WPT) presents a promising solution due to its flexibility in deployment and its ability to provide on-demand energy, especially in scenarios where traditional power delivery methods are challenging. By Leveraging their mobility, UAVs can dynamically adjust their locations to optimize energy transfer, making them particularly effective for remote or hard-to-reach areas where conventional infrastructure is either unavailable or cost-prohibitive.

While UAV-enabled WPT has shown great potential, existing solutions primarily rely on heuristic or locally optimal approaches [21, 22], which often fail to fully exploit the UAV's mobility capabilities. These heuristic approaches, though providing feasible solutions, may not be efficient or robust enough for all scenarios. Moreover, these methods usually involve high computational complexity or lack in guaranteeing optimal energy distribution among ground nodes, which limits the practical application in real-world deployments. Additionally, the lack of a structured approach to determine the optimal hovering positions and flight paths of UAVs further constrains the effectiveness of these methods. Therefore, there is a pressing need for a trajectory design that not only optimizes the energy transfer performance but also remains computationally feasible for practical implementation, ensuring fair and efficient energy distribution to all ground nodes in diverse network settings.

1.1.3 Main Contributions

The main contributions of this work can be summarized as follows:

(1) **Optimal 1D UAV Trajectory Solution**: We present the globally optimal trajectory solution for a one-dimensional (1D) UAV-enabled WPT network with a linear topology. By transforming the speed-constrained problem into a speed-free one and solving it using the Lagrange dual method, we establish that the optimal trajectory follows a multi-location hovering structure, where the UAV flies at maximum speed between hovering points.

(2) **Efficient Low-Complexity Trajectory Design**: Leveraging the structure of the optimal solution, we propose an efficient and low-complexity trajectory design using the successive convex approximation (SCA) technique. This design reduces computational complexity while maintaining high performance, making it suitable for practical implementation.

(3) **Complexity Analysis**: We provide an in-depth complexity analysis of both the proposed optimal and low-complexity trajectory designs. Compared to existing benchmark algorithms, such as the heuristic successive hover-and-fly (SHF) and SCP-based designs, our suboptimal approach demonstrates a constant complexity irrespective of geographical network size, which is an advantage for practical scenarios.
(4) **Numerical Results**: We validate the performance of the proposed trajectory designs through numerical simulations, showing that both the optimal and low-complexity solutions outperform existing methods in terms of energy efficiency under different scenarios. Furthermore, we emphasize the contrast between our proposed SHF and the heuristic SHF, highlighting the superior performance of our approach due to accounting for received energy during UAV flying.

1.1.4 Structural Organization

The rest of this chapter is organized as follows. In Sect. 1.2, we review related works, focusing on UAV-enabled wireless communication and UAV-enabled wireless power transfer. Section 1.3 presents the system model and the problem formulation for UAV trajectory optimization. In Sect. 1.4, we characterize the optimal UAV trajectory structure for the energy harvesting problem. Section 1.5 details the proposed trajectory design and proposes an efficient low-complexity solution, including convex approximation and an iterative algorithm. In Sect. 1.6, we provide a complexity level analysis of the proposed and benchmark trajectory designs. Section 1.7 presents numerical results to validate the proposed methods. Finally, Sect. 1.8 summarizes the conclusions of this work.

1.2 Related Works

1.2.1 UAV-Enabled Wireless Communication

Several studies have focused on optimizing UAV deployment and trajectory to improve wireless communication performance. For instance, UAVs acting as mobile base stations have been employed to maximize the coverage area for ground users by optimizing their quasi-static positions [4–6]. In addition, UAV-enabled relaying systems have been proposed, where UAV trajectories are optimized jointly with communication resources to enhance data throughput [7–11] or energy efficiency [12]. For example, the authors of [10, 11] investigated the way to improve data rates through optimizing in a mobile relaying setup, while [12] focused on enhancing the energy efficiency of UAV-assisted communication.

UAV-enabled wireless communication has shown promising results in enhancing system performance, particularly in scenarios where terrestrial infrastructure is limited. UAVs provide an effective way to dynamically adapt network configurations, thereby improving the overall efficiency and flexibility of wireless communication systems.

1.2.2 UAV-Enabled Wireless Power Transfer

UAV-enabled WPT is an emerging technology designed to wirelessly charge low-power ground nodes, such as sensors and IoT devices, using UAVs as mobile energy transmitters. Unlike terrestrial energy transmitters, UAVs can dynamically adjust their position and trajectory, providing additional degrees of freedom for enhancing the efficiency of energy transmission. Recent studies have focused on optimizing UAV trajectories to maximize energy transfer performance under various operational constraints. Specifically, assuming UAV flying at a fixed altitude, prior works optimized both 1D and two-dimensional (2D) UAV trajectories for maximizing energy transfer to ground nodes. For example, Xu et al. [21] considered a two-user scenario with a linear topology and optimized the 1D trajectory of a UAV to characterize the Pareto boundary of the achievable energy region for the two users. This study provided a basis for understanding energy-efficient UAV movement and charging strategies.

Subsequently, this idea was extended in [22] to a more general multi-user scenario, where the UAV trajectory was optimized in a 2D topology to maximize the minimum received energy among the users. The approach considered a relaxed version of the problem by ignoring the UAV speed constraints, leading to a trajectory where the UAV follows a multi-location-hovering structure. This structure allows the UAV to hover at a finite number of optimized locations, with a specified duration, while flying at maximum speed between these locations. This was further refined by applying the heuristic SHF trajectory design, ensuring the UAV could successively hover over these locations and fly among them while adhering to the speed constraint.

In addition to SHF, the SCA method was adopted to refine the UAV trajectory by quantizing the continuous UAV path into discrete waypoints. While these approaches resulted in locally optimal or heuristic solutions, they laid a foundational framework for UAV trajectory design in WPT applications. However, they faced challenges in achieving globally optimal solutions, especially when considering stringent speed constraints and high complexity introduced by fine-grained waypoint quantization.

1.3 System Model and Problem Statement

1.3.1 System Model

In this chapter, we focus on a UAV-enabled WPT network, as depicted in Fig. 1.1, where the UAV operates at a fixed altitude H, providing energy to a set of K ground nodes, such as IoT devices or sensors, that are distributed in a linear topology. The nodes are located along a straight line, often representing specific real-world scenarios such as roads, rivers, or tunnels. We define the horizontal coordinates of the ground nodes as w_k, where $k = \mathcal{K} \triangleq 1, 2, \ldots, K$, and assume that the nodes are ordered such that $w_1 \leq w_2 \leq \ldots \leq w_K$.

To ensure efficient energy delivery, the UAV's horizontal trajectory is optimized over a time period T, during which the UAV's horizontal location is represented by $x(t), t \in \mathcal{T} \triangleq [0, T]$. The UAV's movement is constrained by a maximum speed V, which is captured by the condition $|\dot{x}(t)| \leq V$, where $\dot{x}(t)$ is the derivative of the UAV's position.

Given that the channels between UAV and the ground nodes primarily consist of LoS links, we employ the free-space path loss model commonly used in UAV communication and WPT systems. At any time t, the channel power gain between the UAV at position $x(t)$ and ground node k is denoted as $h_k(x(t)) = \frac{\beta_0}{(x(t)-w_k)^2+H^2}$, where β_0 represents the reference channel power gain at a distance of 1 m. The corresponding received power at node k at time t is given as:

$$Q_k(x(t)) = \frac{\beta_0 P}{(x(t) - w_k)^2 + H^2}, \tag{1.1}$$

where P denotes the constant transmit power of the UAV. Although in practical scenarios, the received RF signal at each ground node needs to be converted to direct current (DC) for energy storage, this process involves non-linearities in the RF-to-DC conversion [23]. For simplicity, in this model, we focus solely on the received RF power, following similar assumptions made in related studies [21, 22].

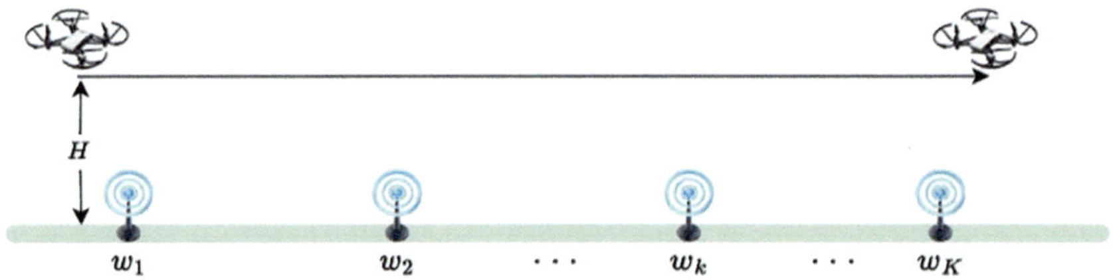

Fig. 1.1 Illustration of the UAV-enabled WPT network with a linear topology

Since the UAV's transmission is broadcasting, all ground nodes simultaneously receive energy during the charging period T. The total energy harvested by node k over the entire charging period is expressed as:

$$E_k(\{x(t)\}) = \int_0^T Q_k(x(t))dt, \tag{1.2}$$

which captures the energy accumulation at each ground node based on the UAV trajectory throughout the charging period.

1.3.2 Problem Statement

Our objective is maximizing the minimum received energy among all K ground nodes during the charging period T through designing the UAV trajectory. Specifically, we assume that the UAV can be freely deployed, with both its initial and final positions optimized to achieve the desired objective. The problem of interest can be formulated as follows:

$$\begin{aligned} \text{(OP)}: \max_{\{x(t)\}} \min_{k\in\mathcal{K}} & \int_0^T Q_k(x(t))dt. \\ \text{s.t.} \quad & |\dot{x}(t)| \le V, \forall t \in \mathcal{T}. \end{aligned} \tag{1.3}$$

To facilitate the solution, we introduce an auxiliary variable E, transforming the original problem (OP) into an equivalent reformulation:

$$\begin{aligned} \text{(P1)}: \max_{\{x(t)\},E} \; & E \\ \text{s.t.} \quad & \int_0^T Q_k(x(t))dt \ge E, \forall k \in \mathcal{K}, \\ & |\dot{x}(t)| \le V, \forall t \in \mathcal{T}. \end{aligned} \tag{1.4}$$

Both the original problem (OP) and the reformulated problem (P1) are non-convex for the non-concave throughput $\int_0^T Q_k(x(t))dt$. Additionally, both problems involve an infinite number of variables $\{x(t)\}$ over continuous time, making the solution to the min-energy maximization problem challenging.

In prior research, several approaches have been proposed, such as the heuristic SHF and the SCP-based trajectory designs in [22], which can be used to solve the problem sub-optimally. However, the heuristic SHF method provides a sub-optimal solution without guaranteeing the global optimality, while the SCP-based approach can achieve higher accuracy but requires cost of significant computational complexity, especially when high quantization accuracy is required. Therefore, developing an optimal or well-structured trajectory design for the UAV that can solve the min-energy maximization problem under speed constraints remains an open challenge.

1.4 Theoretical Design of Optimal Trajectory

In this section, we focus on problem (P1) to characterize the optimal trajectory solution for the UAV-enabled wireless power transfer problem. Our aim is to derive an effective trajectory that maximizes the minimum received energy among all ground nodes, subject to practical UAV mobility constraints. One key finding is the existence of the SHF structure, where the UAV hovers at certain points and moves with maximum speed between those points, which can simplify the trajectory optimization significantly.

First, it is important to note that there always exists a unidirectional trajectory that is optimal for problem (P1), meaning that the UAV moves monotonically in one direction without revisiting previous positions. Therefore, in this section, we focus on such unidirectional trajectories without loss of optimality. To further simplify the problem (P1), we consider a special case, denoted as sub-problem (P1.1), where the initial and final positions of the UAV are given as x_1 and x_F, i.e., $x(0) = x_1$ and $x(T) = x_F$. The formulation of sub-problem (P1.1) is as follows:

$$
\begin{aligned}
\text{(P1.1)}: \max_{\{x(t)\},E} \; & E \\
\text{s.t.} \; & \int_0^T Q_k(x(t))dt \geq E, \forall k \in \mathcal{K}, \\
& |\dot{x}(t)| \leq V, \forall t \in \mathcal{T}, \\
& x_I \leq x(t) \leq x_F, \forall t \in \mathcal{T}.
\end{aligned}
\tag{1.5}
$$

As the solution to (P1) can be viewed as a special case of (P1.1) (with optimal initial and final locations), understanding the optimal trajectory structure of (P1.1) provides key insights into the solution for (P1). In the subsequent sections, we demonstrate how any speed-constrained trajectory to problem (P1.1) can be equivalently represented as a combination of maximum-speed and speed-free segments, and how this representation helps in deriving optimal trajectory solution for (P1).

1.4.1 Reformulation via Equivalent Trajectories

In order to effectively solve the UAV-enabled WPT trajectory optimization problem, it is essential to simplify the original problem. The initial formulation, referred to as problem (P1.1), involves optimizing the UAV trajectory with specific initial and final positions while adhering to speed constraints. All of the above make the problem non-trivial due to the complexity introduced by these constraints, which couple the UAV's location over the entire time duration. Therefore, in this subsection, we introduce a reformulation approach that relies on constructing equivalent trajectories to make the problem more tractable.

To start, we consider a unidirectional trajectory that satisfies the speed constraint $|\dot{x}(t)| \leq V$ for all $t \in [0, T]$. We establish that for any such unidirectional trajectory, it is possible to construct two equivalent trajectories that together achieve the same WPT performance. For a given initial position $x(0) = x_I$ and final position $x(T) = x_F$, there exist two trajectories, denoted as $\overline{x}(t)$ and $\hat{x}(t)$, that can be combined to achieve the same energy transfer as the original trajectory. The first trajectory, $\overline{x}(t)$, represents the UAV flying from x_I to x_F at maximum speed V. The duration of this maximum-speed trajectory is $\overline{T} = (x_F - x_I)/V$, during which the UAV moves in a straight line without stopping. The second trajectory, $\hat{x}(t)$, represents the UAV during the remaining time period $\hat{T} = T - (x_F - x_I)/V$, where it is free to adjust its position without any speed constraints. The key idea here is that the UAV can use these two trajectories in combination, effectively hovering or moving freely during the extra time, to ensure that the energy transfer to each ground node remains optimal. Mathematically, we can express the energy received by each ground node k as follows:

$$\int_0^T Q_k(x(t))dt = \int_0^{\overline{T}} Q_k(\overline{x}(t))dt + \int_0^{\hat{T}} Q_k(\hat{x}(t))dt, \forall k \in \mathcal{K}, \tag{1.6}$$

where $Q_k(x(t))$ represents the received power by node k when the UAV is at position $x(t)$. The equivalence ensures that the energy received by each ground node is identical to that achieved by the original trajectory, thus preserving the performance of the WPT system.

Based on this equivalence, we reformulate problem (P1.1) into a new optimization problem, referred to as (P2). In problem (P2), the UAV trajectory $\hat{x}(t)$ is optimized over the remaining time period $\left[0, \hat{T}\right]$, and an additional constant energy term $\overline{E}_k$ is introduced to account for the energy contribution from the maximum-speed trajectory $\overline{x}(t)$. The reformulated problem (P2) is given by:

$$
\begin{aligned}
\text{(P2)}: & \max_{\{\hat{x}(t)\},E} \ \mathrm{E} \\
& \text{s.t} \int_0^{\hat{T}} Q_k(\hat{x}(t))dt + \overline{E}_k \geq E, \forall k \in \mathcal{K}, \\
& x_I \leq \hat{x}(t) \leq x_F, \forall t \in \hat{\mathcal{T}}.
\end{aligned} \tag{1.7}
$$

In this reformulation, $\overline{E}_k$ is a constant defined based on the maximum-speed segment, and it represents the energy received by node k during the initial period $\overline{T}$. The reformulated problem (P2) effectively removes the speed constraint, allowing the UAV to freely adjust its position within the given bounds, thereby making the optimization more tractable.

It is worth noting that problems (P1), (P1.1), and (P2) are all non-convex due to the inherent structure of the WPT system. However, they satisfy the so-called time-sharing condition [24], which ensures that strong duality holds between each problem and its Lagrange dual. The primary motivation for reformulating problem (P1.1) into (P2) lies in the fact that removing the speed constraint simplifies the Lagrange dual function. In (P1) and (P1.1), the speed constraint couples the UAV's position over time, resulting in an infinite number of Lagrange multipliers and making the problem non-tractable. In contrast, (P2) eliminates this complexity, allowing efficient solution via the Lagrange dual method [25].

To sum up, by reformulating the original problem using equivalent trajectories, we are able to reduce the complexity of the optimization while preserving the energy transfer performance. This reformulation provides a foundation for deriving efficient solutions to the UAV trajectory optimization problem, which will be further explored in the subsequent sections.

1.4.2 Optimal Solution for Equivalent Problem (P2)

To solve the equivalent problem (P2) introduced in the prior subsection, we employ the Lagrange dual method. The reformulation of problem (P2) effectively removes the speed constraint, making it more tractable. By leveraging the dual approach, we can simplify the optimization process while retaining the essential characteristics of the WPT system.

We begin by constructing the Lagrange function for problem (P2). The Lagrange function $\mathcal{L}$ is defined as follows:

$$\mathcal{L}(\{\hat{x}(t)\}, E, \{\lambda_k\}) = E + \sum_{k\in\mathcal{K}} \lambda_k \left(\int_0^{\hat{T}} Q_k(\hat{x}(t))dt + \overline{E}_k - E \right)$$
$$= \left(1 - \sum_{k\in\mathcal{K}} \lambda_k\right) E + \sum_{k\in\mathcal{K}} \lambda_k \overline{E}_k + \int_0^{\hat{T}} \sum_{k\in\mathcal{K}} \lambda_k Q_k(\hat{x}(t))dt, \quad (1.8)$$

where $\lambda_k \geq 0$ are the Lagrange multipliers associated with the energy constraints for each ground node k. The Lagrange function integrates the original objective with the energy constraints, enabling us to handle the optimization in a unified manner.

The corresponding dual function $f_2(\{\lambda_k\})$ is then defined by maximizing the Lagrange function over the primal variables $\hat{x}(t)$ and E, subject to the position constraints, and given as:

$$f_2(\{\lambda_k\}) = \max_{\{\hat{x}(t)\}, E} \mathcal{L}_2(\{\hat{x}(t)\}, E, \{\lambda_k\})$$
$$\text{s.t } x_I \leq \hat{x}(t) \leq x_F, \forall t \in \hat{\mathcal{T}}. \quad (1.9)$$

To ensure that the dual function $f_2(\{\lambda_k\})$ is bounded from above, the condition $1 - \sum_{k\in\mathcal{K}} \lambda_k = 0$ must be satisfied. Otherwise, if $1 - \sum_{k\in\mathcal{K}} \lambda_k < 0$, the dual function would become unbounded. With this condition, the dual problem (DP2) can be formulated as follows

$$\text{(DP2)}: \min_{\{\lambda_k\}} f_2(\{\lambda_k\}\})$$
$$\text{s.t. } 1 - \sum_{k\in\mathcal{K}} \lambda_k = 0, \lambda_k \geq 0, \forall k \in \mathcal{K}. \quad (1.10)$$

The dual problem aims to find the optimal values of the Lagrange multipliers that minimize the dual function $f_2(\{\lambda_k\})$, while satisfying the equality and non-negativity constraints. By solving the dual problem, we can derive conditions for the optimal primal solution, which significantly simplifies the original optimization problem.

Once the dual problem is solved, we proceed to determine the optimal hover times at specific locations along the trajectory. This is done through linear programming, where the goal is to allocate the hovering time at each location such that the energy received by all ground nodes is maximized while ensuring fairness. The resulting optimal trajectory follows the SHF structure, where the UAV moves at maximum speed between selected locations and hovers at these locations for specific durations.

The optimal solution to problem (P2) can be characterized by the following key properties:

(1) The UAV follows a piecewise trajectory consisting of multiple segments, each either at maximum speed or at a hover.
(2) The number of hover locations is at most $K + 2$, where K is the number of ground nodes.
(3) The hovering times are allocated in such a way that the minimum energy received by any ground node is maximized, ensuring equitable energy distribution.

In summary, by employing the Lagrange dual method and linear programming, we derive the optimal solution for problem (P2). The resulting trajectory exhibits the SHF structure, which not only simplifies the UAV's movement but also ensures optimal energy transfer to all ground nodes. In the next subsection, we will further analyse the properties of this trajectory and its implications for practical UAV-enabled WPT systems.

1.4.3 Optimal Trajectory for Constrained Problem (P1.1)

The optimal solution obtained for the equivalent problem (P2) can be used to construct the optimal trajectory for the original constrained problem (P1.1). Since (P1.1) is a more constrained version of the trajectory optimization problem, involving specific initial and final positions, we leverage the solution to (P2) to derive the optimal trajectory.

Suppose that the optimal solution to problem (P2) is $\hat{x}(t)$ with optimal hovering points $\hat{x}_1, \hat{x}_2, \ldots, \hat{x}_N$ and corresponding optimal hovering durations $\hat{\tau}_1, \hat{\tau}_2, \ldots, \hat{\tau}_N$. To construct the optimal solution for (P1.1), we combine this solution with a maximum-speed flying segment from x_I to x_F. Specifically, the UAV moves from x_I to x_F at maximum speed V, while stopping/hovering at the N hovering points along the way with the corresponding optimal hovering durations.

To clearly define the trajectory, we introduce boundary conditions for the hovering points: $x_0^* = x_I, x_{N+1}^* = x_F$, and the corresponding hovering times $\hat{\tau}_0 = \hat{\tau}_{N+1} = 0$. Thus, the complete optimal trajectory $\{x^*(t)\}$ for problem (P1.1) is composed of flying segments between these hovering points and hovering at the points for the specified durations. This trajectory inherits the SHF structure from (P2), ensuring that the UAV flies at maximum speed between selected locations and hovers at those points for optimal durations. The resulting trajectory not only satisfies the initial and final position constraints but also maximizes the energy transfer to all ground nodes, achieving the best possible energy distribution.

This structured approach provides a clear and rigorous way to derive the optimal trajectory for (P1.1) based on the solution of (P2), ensuring both mathematical precision and practical feasibility.

1.4.4 Structure of Optimal Trajectory Solution to Problem (P1)

In this section, we describe the structure of the optimal trajectory solution to problem (P1). The optimal trajectory solution to problem (P1.1) or problem (P1) follows the SHF structure, consisting of N hovering locations, where N satisfies $N \leq K + 2$. At each hovering location, the UAV flies at maximum speed from one point to another and then hovers for a specified duration to maximize energy delivery.

This result can be explained by considering the characteristics of the SHF structure. The UAV trajectory consists of maximum-speed flying segments and hovering segments, which jointly achieve optimal energy transfer. There exists an optimal trajectory with no more than $K + 2$ hovering locations, ensuring the best possible energy distribution. The conclusions for problem (P1.1) extend to (P1) since the solution to (P1) is obtained by optimizing over all possible pairs of initial and final positions (x_I, x_F), ensuring the global optimality of the resulting trajectory.

In particular, the optimal trajectory starts from the initial position x_I and ends at the final position x_F, with N intermediate hovering locations that are selected to maximize the energy transfer to all ground nodes. The UAV flies at maximum speed between these hovering points, and the number of hovering points is at most $K + 2$. This upper bound includes the possibility that both the initial and final positions may be considered as hovering points, depending on the optimal trajectory design.

Thus, the optimal trajectory for problem (P1) follows the SHF structure, ensuring that the UAV hovers at specific locations to deliver energy effectively, and flies at maximum speed between these locations to minimize travel time. This structure guarantees that the energy received by the ground nodes is maximized while maintaining the efficiency of UAV movement.

1.5 Efficient Low-Complexity Trajectory Design

Building on the optimal trajectory characterization established in the previous sections, we now focus on deriving both an optimal solution and a low-complexity suboptimal solution to problem (P1). The aim of this section is to propose efficient approaches that retain high performance while significantly reducing computational complexity.

1.5.1 Optimal Solution to Problem (P1)

In Sect. 1.4, we demonstrated that the global optimal trajectory for problem (P1.1) can be derived, which provides a basis for solving problem (P1) under specified initial and final positions (x_I, x_F). The approach involves performing a two-dimensional exhaustive search over all possible pairs of initial and final positions while solving problem (P1.1) for each pair, ultimately yielding the global optimal trajectory solution to problem (P1).

It is evident that if the UAV hovers at a position outside the region of ground nodes, it does not contribute to energy transfer. Therefore, the feasible set for the initial position x_I is $[w_1, w_K]$, while the feasible set for the final position x_F is $[x_I, w_K]$. To conduct an exhaustive search over x_I or x_F within their respective feasible sets, we introduce d_{min} as the distance resolution. We assume d_{min} to be sufficiently small and ensure that $(w_K - w_1)/d_{min}$ is an integer.

As a result, the feasible locations for x_I are $\{w_1, w_1 + d_{min}, w_1 + 2d_{min}, \ldots, w_K\}$, and for x_F are $\{x_I, x_I + d_{min}, x_I + 2d_{min}, \ldots, w_K\}$. This discretization allows for efficient implementation of the exhaustive search approach.

1.5.2 An Efficient Low-Complexity Trajectory Design to (P1)

In the following subsections, we propose a low-complexity approach to solve problem (P1) efficiently. By leveraging the SHF structure and applying suitable approximations, we aim to propose a trajectory design that significantly reduces computational requirements while maintaining performance close to the optimal solution.

1.5.3 Problem Reformulation Based on SHF Structure

According to the SHF structure, in order to solve problem (P1), we need to determine up to $K+2$ hovering points, denoted as $\{x_1, \ldots, x_{K+2}\}$, along with their corresponding hovering durations, denoted as $\{\tau_1, \ldots, \tau_{K+2}\}$, where these points satisfy $w_1 \le x_1 \le x_2 \le \cdots \le x_{K+2} \le w_K$. These $K+2$ hovering points include the initial and final locations, which are also potential optimal hovering points for problem (P1). In cases where the optimal trajectory solution contains fewer than $K+2$ hovering points, this corresponds to scenarios in which certain hovering durations are zero, effectively reducing the number of active hovering locations.

Thus, this formulation with up to $K+2$ hovering points and corresponding hovering durations serves as a general representation to characterize the optimal trajectory. Following the optimal solution expressed in (1.11a), the trajectory with $K+2$ hovering points is expressed in (1.11b), as shown below:

$$x^*(t) = \begin{cases} \hat{x}_i^*, \text{ if } \sum_{j=1}^{i} \hat{\tau}_j^* - \hat{\tau}_i^* + \frac{x_i^* - x_0^*}{V} \le t \le \sum_{j=1}^{i} \hat{\tau}_j^* + \frac{x_i^* - x_1^*}{V}, 1 \le i \le N+1 \\ x_0^* + V\left(t - \sum_{j=1}^{i} \hat{\tau}_j^*\right), \text{ if } \sum_{j=1}^{i} \hat{\tau}_j^* + \frac{x_i^* - x_0^*}{V} \le t \le \sum_{j=1}^{i} \hat{\tau}_j^* + \frac{x_{i+1}^* - x_0^*}{V}, 1 \le i \le N, \end{cases} \quad (1.11a)$$

$$x(t) = \begin{cases} x_i, \text{ if } \sum_{j=1}^{i} \tau_j - \tau_i + \frac{x_i - x_1}{V} \le t \le \sum_{j=1}^{i} \tau_j + \frac{x_i - x_1}{V}, 1 \le i \le K+2 \\ x_1 + V\left(t - \sum_{j=1}^{i} \tau_j\right), \text{ if } \sum_{j=1}^{i} \tau_j + \frac{x_i - x_1}{V} \le t \le \sum_{j=1}^{i} \tau_j + \frac{x_{i+1} - x_1}{V}, 1 \le i \le K+1. \end{cases} \tag{1.11b}$$

Thus, the total received energy can be expressed as:

$$\begin{aligned} E_k(\mathbf{x}, \mathbf{t}) &= E_{k,\text{ Hov}}(\mathbf{x}, \mathbf{t}) + E_{k,\text{Fly}}(x_1, x_{K+2}) = \sum_{i=1}^{K+2} E_{k,i,\text{Hov}}(x_i, \tau_i) + E_{k,\text{Fly}}(x_1, x_{K+2}) \\ &= \sum_{i=1}^{K+2} \frac{\beta_0 P \tau_i}{(x_i - w_k)^2 + H^2} + \frac{\beta_0 P}{VH} \arctan\left(\frac{x_{K+2} - w_k}{H}\right) \\ &\quad - \frac{\beta_0 P}{VH} \arctan\left(\frac{x_1 - w_k}{H}\right), \end{aligned} \tag{1.12}$$

where $E_{k,I,\text{Hov}}(x_I, \tau_i) = \frac{\beta_0 P \tau_i}{(x_i - w_k)^2 + H^2}$ denotes the received energy at node k during UAV hovering at location x_i, and $E_{k,\text{Hov}}(\mathbf{x}, \mathbf{t})$ represents the sum of received energy of node k when the UAV hovers at all $K + 2$ hovering points. Additionally, $E_{k,\text{Fly}}(x_1, x_{K+2})$ denotes the received energy at node k when the UAV is flying between x_1 and x_{K+2}.

Based on this reformulation, problem (P1) is equivalently reformulated into:

$$\begin{aligned} \text{(P3): } & \max_{x,t,E} E \\ & \text{s.t.} E_k(\boldsymbol{x}, \boldsymbol{t}) \ge E, \forall k \in \mathcal{K}, \\ & \sum_{i=1}^{K+2} \tau_i + \frac{x_{K+2} - x_1}{V} = T, \\ & w_1 \le x_1 \le x_2 \le \cdots \le x_{K+2} \le w_K, \\ & \tau_i \ge 0, \quad \forall i \in \{1, 2, \ldots, K+2\}, \end{aligned} \tag{1.13}$$

where the second constraint ensures that the total charging duration T is fully utilized, including both hovering and flying time.

Although problem (P3) is non-convex as the first constraint is non-convex, we propose an efficient solution using the SCA method. Unlike conventional alternating optimization approaches, which have been widely used in UAV trajectory design to tackle non-convex problems, the alternating method is not suitable for problem (P3). Specifically, alternating optimization would involve iteratively optimizing $\mathbf{t}$ and $\mathbf{x}$; while $\mathbf{t}$ is a linear problem, $\mathbf{x}$ can be approximated as a convex problem using Taylor expansions. However, during alternating optimization, it can be observed that the sum of hovering durations $\sum_{i=1}^{K+2} \tau_i$ remains constant for any given $\mathbf{t}$, and similarly for $\mathbf{x}$ based on the constraint $\sum_{i=1}^{K+2} \tau_i + \frac{x_{K+2}-x_1}{V} = T$. This implies that the total hovering and flying time is fixed regardless of the optimization iteration, leading to an unchanged objective value and resulting in a potentially poor solution if the initial point is not well chosen.

1.5.4 Convex Approximation

In this subsection, we develop a convex approximation for problem (P3). The primary goal is to construct a concave function for each iteration, denoted as $E_k^{(r)}$ for the rth iteration, ensuring that $E_k \geq E_k^{(r)}$ and the equality holds at the local point $\left(\mathbf{x}^{(r)}, \mathbf{t}^{(r)}\right)$ during the iteration.

According to (1.12), the energy E_k received by ground node k consists of two components. In the following, we present concave functions to approximate these two components separately. We begin with the part of energy received when the UAV is hovering, distinguishing between $\tau_i^{(r)} > 0$ and $\tau_i^{(r)} = 0$:

(1) **Case** $\tau_i^{(r)} > 0$: The approximation $f_{i,k}^{(r)}(x_i, \tau_i)$ is obtained as follows:

The function $\frac{1}{(x_i-w_k)^2+H^2}$ is convex in $(x_i - w_k)^2$. By applying the first-order condition, it holds for the two points $(x_i - w_k)^2$ and $\left(x_i^{(r)} - w_k\right)^2$ that:

$$\frac{\beta_0 P \tau_i}{(x_i - w_k)^2 + H^2} \geq -A_{i,k}^{(r)}\left((x_i - w_k)^2 - \left(x_i^{(r)} - w_k\right)^2\right)\tau_i + B_{i,k}^{(r)}\tau_i, \quad (1.14)$$

where $A_{i,k}^{(r)} = \frac{\beta_0 P}{\left(\left(x_i^{(r)}-w_k\right)^2+H^2\right)^2}$ and $B_{i,k}^{(r)} = \frac{\beta_0 P}{\left(x_i^{(r)}-w_k\right)^2+H^2}$. The equality holds when $x_i = x_i^{(r)}$.

Furthermore, by introducing two positive constants $C_{i,k}^{(r)} > 0$ and $D_{i,k}^{(r)} > 0$, we obtain the following inequality:

$$\begin{aligned}(x_i - w_k)^2\tau_i &= \underline{\left((x_i - w_k)^2 + C_{i,k}^{(r)}\right)\left(\tau_i + D_{i,k}^{(r)}\right)} \\ &\quad - C_{i,k}^{(r)}\tau_i - D_{i,k}^{(r)}(x_i - w_k)^2 - C_{i,k}^{(r)}D_{i,k}^{(r)} \\ &\leq \underline{\frac{F_{i,k}^{(r)}}{2}\left((x_i - w_k)^2 + C_{i,k}^{(r)}\right)^2 + \frac{1}{2F_{i,k}^{(r)}}\left(\tau_i + D_{i,k}^{(r)}\right)^2} \\ &\quad - C_{i,k}^{(r)}\tau_i - D_{i,k}^{(r)}(x_i - w_k)^2 - C_{i,k}^{(r)}D_{i,k}^{(r)},\end{aligned} \quad (1.15)$$

where $F_{i,k}^{(r)} = \frac{\tau_i^{(r)}+D_{i,k}^{(r)}}{\left(x_i^{(r)}-w_k\right)^2+C_{i,k}^{(r)}}$, and the equality holds at local point. Combining the results from (1.14) to (1.15), we derive the following lower-bound approximation of $E_{k,i,\mathrm{Hov}}$:

$$
\begin{aligned}
E_{k,i,\mathrm{Hov}}(x_i, \tau_i) &\geq -A_{i,k}^{(r)}\left((x_i - w_k)^2 - \left(x_i^{(r)} - w_k\right)^2\right)\tau_i + B_{i,k}^{(r)}\tau_i \\
&\geq -\frac{A_{i,k}^{(r)}F_{i,k}^{(r)}}{2}\left((x_i - w_k)^2 + C_{i,k}^{(r)}\right)^2 - \frac{A_{i,k}^{(r)}}{2F_{i,k}^{(r)}}\left(\tau_i + D_{i,k}^{(r)}\right)^2 \\
&\quad + A_{i,k}^{(r)}\left(x_i^{(r)} - w_k\right)^2\tau_i + B_{i,k}^{(r)}\tau_i + A_{i,k}^{(r)}C_{i,k}^{(r)}\tau_i \\
&\quad + A_{i,k}^{(r)}D_{i,k}^{(r)}(x_i - w_k)^2 + A_{i,k}^{(r)}C_{i,k}^{(r)}D_{i,k}^{(r)} \\
&\triangleq f_{i,k}^{(r)}(x_i, \tau_i),
\end{aligned}
\tag{1.16}
$$

which holds at $x_i = x_i^{(r)}$, as the inequalities have been shown to be satisfied under the given conditions. To guarantee the concavity of $f_{i,k}^{(r)}(x_i, \tau_i)$, $F_{i,k}^{(r)}$ is required to be limited by $F_{i,k}^{(r)} \in \left(0, \frac{\tau_i^{(r)}}{\left(x_i^{(r)} - w_k\right)^2}\right)$. In other words, the choices of $C_{i,k}^{(r)}, D_{i,k}^{(r)}$ should satisfy $0 < \frac{\tau_i^{(r)} + D_{i,k}^{(r)}}{\left(x_i^{(r)} - w_k\right)^2 + C_{i,k}^{(r)}} \leq \frac{\tau_i^{(r)}}{\left(x_i^{(r)} - w_k\right)^2}$.

(2) **Case** $\tau_i^{(r)} = 0$: The approximation $f_{i,k}^{(r)}(x_i, \tau_i)$ is given by $f_{i,k}^{(r)}(x_i, \tau_i) \triangleq \frac{\beta_0 P \tau_i}{\max\{(w_1 - w_k)^2, (w_K - w_k)^2\} + H^2} \leq \frac{\beta_0 P \tau_i}{(x_i - w_k)^2 + H^2}$.

The inequality holds because $\max\{(w_1 - w_k)^2, (w_K - w_k)^2\} \geq (x_i - w_k)^2$. Additionally, the equality holds at $x_i = x_i^{(r)}$ since $\tau_i = 0$ makes both sides of the inequality equal to zero. Moreover, $f_{i,k}^{(r)}(x_i, \tau_i)$ is linear and therefore concave.

Next, we consider the approximation for $E_{k,\mathrm{Fly}} = \frac{\beta_0 P}{VH}\arctan\left(\frac{x_{K+2} - w_k}{H}\right) - \frac{\beta_0 P}{VH}\arctan\left(\frac{x_1 - w_k}{H}\right)$, representing the difference between two arctangent functions. Despite the well-known convexity/concavity properties of the arctangent function, i.e., $\arctan(\mu)$ is convex for $\mu > 0$ and concave for $\mu < 0$, it remains challenging to construct a concave function to approximate $E_{k,\mathrm{Fly}}$. This difficulty arises because the terms $x_{K+2} - w_k$ and $x_1 - w_k$ may take both positive and negative values, depending on whether $\min\{\boldsymbol{w}\} \leq x_1 \leq x_2 \leq \cdots \leq x_{K+2} \leq \max\{\boldsymbol{w}\}$, making the approximation nontrivial.

In our work, we derive the approximation $g_k^{(r)}(x_1, x_N)$ as follows:

(1) Reformulation of $E_{k,\mathrm{Fly}}$: First, we reformulate $E_{k,\mathrm{Fly}}$ as the sum of two arctangent functions, i.e., $E_{k,\mathrm{Fly}}(x_1, x_N) = \frac{\beta_0 P}{VH}\arctan\left(\frac{x_{K+2} - w_k}{H}\right) + \frac{\beta_0 P}{VH}\arctan\left(\frac{w_k - x_1}{H}\right)$.
(2) Concave Approximation: We approximate each arctangent function by a concave quadratic function. This quadratic function is a lower bound to the original arctangent function, and the two functions share the same value at the local point $\left(x_1^{(r)}, x_{K+2}^{(r)}\right) = (x_1, x_{K+2})$. Consequently, the sum of the concave approximations serves as a lower bound to $E_{k,\mathrm{Fly}}$ while maintaining the same value as $E_{k,\mathrm{Fly}}$ at $\left(x_1^{(r)}, x_{K+2}^{(r)}\right) = (x_1, x_{K+2})$.

Following this approach, we establish a lower bound concave approximation for $g_k^{(r)}(x_1, x_N)$, ensuring the concavity of the resulting function. To achieve this, we derive specific parameters for the quadratic function, ensuring that the concavity condition is satisfied and that both functions share the same value at a local point $z = z_0$.

In particular, for a local hovering point $x^{(r)}$ in the rth iteration, we determine the corresponding parameters $\left(a_{K+2,k}^{(r)}, b_{K+2,k}^{(r)}, c_{K+2,k}^{(r)}\right)$ and $\left(a_{1,k}^{(r)}, b_{1,k}^{(r)}, c_{1,k}^{(r)}\right)$ to approximate $\arctan\left(\frac{x_{K+2}-w_k}{H}\right)$ and $\arctan\left(\frac{w_k-x_1}{H}\right)$ by $-a_{K+2,k}^{(r)}\left(\frac{x_{K+2}-w_k}{H}\right)^2 + b_{K+2,k}^{(r)}\frac{x_{K+2}-w_k}{H} + c_{K+2,k}^{(r)}$ and $-a_{1,k}^{(r)}\left(\frac{w_k-x_1}{H}\right)^2 + b_{1,k}^{(r)}\frac{w_k-x_1}{H} + c_{1,k}^{(r)}$ respectively. These parameters are chosen to guarantee that the approximations match the original arctangent functions at specific points, ensuring concavity and a lower bound property.

Define $g_k^{(r)}(x_1, x_N)$ as the approximation of $E_{k,\text{Fly}}$ in the r-th iteration. Hence, for any feasible point x in problem (P3), $g_k^{(r)}(x_1, x_N)$ is given by:

$$\begin{aligned} g_k^{(r)}(x_1, x_{K+2}) &\triangleq \frac{\beta_0 P}{VH}\Bigg(-a_{K+2,k}^{(r)}\left(\frac{x_{K+2}-w_k}{H}\right)^2 + b_{K+2,k}^{(r)}\frac{x_{K+2}-w_k}{H} \\ &\quad + c_{K+2,k}^{(r)} - a_{1,k}^{(r)}\left(\frac{w_k-x_1}{H}\right)^2 + b_{1,k}^{(r)}\frac{w_k-x_1}{H} + c_{1,k}^{(r)}\Bigg) \\ &\le \frac{\beta_0 P}{VH}\arctan\left(\frac{x_{K+2}-w_k}{H}\right) + \frac{\beta_0 P}{VH}\arctan\left(\frac{w_k-x_1}{H}\right) \\ &= \frac{\beta_0 P}{VH}\arctan\left(\frac{x_{K+2}-w_k}{H}\right) - \frac{\beta_0 P}{VH}\arctan\left(\frac{x_1-w_k}{H}\right) \\ &= E_{k,\text{ Fly}}(x_1, x_{K+2}), \end{aligned} \tag{1.17}$$

which serves as a concave lower bound for $E_{k,\text{Fly}}(x_1, x_{K+2})$, ensuring that the approximation matches the original function at (x_1, x_{K+2}).

In the next step, we establish the combined lower-bounded approximation for $E_k(x, t)$ by combining the individual lower bounds derived for $E_{k,\text{Hov}}(x_i, t_i)$ and $E_{k,\text{Fly}}(x_1, x_{K+2})$. Specifically, for any feasible point or trajectory x in problem (P3), we define the combined approximation $E_k^{(r)}(\mathbf{x}, \mathbf{t})$ as follows:

$$E_k^{(r)}(\mathbf{x}, \mathbf{t}) = \sum_{i=1}^{N} f_{i,k}^{(r)}(x_i, \tau_i) + g_k^{(r)}(x_1, x_{K+2}) \le E_k(\mathbf{x}, \mathbf{t}), \tag{1.18}$$

where $f_{i,k}^{(r)}(x_i, t_i)$ and $g_k^{(r)}(x_1, x_{K+2})$ are concave functions derived previously that approximate $E_{k,\text{Hov}}(x_i, t_i)$ and $E_{k,\text{Fly}}(x_1, x_{K+2})$, respectively. These functions serve as lower-bounded approximations, ensuring the concavity of $E_k^{(r)}(\mathbf{x}, \mathbf{t})$ in each iteration of the optimization.

1.5.5 *Iterative Algorithm*

Based on the previously established convex approximation, this subsection introduces an iterative algorithm to solve problem (P3).

In the initialization phase, a feasible initial local point $\left(\mathbf{x}^0, \mathbf{t}^0\right)$ is set by uniformly distributing potential hovering points along the line segment within the interval $[w_1, w_K]$, and assigning an initial hovering time at each point. This ensures that the total length of all segments lies within $[w_1, w_K]$ and that the cumulative segment length does not exceed VT.

After initialization, the algorithm proceeds iteratively. During each iteration r, a convex approximation of problem (P3) is developed at the local point $\left(\mathbf{x}^{(r)}, \mathbf{t}^{(r)}\right)$. Using the convex approximation derived in the previous subsection, an approximated problem is formulated as follows in (P4):

$$
\begin{aligned}
\text{(P4)}: \max_{x,t,E}\ & E \\
\text{s.t.}\ & (\mathbf{x}, \mathbf{t}) \geq E, \forall k \in \mathcal{K}, \\
& \sum_{i=1}^{K+2} \tau_i + \frac{x_{K+2} - x_1}{V} = T, \\
& \tau_i \geq 0, \quad \forall i \in \{1, 2, \ldots, K+2\}.
\end{aligned} \tag{1.19}
$$

Note that $E_k^{(r)}(\mathbf{x}, \mathbf{t})$ is a concave function. Hence, problem (P4) is a convex optimization problem and can be solved efficiently. The solution of problem (P4) is then used as the local point for the next iteration.

1.6 Complexity Level Analysis

In this section, we analyse the complexity of the proposed trajectory designs, including the **proposed optimal SHF** in Sect. 1.5.1 and the **proposed low-complexity SHF** in Sect. 1.5.2, along with two benchmark solutions presented in [22], namely, **heuristic SHF** and **SCP with time quantization**, which were discussed after Problem (P1) in Sect. 1.3.

One key feature of these four trajectory designs is that they use iterative methods to minimize convex functions. It is worth noting that there are different iterative methods available, which leads to variations in computational complexity. To ensure a fair comparison of complexity, we assume all three trajectory designs adopt the same quantization size for iteration, and iterations terminate at the same error threshold $\varepsilon > 0$. Specifically, we assume that the ellipsoid method is applied for the iterations, motivated by its properties [26]: the volume of the ellipsoids decreases uniformly at each step, and the method converges to a minimizer of a convex function within finite iterations.

The complexity of the ellipsoid method depends on both the number of variables and the calculations required for objective and constraint functions [27]. Using the complexity analysis in [27], applying the ellipsoid method to solve a convex subproblem with N variables involve $\mathcal{O}(N^2\log(\frac{1}{\varepsilon}))$ calls, representing the number of ellipsoid iterations required. Each call includes computational complexity resulting from evaluation of objective and constraint functions. For computational analysis per call, we calculate the number of floating-point operations following the analysis in [28].

The results of the complexity comparison are summarized in Table 1.1, where φ denotes the number of iterations required by the two iterative algorithms to reach convergence, i.e., when the error threshold ε is satisfied. Moreover, $d_{\min}$ represents the resolution in distance for the exhaustive search, and $t_{\min}$ is the quantization size for the SCP with time quantization.

In the proposed optimal SHF, the total number of possible (x_I, x_F) pairs in the exhaustive search is $\mathcal{O}\left(\left(\frac{D}{d_{\min}}\right)^2\right)$. For each pair (x_I, x_F), a dual problem with K variables (i.e., K Lagrange multipliers) is solved, which requires $\mathcal{O}(K^2\log(\frac{1}{\varepsilon}))$ calls. Each call introduces computational complexities of $\mathcal{O}(K^2)$ and $\mathcal{O}\left(K\frac{D}{d_{\min}}\right)$ due to the ellipsoid updates [26] and solving (13) based on the analysis in [28]. Therefore, the overall complexity level of the optimal SHF is given by $\mathcal{O}\left(\left(\frac{D}{d_{\min}}\right)^2\right) \cdot \mathcal{O}(K^2\log(\frac{1}{\varepsilon})) \cdot \left(\mathcal{O}(K^2) + \mathcal{O}\left(K\frac{D}{d_{\min}}\right)\right) = \mathcal{O}\left(\max\left(K, \frac{D}{d_{\min}}\right)K^3\left(\frac{D}{d_{\min}}\right)^2\log(\frac{1}{\varepsilon})\right)$. For the proposed low-complexity SHF, in each iteration, a convex problem with $2N$ variables is solved using the ellipsoid method, which requires $\mathcal{O}((2N)^2\log(\frac{1}{\varepsilon}))$ calls. Each call has complexities of $\mathcal{O}((2N)^2)$ for the ellipsoid updates and $\mathcal{O}(KN)$ for the objective function evaluation. Since the number of iterations of this method is φ and the number of hovering points N is given by $N = K + 2$, the total complexity level for this design is given by $\mathcal{O}(\varphi(2N)^2\log(\frac{1}{\varepsilon})) \cdot (\mathcal{O}((2N)^2) + \mathcal{O}(KN)) = \mathcal{O}(\varphi K^4\log(\frac{1}{\varepsilon}))$. The heuristic SHF has the same complexity as the optimal SHF for a given (x_I, x_F) pair. Therefore, the overall complexity of the heuristic SHF

Table 1.1 Complexity level comparison

Algorithm	Complexity level
Proposed optimal SHF	$\mathcal{O}\left(\max\left(K, \frac{D}{d_{\min}}\right)K^3\left(\frac{D}{d_{\min}}\right)^2\log(\frac{1}{\varepsilon})\right)$
Proposed low-complexity SHF	$\mathcal{O}(\varphi K^4\log(\frac{1}{\varepsilon}))$
Heuristic SHF [22]	$\mathcal{O}\left(\max\left(K, \frac{D}{d_{\min}}\right)K^3\log(\frac{1}{\varepsilon})\right)$
SCP with time quantization [22]	$\mathcal{O}\left(\max\left(K, \frac{D}{d_{\min}}\right)K^3\log(\frac{1}{\varepsilon})\right) +$ $\mathcal{O}\left(\varphi\left(\frac{T}{t_{\min}}\right)^3\max\left(\frac{T}{t_{\min}}, K\right)\log(\frac{1}{\varepsilon})\right)$

is $\mathcal{O}\left(\max\left(K, \frac{D}{d_{\min}}\right)K^3 \log\left(\frac{1}{\varepsilon}\right)\right)$. Finally, the SCP with time quantization improves upon the heuristic SHF by adding an SCP process with $\frac{T}{t_{\min}}$ variables, resulting in a complexity of $\mathcal{O}\left(\varphi\left(\frac{T}{t_{\min}}\right)^3 \max\left(\frac{T}{t_{\min}}, K\right) \log\left(\frac{1}{\varepsilon}\right)\right)$.

Clearly, to obtain an accurate trajectory solution, d_{min} and t_{min} need to be sufficiently small, making $\frac{D}{d_{min}}$ and $\frac{T}{t_{min}}$ significantly larger than other terms in the complexity expressions for all solutions. In such cases, SCP with time quantization is expected to exhibit the highest complexity, with a complexity level on the order of $\left(\frac{T}{t_{min}}\right)^4$. Furthermore, the proposed optimal SHF and heuristic SHF also have high complexities, which are of the orders $\left(\frac{D}{d_{min}}\right)^3$ and $\frac{D}{d_{min}}$, respectively. Thus, the complexities of these three algorithms decrease as d_{min} or t_{min} decrease, while they increase with the geographical network size D. Meanwhile, the complexity level of the proposed low-complexity SHF is not influenced by the resolutions d_{min} and t_{min}, meaning it remains unaffected by the geographical network size D given the specified resolutions. Therefore, the low-complexity SHF is expected to introduce significantly lower computational complexity while maintaining efficient WPT performance.

1.7 Numerical Simulation

In this section, we assess the performance of the proposed optimal SHF and the low-complexity SHF, in comparison to the heuristic SHF [22] and SCP with time quantization [22], based on the complexity analysis presented earlier. To evaluate the WPT performance, we randomly generate 20 different topologies by placing ground nodes in various locations. Each of the four algorithms is then applied to each topology, and the results are averaged to determine the max–min received power across all ground nodes for these random configurations. In the simulation, we have the following default setups of parameters: $\beta_0 = -30$dB, $P = 40$ dBm, $K = 3$, $N = K + 2$, $H = 5$ m, $T = 20s$, $V = 1$ m/s and $D = 20$m. In addition, we set the quantization size of distance for the exhaustive search to $d_{\min} = 0.01$m and set accordingly the time quantization size in the reference algorithm SCP with time quantization to $t_{\min} = d_{\min}/V$.

Figure 1.2 illustrates the impact of the charging duration T on WPT performance, while also showing the upper bound for the ideal scenario where UAV speed constraints are ignored. As the charging duration T increases, it can be seen that the performance of all four algorithms improves, converging toward the upper bound. Additionally, the SCP with time quantization outperforms the heuristic SHF due to the additional SCP processing, which aligns with the findings of [22]. Moreover, both proposed algorithms (optimal SHF and low-complexity SHF) consistently outperform the two reference algorithms throughout the entire charging duration range. Specifically, the proposed low-complexity SHF achieves nearly identical

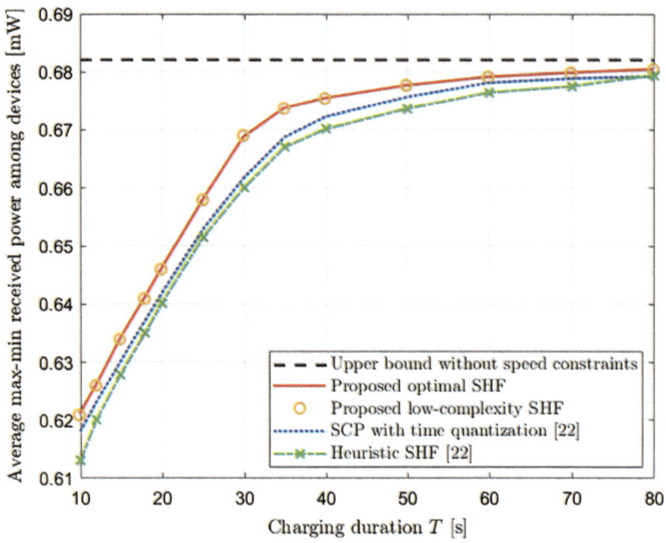

Fig. 1.2 Average WPT performance (over 20 realizations of random ground node topologies) versus the charging duration

performance to the optimal SHF, despite having significantly lower computational complexity compared to the other three algorithms. This result underscores the efficiency of the proposed low-complexity approach. Moreover, similar trends among the algorithms have been observed in additional simulations, even for scenarios with shorter charging durations T.

Figure 1.3 illustrates the relationship between the WPT performance and the UAV speed. It is evident that, across all algorithms, the max–min received power increases as V grows. As the UAV speed becomes sufficiently high, the performance of all four methods approaches the upper bound. Again, the proposed optimal SHF and low-complexity SHF outperform the heuristic SHF and SCP with time quantization. While similar trends are observed even for higher UAV speeds, it is noteworthy that the performance gap between the proposed and reference algorithms narrows as the UAV speed approaches infinity. Furthermore, recall that the heuristic SHF has two steps: first solving the relaxed problem assuming an infinite UAV speed and then based on it obtaining a new trajectory by using TSP or SCP to satisfy the speed constraint. Consequently, there is some inherent performance loss in the heuristic SHF compared to the optimal solution. However, as the UAV speed increases, this performance gap diminishes. On the other hand, the proposed low-complexity SHF closely matches the optimal solution's performance, which suggests that the relative advantage of the proposed designs decreases when the UAV speed becomes very large.

Next, we present a comparison of the computational complexity among the four algorithms based on their average computing time in Fig. 1.4. In the comparison, In

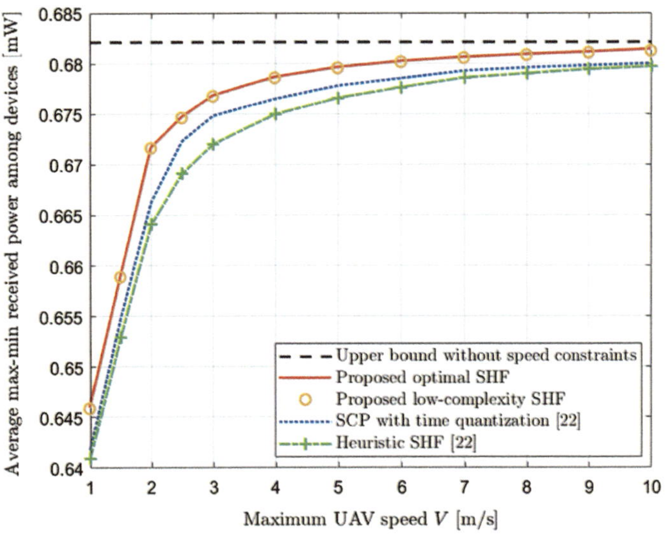

Fig. 1.3 Average WPT performance (over 20 realizations of random ground node topologies) versus the maximum UAV speed

this comparison, we scale both geographical network size D and the UAV speed V while ensuring $D/V = T$, which maintains the scenario's validity. This condition guarantees that if the UAV's speed is too low relative to the network size, there will not be sufficient time for hovering, thereby minimizing the benefits from any trajectory optimization. The results indicate that, with a fixed resolution in distance or time, the computing time for both the proposed optimal SHF and the SCP with time quantization becomes excessively large when D is increased. Due to time constraints, we only obtained partial results for larger geographical network sizes for these two algorithms. It is evident that the proposed optimal SHF requires comparatively less computational time than the SCP with time quantization, which aligns with the complexity analysis provided at the end of Sect. 1.6. The complexities of the optimal SHF and SCP with time quantization scale with orders of $\left(\frac{D}{d_{min}}\right)^3$ and $\left(\frac{T}{t_{min}}\right)^4$ respectively. Since both $\frac{D}{d_{min}}$ and $\frac{T}{t_{min}}$ increase linearly with D, the computational demands for these methods rise polynomially. Besides, Fig. 1.4 also highlights the advantages of the proposed low-complexity SHF, as its computational cost remains unaffected by the geographical network size. This matches our theoretical analysis, confirming the efficiency and practicality of the proposed low-complexity SHF in real-world applications. The results also validate the benefits of the proposed low-complexity SHF structure, particularly in terms of its balance between WPT performance and computational efficiency.

Figure 1.5 illustrates the impact of resolution on the WPT performance for the four considered algorithms. Notably, we study how the resolution parameters d_{min} and t_{min} influence the performance of each method. It can be observed that different

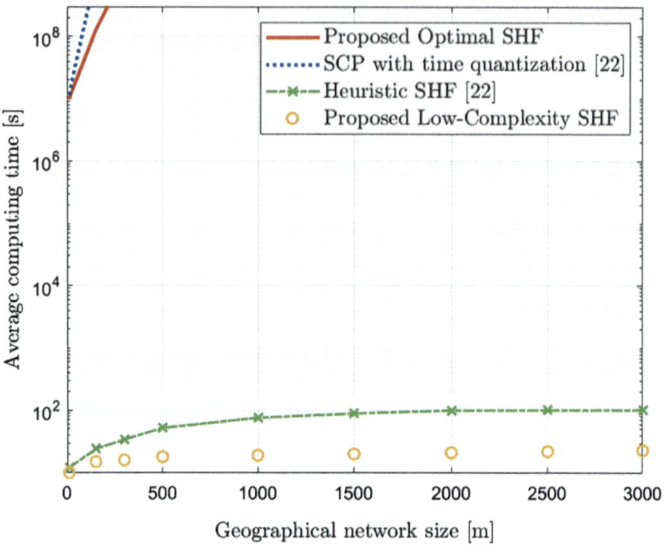

Fig. 1.4 The comparison in terms of computing time

algorithms respond differently to changes in resolution values. First, as expected, the proposed low-complexity SHF remains unaffected by these resolutions and consistently delivers excellent WPT performance. In contrast, the SCP approach with time quantization heavily relies on the resolution level. When t_{min} takes relatively large values, the performance of SCP is observed to degrade significantly. Notably, it delivers competitive results only under high complexity conditions, specifically when $t_{min} \leq 10^{-2}$ s. This effect arises because SCP approximates the UAV's flying trajectory within a time slot as a hovering point, treating the flying time as an equivalent hovering duration. Therefore, when t_{min} is not sufficiently small, the actual SCP performance tends to be lower than the ideal optimized results of the approximation. Both the proposed optimal SHF and the heuristic SHF demonstrate some insensitivity to d_{min}, particularly when d_{min} is not excessively large. For heuristic SHF, exhaustive search is employed to determine hovering points, followed by allocating hovering times based on those points, which mitigates the negative influence of imprecise hovering points. For example, poor hovering points could simply be assigned zero-time allocation, thereby reducing their impact. Interestingly, the proposed optimal SHF consistently delivers robust performance even with larger values of d_{min}. In this situation, inaccuracies arise in determining locations (x_I, x_F), yet the algorithm effectively compensates by further optimizing the hovering points and their durations. This adaptability ensures that the overall performance remains largely unaffected by slight errors in the initial location estimates (x_I, x_F).

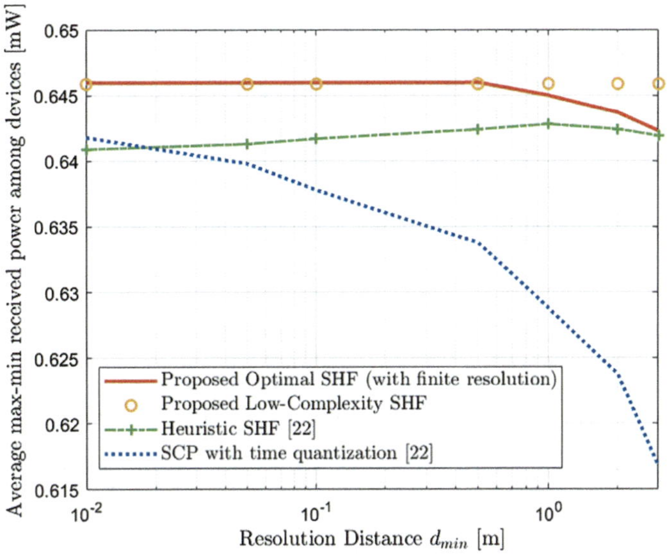

Fig. 1.5 The WPT performance comparison in terms of the resolution in exhaustive search

1.8 Conclusion

In this section, we conclude our work on UAV-enabled WPT networks with a linear topology. We studied the 1D UAV trajectory design problem with the aim of maximizing the minimum received power among all ground nodes, while respecting the maximum UAV speed constraint. Unlike previous studies that offered heuristic and locally optimal solutions, we proposed a globally optimal 1D UAV trajectory solution for the WPT problem in an SHF structure. Building on this SHF framework, we also developed an efficient, low-complexity trajectory design using the SCA technique. Our analysis and simulation results have demonstrated that the proposed low-complexity design maintains constant complexity, independent of the geographical network size, which makes it promising for practical implementation. Additionally, numerical results have confirmed the superior performance of both the optimal and low-complexity SHF trajectory designs, compared to benchmark schemes found in the literature.

The principles of our proposed trajectory design can be extended to UAV-enabled wireless communication scenarios, such as information broadcasting and multicasting in 1D network topologies, where communication users are positioned in a line on the ground. Our proof can also be employed to demonstrate that the optimal trajectory in such networks continues to follow the SHF structure. Determining the optimal hovering locations together with resource allocation is an interesting area for future work. Our 1D trajectory design also offers insights for more general 2D/3D trajectory designs. For example, in 2D/3D settings, the UAV is likely to follow an SHF trajectory structure, but optimizing the maximum-speed flying path becomes crucial for achieving the best performance. Designing such a path presents a non-trivial challenge, which we plan to explore in future research.

References

1. Zeng, Y., Zhang, R., & Lim, T. J. (2016). Wireless communications with unmanned aerial vehicles: Opportunities and challenges. *IEEE Communications Magazine, 54*, 36–42.
2. Mozaffari, M., Saad, W., Bennis, M., & Debbah, M. (2016). Efficient deployment of multiple unmanned aerial vehicles for optimal wireless coverage. *IEEE Communications Letters, 20*, 1647–1650.
3. Gupta, L., Jain, R., & Vaszkun, G. (2016). Survey of important issues in UAV communication networks. *IEEE Communications Surveys & Tutorials, 18*, 1123–1152.
4. Al-Hourani, A., Kandeepan, S., & Lardner, S. (2014). Optimal LAP altitude for maximum coverage. *IEEE Wireless Communications Letters, 3*, 569–572.
5. Alzenad, M., El-Keyi, A., & Yanikomeroglu, H. (2018). 3-D placement of an unmanned aerial vehicle base station for maximum coverage of users with different QoS requirements. *IEEE Wireless Communications Letters, 7*, 38–41.
6. Lyu, J., Zeng, Y., Zhang, R., & Lim, T. J. (2017). Placement optimization of UAV-mounted mobile base stations. *IEEE Communications Letters, 21*, 604–607.
7. Wu, Q., Zeng, Y., & Zhang, R. (2018). Joint trajectory and communication design for multi-UAV enabled wireless networks. *IEEE Transactions on Wireless Communications, 17*, 2109–2121.
8. Mozaffari, M., Saad, W., Bennis, M., & Debbah, M. (2016). Unmanned aerial vehicle with underlaid device-to-device communications: Performance and tradeoffs. *IEEE Transactions on Wireless Communications, 15*, 3949–3963.
9. Fan, R., Cui, J., Jin, S., Yang, K., & An, J. (2018). Optimal node placement and resource allocation for UAV relaying network. *IEEE Communications Letters, 22*, 808–811.
10. Zeng, Y., Zhang, R., & Lim, T. J. (2016). Throughput maximization for UAV-enabled mobile relaying systems. *IEEE Transactions on Communications, 64*, 4983–4996.
11. Sharma, V., Bennis, M., & Kumar, R. (2016). UAV-assisted heterogeneous networks for capacity enhancement. *IEEE Communications Letters, 20*, 1207–1210.
12. Zhan, C., Zeng, Y., & Zhang, R. (2018). Energy-efficient data collection in UAV enabled wireless sensor network. *IEEE Wireless Communications Letters, 7*, 328–331.
13. Cheng, F., et al. (2018). UAV trajectory optimization for data offloading at the edge of multiple cells. *IEEE Transactions on Vehicular Technology, 67*, 6732–6736.
14. Wu, Q., Xu, J., & Zhang, R. (2018). Capacity characterization of UAV-enabled two-user broadcast channel. *IEEE Journal on Selected Areas in Communications, 36*, 1955–1971.
15. Xie, L., Xu, J., & Zhang, R. (2019). Throughput maximization for UAV-enabled wireless powered communication networks. *IEEE Internet of Things Journal, 6*, 1690–1703.
16. Ju, H., & Zhang, R. (2014). Optimal resource allocation in full-duplex wireless powered communication network. *IEEE Transactions on Communications, 62*, 3528–3540.
17. Bi, S., Ho, C. K., & Zhang, R. (2015). Wireless powered communication: Opportunities and challenges. *IEEE Communications Magazine, 53*, 117–125.
18. Lu, X., Wang, P., Niyato, D., Kim, D. I., & Han, Z. (2015). Wireless networks with RF energy harvesting: A contemporary survey. *IEEE Communications Surveys & Tutorials, 17*, 757–789.
19. Wang, F., Xu, J., Wang, X., & Cui, S. (2018). Joint offloading and computing optimization in wireless powered mobile-edge computing systems. *IEEE Transactions on Wireless Communications, 17*, 1784–1797.

20. Zhou, F., Wu, Y., Hu, R. Q., & Qian, Y. (2018). Computation rate maximization in UAV-enabled wireless-powered mobile-edge computing systems. *IEEE Journal on Selected Areas in Communications, 36*, 1927–1941.
21. Xu, J., Zeng, Y., & Zhang, R. (2017). UAV-enabled wireless power transfer: Trajectory design and energy region characterization. In *2017 IEEE Globecom Workshops* (pp. 1–7). Singapore. https://doi.org/10.1109/GLOCOMW.2017.8269097
22. Xu, J., Zeng, Y., & Zhang, R. (2018). UAV-enabled wireless power transfer: Trajectory design and energy optimization. *IEEE Transactions on Wireless Communications, 17*, 5092–5106.
23. Clerckx, B., Zhang, R., Schober, R., Ng, D. W. K., Kim, D. I., & Poor, H. V. *Fundamentals of wireless information and power transfer: From RF energy harvester models to signal and system designs*. https://arxiv.org/abs/1803.07123
24. Yu, W., & Lui, R. (2006). Dual methods for nonconvex spectrum optimization of multicarrier systems. *IEEE Transactions on Communications, 54*, 1310–1322.
25. Boyd, S., & Vandenberghe, L. (2004). *Convex optimization*. Cambridge University Press.
26. Boyd, S. (2017). *EE364b convex optimization II, Course notes*. Stanford University. http://www.stanford.edu/class/ee364b/
27. Bubeck, S. (2015). Convex optimization: Algorithms complexity. *Foundations and Trends in Machine Learning, 8*, 231–357. https://arxiv.org/pdf/1405.4980v2.pdf
28. Hunger, R. (2005). *Floating point operations in matrix-vector calculus*. Munich University of Technology.

Chapter 2
Efficient UAV Trajectory Design for Multiuser Wireless Power Transfer

Xinran Wang, Xiaopeng Yuan, and Yulin Hu

Abstract In this chapter, we explore a UAV-assisted multi-user wireless power transfer (WPT) network, where a UAV is deployed to wirelessly transfer energy to multiple ground devices (GDs) distributed across a defined region. Our focus is on optimizing the UAV trajectory under maximum flying speed constraints to maximize the minimum energy harvested by the GDs over a defined charging duration. Unlike prior works that employed simplified linear energy harvesting (EH) models, this study introduces a realistic nonlinear EH model to address UAV trajectory design, marking its first application in this context. The formulated trajectory design problem is highly non-convex, involving infinitely many variables, which poses significant challenges for achieving an optimal solution. To address this, we propose a systematic three-step approach to derive an efficient solution. First, we rigorously characterize the optimal trajectory, which adopts a successive-hover-and-fly (SHF) structure, where the UAV hovers at designated points for efficient energy transfer and moves at maximum speed along arcs between these points. Leveraging this SHF structure, we reformulate the original problem to identify turning points that allow the UAV to change direction during maximum-speed flight without additional hovering. Finally, convex approximation techniques are employed to solve the reformulated problem. By utilizing the convex properties of the nonlinear EH model, an iterative approach is applied to solve a series of convex optimization problems, progressively refining the UAV trajectory towards a high-quality solution. Numerical results confirm the convergence of the proposed approach and substantiate its performance advantages over the benchmarks.

X. Wang · Y. Hu (✉)
School of Electronic Information, Wuhan University, Wuhan, China
e-mail: yulin.hu@whu.edu.cn

X. Wang
e-mail: xinran.wang@whu.edu.cn

X. Yuan
Chair of Information Theory and Data Analytics, RWTH Aachen University, Aachen, Germany
e-mail: yuan@inda.rwth-aachen.de

Y. Hu et al. (eds.), *Reliable and Efficient Trajectory Design in UAV-Aided Wireless Networks*, https://doi.org/10.1007/978-981-96-5988-3_2

Keywords Unmanned aerial vehicle · Nonlinear energy harvesting · Multiple users · Energy fairness · Trajectory optimization · Successive-hover-and-fly

Abbreviations

1D	One Dimension
2D	Two Dimension
BS	Base Stations
DC	Direct Current
GD	Ground Device
EH	Energy Harvesting
RF	Radio-Frequency
SHF	Successive Hover-and-Fly
SCA	Successive Convex Approximation
UAV	Unmanned Aerial Vehicle
WPT	Wireless Power Transfer
WPCN	Wireless Powered Communication Networks

2.1 Introduction

2.1.1 Background

The rapid advancement of unmanned aerial vehicles (UAVs) has significantly transformed various fields, particularly wireless communication networks. UAVs, with their high mobility, flexible deployment, and ability to operate in three-dimensional (3D) space, offer substantial benefits for enhancing the efficiency and coverage of modern communication systems [1]. Compared to conventional terrestrial networks, UAV-enabled systems exhibit superior adaptability and dynamic capabilities, making them particularly advantageous in environments where fixed infrastructure is either impractical or prohibitively expensive to implement [2].

In the context of wireless communication, UAVs have been deployed in numerous applications, including serving as aerial base stations or acting as mobile relays to support ground devices (GDs) [3]. By leveraging their altitude and mobility, UAVs are capable of establishing line-of-sight (LoS) communication links, thereby mitigating interference and enhancing signal quality. Such characteristics make UAVs suitable for diverse scenarios, such as disaster recovery, rural area coverage, and temporary hotspot provisioning. Furthermore, the flexibility in UAV deployment facilitates dynamic optimization of their trajectories, thereby enabling improvements in network performance metrics, including coverage [4–6], capacity [7], and energy efficiency [8].

A particularly promising application of UAVs lies in their integration with wireless power transfer (WPT) systems [9]. In these systems, UAVs function as energy transmitters to wirelessly charge distributed ground devices. This UAV-WPT integration is highly promising in scenarios where GDs lack direct access to traditional power sources, such as remote or hard-to-reach areas [10]. To fully realize the potential of UAV-enabled WPT networks, however, efficient trajectory planning is imperative. The UAV's trajectory has a direct impact on the amount of energy transferred to each GD, and optimizing this trajectory presents considerable challenges due to the complex interplay between UAV movement, energy harvesting efficiency, and the inherent nonlinear characteristics of the energy harvesting (EH) model [11].

2.1.2 Motivation

The increasing demand for efficient energy solutions in UAV-enabled WPT networks, particularly in scenarios where GDs are distributed across challenging environments, motivates the need for a detailed exploration of this topic. Existing research on UAV trajectory design for WPT has often relied on a linear EH model, which simplifies the analysis considerably [12–14]. However, in practical applications, the energy harvesting process is inherently nonlinear, resulting in inefficiencies when linear models are applied.

Addressing the nonlinearity of the EH model is crucial for achieving optimal energy transfer. The nonlinear characteristics of energy harvesting introduce complexities that linear approximations fail to capture, resulting in suboptimal system performance. Specifically, ignoring the nonlinearity can lead to inaccurate predictions of the energy harvested by GDs, thereby compromising overall system efficiency. Therefore, this chapter seeks to provide a comprehensive approach that takes into account the realistic nonlinear EH behavior, ensuring more accurate and effective energy delivery.

Moreover, optimizing the UAV trajectory under the nonlinear EH model presents a challenging and highly non-convex problem involving numerous variables. This chapter aims to address this challenge by characterizing the optimal trajectory structure, reducing the complexity of the problem, and proposing an effective method that enables the UAV to maximize the minimum energy harvested by the GDs.

2.1.3 Main Contributions

The main contributions of this chapter are outlined as follows:

(1) Characterization of Optimal UAV Trajectory Structure: We rigorously prove that the optimal UAV trajectory in a two-dimensional (2D) setup follows a successive-hover-and-fly (SHF) structure. In this structure, the UAV hovers at

specific positions to efficiently transfer energy and then moves along curved paths at maximum speed between these positions. Furthermore, the maximal number of hovering points is derived.

(2) Problem Reformulation with Turning Points: To facilitate problem-solving, we introduce a series of turning points into the UAV max-speed flight period, approximating curved paths as a series of line segments. Utilizing the SHF structure along with these turning points, the original problem is transformed into a simplified version involving fewer variables, including hovering positions, durations, and turning points.

(3) Solution via Convex Approximation: We apply convex approximation techniques to iteratively solve the reformulated problem. In each iteration, the harvested energy is approximated as a concave function of the UAV positions at the local level. By solving a series of convex approximations, UAV positions are updated iteratively towards convergence.

(4) Performance Evaluation through Numerical Results: We present numerical simulation results to evaluate the proposed trajectory design. These results show that the proposed design significantly outperforms conventional designs that assume linear EH models, achieving performance close to the theoretical upper bound, especially in scenarios where UAV flight speed constraints are disregarded.

2.1.4 Structural Organization

The rest of this chapter is organized as follows. In Sect. 2.2, we review related works, focusing on UAV-enabled WPT technology and energy harvesting models. Section 2.3 presents the system model and the problem formulation for UAV trajectory optimization. In Sect. 2.4, we characterize the optimal UAV trajectory structure. Section 2.5 details the proposed trajectory design, including problem reformulation based on SHF structure, convex approximation, and an iterative solution approach. Numerical results validating the proposed methods are provided in Sect. 2.6, and the conclusions are summarized in Sect. 2.7.

2.2 Related Works

2.2.1 UAV-Enabled Wireless Power Transfer

Integrating WPT technologies into communication networks has become a prominent approach to extending the lifetime of low-power rechargeable sensors and devices [9–11]. UAV-enabled WPT has emerged as a promising solution, leveraging the mobility of UAVs to efficiently deliver power to GDs [15–19]. UAVs can serve as mobile energy transmitters, providing dynamic power supply solutions in challenging environments where traditional power infrastructure may be unavailable or impractical. Previous research has explored the optimization of UAV trajectories to enhance energy transmit efficiency, focusing on various strategies to maximize harvested energy. In [15], the authors investigated the UAV trajectory optimization problem with the objective of maximizing the minimum radio-frequency (RF) energy transferred to a set of GDs, subject to UAV flight speed constraint. To address this problem, they first derived a multi-location-hovering solution for the ideal case without speed constraints and then proposed a suboptimal SHF trajectory structure, where the UAV hovers above specific points for a certain period and then flies straight at maximum speed between these points, balancing practical flight constraints with energy transfer efficiency. Building on [15, 16] extended the analysis to a simplified one-dimensional (1D) setup, where GDs were arranged linearly. In this 1D scenario, the authors characterized the optimal UAV trajectory structure and demonstrated that it must follow the SHF structure to achieve energy maximization. They then proposed the optimal solution for the energy maximization problem based on this structure. However, the solution in [16] is applicable to 1D scenarios, and extending the characterization of the optimal UAV trajectory to a more general 2D topology remains an unresolved challenge in the previous literature.

2.2.2 Energy Harvesting

Energy harvesting plays a crucial role in UAV-enabled WPT networks, as it directly impacts the efficiency of energy transfer to GDs. In the existing literature on UAV-enabled WPT and wireless powered communication networks (WPCN) [18–20], RF energy is commonly used as a performance metric while often neglecting the specifics of the EH process [21], or assuming simplified linear EH models at the GDs [12–14]. It has been demonstrated that the efficiency of EH, specifically the RF-to-direct current (DC) conversion, can vary significantly depending on the input power and signal waveforms [22]. In particular, studies have shown that EH is inherently a nonlinear process [23–25]. The authors in [20, 26] have made efforts to model and incorporate nonlinearity of EH process in their studies. However, the EH models they introduced relied on curve fitting based on statistical results, lacking essential analytical proof or guarantees. In contrast, Boshkovska et al. [24] rigorously derived the nonlinear EH process, showing that accurately modelling the nonlinearity in EH leads to better system performance compared to traditional linear models.

To the best of our knowledge, existing studies have not yet fully explored UAV trajectory design with the realistic accurate nonlinear energy harvesting (EH) model, which motivates the investigation presented in this chapter.

2.3 System Model and Problem Formulation

2.3.1 Description of UAV-Enabled WPT Network

Considered a UAV-enabled WPT system, where K GDs are deployed within a given area. The horizontal position of each GD, indexed by $k \in \mathcal{K} \triangleq \{1, 2, \ldots, K\}$, is denoted by $\left(w_{x,k}, w_{y,k}\right)$, and it is assumed that these locations are known beforehand by the UAV. Additionally, the UAV operates at a fixed altitude $H > 0$, maintaining a steady height throughout the energy transmission process. The power transfer occurs over a fixed time period $\mathcal{T} \triangleq [0, T]$, which is determined based on the available battery energy and maximum allowable power consumption of the UAV. This scheduling ensures that the UAV can complete its energy transfer mission and return for recharging before its battery is depleted.

The horizontal location of UAV at any time t is represented by the coordinates $(x(t), y(t))$, which will be optimized for the power transfer operation. The UAV is subject to the speed constraint, with its maximum flight speed denoted by V. This constraint imposes the following condition on its speed: $(\dot{x}(t))^2 + (\dot{y}(t))^2 \leq V^2, \forall t \in \mathcal{T}$, where $\dot{x}(t)$ and $\dot{y}(t)$ denote the time derivatives of $x(t)$ and $y(t)$, respectively.

2.3.2 Channel Model

In a practical setting, the wireless communication between the UAV and the GDs predominantly experiences LoS propagation. To simplify the analysis, we adopt a free-space channel model, where the path loss exponent is set to two for all the communication links between the UAV and the GDs. This assumption is appropriate given the typically unobstructed aerial-to-ground signal paths. At any given time t, the channel gain from the UAV to the k th GD, denoted by $h_k(x(t), y(t))$, can be expressed as $h_k(x(t), y(t)) = \frac{\beta_0}{\left(x(t)-w_{x,k}\right)^2+\left(y(t)-w_{y,k}\right)^2+H^2}$, where β_0 represents the channel power gain at a reference distance of one meter. The term $\sqrt{\left(x(t) - w_{x,k}\right)^2 + \left(y(t) - w_{y,k}\right)^2 + H^2}$ denotes the distance between the UAV and the k th GD, with H being the fixed altitude of the UAV. As a result, the RF power received by the k th GD at time t, denoted by $Q_k(x(t), y(t))$, is given by:

$$Q_k(x(t), y(t)) = \frac{\beta_0 P}{\left(x(t) - w_{x,k}\right)^2 + \left(y(t) - w_{y,k}\right)^2 + H^2}, \tag{2.1}$$

where P represents the constant transmit power of the UAV used for wireless power transfer.

2.3.3 Nonlinear Energy Harvesting Model

In the context of WPT, the EH process at each GD is a critical factor influencing the system overall performance. Most of previous works generally assume a linear relationship between the received RF power and the harvested energy, which simplifies analysis but fails to capture the actual behaviour of energy harvesters in practical applications [15–19]. Recent studies have demonstrated that the EH process exhibits nonlinear characteristics, especially under varying input power conditions [23]. Therefore, it is essential to employ a more accurate nonlinear EH model in our design.

The received RF signal at each GD is first converted into the DC signal, which is subsequently used to charge the device's battery. Let $P_{ch,k}(t)$ denote the harvested DC power at the k th GD at time t. In contrast to simplified linear models, the EH process is inherently nonlinear due to the rectification and conversion process within the EH circuit. This nonlinear relationship can be modelled as a function of the received RF power $Q_k(x(t), y(t))$. Specifically, we denote this nonlinear mapping by $\mathcal{F}_{nl}(Q_k(x(t), y(t)))$, which relates the received RF power to the harvested DC power.

The nonlinear EH process can be approximated by an implicit function derived from the physical properties of the rectifier circuit. Let $I_{\text{out}}(Q_k(t))$ represent the output current of the energy harvester, which is a function of the received RF power $Q_k(t)$. Following the work [27], this relationship is expressed as:

$$e^{\frac{R_L I_{\text{out}}(Q_k(t))}{n v_t}} (I_{\text{out}}(Q_k(t)) + I_s) \approx \sum_{j=0}^{n_o} \alpha_j Q_k^j(t), \tag{2.2}$$

where R_L represents the load resistance, n represents the ideality factor of the diode, v_t represents the thermal voltage, and I_s represents the reverse bias saturation current of the diode. The coefficients α_j are determined based on a Taylor series expansion and are positive constants that depend on the specific characteristics of the energy harvester. Then, the harvested DC power at GD k at time t can be written as:

$$P_{ch,k}(t) = \mathcal{F}_{nl}(Q_k(t)) = (I_{\text{out}}(Q_k(t)))^2 R_L, \tag{2.3}$$

where the nonlinear function $\mathcal{F}_{nl}(Q_k(t))$ captures the relationship between the received RF power and the harvested DC power. Notably, this function is convex in terms of the received power $\frac{1}{Q_k(t)}$, implying that the EH efficiency increases with higher input power but at a diminishing rate [27].

Thus, the total harvested energy at GD k over the entire charging period T is given by the integral of the harvested power over time as:

$$E_k(x(t), y(t)) = \int_0^T \mathcal{F}_{nl}(Q_k(x(t), y(t)))dt, \tag{2.4}$$

which serves as a key metric for evaluating the performance of the WPT system.

2.3.4 Problem Formulation

To ensure the fairness in the WPT network, our goal is to design the UAV trajectory for maximizing the minimum harvested energy across all GDs within the charging period T. This objective ensures that no GD is left with significantly less energy than others, balancing the energy distribution among all devices. The optimization problem can be formulated as follows:

$$\begin{aligned}\text{(OP)}: \max_{\{x(t),y(t)\}} \min_{k\in\mathcal{K}} & \int_0^T \mathcal{F}_{nl}(Q_k(x(t), y(t)))dt \\ \text{s.t.} \quad & (\dot{x}(t))^2 + (\dot{y}(t))^2 \le V^2, \forall t \in \mathcal{T}. \end{aligned} \tag{2.5}$$

Here, the UAV trajectory is subject to the maximum speed limit V. The integral term corresponds to the total harvested energy by GD k during the charging period. To simplify the optimization problem, we introduce an auxiliary variable E, , which represents the minimum energy harvested by any GD. The original problem (OP) can thus be reformulated as:

$$\begin{aligned}\text{(P1)}: \max_{\{x(t),y(t)\},E} \quad & E \\ \text{s.t.} \quad & \int_0^T \mathcal{F}_{nl}(Q_k(x(t), y(t)))dt \ge E, \forall k \in \mathcal{K}, \\ & (\dot{x}(t))^2 + (\dot{y}(t))^2 \le V^2, \forall t \in \mathcal{T}. \end{aligned} \tag{2.6}$$

Both the original problem (OP) and its reformulated version (P1) are inherently non-convex. The objective function in (OP) is non-concave, and the constraint in (P1) is non-convex due to the nonlinear EH function $\mathcal{F}_{nl}$. Additionally, both problems involve continuous-time variables $\{x(t), y(t)\}$, which introduces an infinite-dimensional optimization space, making it challenging to solve these problems optimally. Moreover, the nonlinearity in the EH function $\mathcal{F}_{nl}(Q_k(x(t), y(t)))$ adds further complexity, as it is an implicit function that depends on the received RF power at each GD. Solving this optimization problem, therefore, requires sophisticated techniques that can handle both non-convexity and the continuous nature of the UAV trajectory.

2.4 Optimal Trajectory Structure

In this section, we characterize the structure of the optimal trajectory for solving problem (OP). It is worth highlighting that the trajectory structure outlined in [16] can be seen as a specific case of the approach we present here, particularly when considering linear EH in a 1D scenario. To thoroughly understand the optimal trajectory of problem (P1), we will discuss three fundamental features that describe its structure.

First, consider a trajectory $x(t), y(t)$ that satisfies the maximum speed constraint V and is executed over a time duration T with a path or arc length d_p. It is always possible to find two separate trajectories, denoted by $\overline{x}(t), \overline{y}(t)$ and $\hat{x}(t), \hat{y}(t)$, that together achieve the same WPT performance as the original trajectory $x(t), y(t)$. Specifically, the trajectory $\overline{x}(t), \overline{y}(t)$ represents the UAV flying along the original path at maximum speed V during a reduced time $\overline{T} = d_p/V$, while $\hat{x}(t), \hat{y}(t)$ is defined within the original path and operates for the remaining duration $\hat{T} = T - \overline{T}$, without any speed constraint (i.e., speed-free). Essentially, for any GD $k \in \mathcal{K}$, the following equality holds:

$$\int_0^T \mathcal{F}_{nl}(Q_k(x(t), y(t)))dt = \int_0^{\overline{T}} \mathcal{F}_{nl}(Q_k(\overline{x}(t), \overline{y}(t)))dt + \int_0^{\hat{T}} \mathcal{F}_{nl}(Q_k(\hat{x}(t), \hat{y}(t)))dt. \tag{2.7}$$

The decomposition of the trajectory into $\{\overline{x}(t), \overline{y}(t)\}$ and $\{\hat{x}(t), \hat{y}(t)\}$ ensures that the UAV can achieve the same energy transfer performance as the original trajectory, but with more efficient use of time. The UAV first flies at maximum speed along the given path, followed by a speed-free hovering phase to spend the remaining time in the most energy-efficient positions.

For any optimal trajectory $\{x^{(*)}(t), y^{(*)}(t)\}$ of problem (P1), Feature 1 implies that it can be divided into two parts: the maximum-speed segment $\{\overline{x}(t), \overline{y}(t)\}$ and the speed-free segment $\{\hat{x}(t), \hat{y}(t)\}$. During the maximum-speed flight phase, the UAV follows the path of $\{x^{(*)}(t), y^{(*)}(t)\}$ with speed V, and the harvested energy by GD $k \in \mathcal{K}$ can be expressed as:

$$\overline{E}_k^* = \int_0^{\overline{T}} \mathcal{F}_{nl}\left(Q_k\left(\overline{x^*}(t), \overline{y^*}(t)\right)\right)dt. \tag{2.8}$$

Since $\{x^{(*)}(t), y^{(*)}(t)\}$ is an optimal solution to problem (P1), the speed-free trajectory $\{\hat{x}(t), \hat{y}(t)\}$ must also be part of the optimal solution to the following problem (P2). Otherwise, a better solution could be constructed for problem (P1) using the result from the following problem (P2):

$$
\begin{aligned}
(\text{P2}): & \max_{\{\hat{x}(t),\hat{y}(t)\},E} E \\
\text{s.t.} & \int_0^{\hat{T}} \mathcal{F}_{nl}(Q_k(\hat{x}(t),\hat{y}(t)))dt + \overline{\mathrm{E}}_k^* \geq E, \forall k \in \mathcal{K}, \\
& (\hat{x}(t),\hat{y}(t)) \in \{(x(t_0),y(t_0))t_0 \in \mathcal{T}\}, \forall t \in \left[0,\hat{T}\right].
\end{aligned} \tag{2.9}
$$

Then, in the optimal trajectory structure, the second feature is that, at the optimal solution of problem (P2), the UAV hovers at several specific points and then flies between these points at maximum speed. This means that the trajectory $\{x(t), y(t)\}$ is composed of a sequence of hovering points, each associated with a corresponding hovering duration. Therefore, the optimal trajectory for problem (P1) follows SHF structure, where the UAV moves (potentially along a curve) between hovering points at maximum speed V and hovers at each point for an optimal time duration.

Furthermore, the third feature of the optimal trajectory for problem (OP) is that it contains no more than K hovering points, including both the starting and ending points, which shows that the number of hovering points is limited by the total number of GDs in the system. This constraint ensures that the UAV makes efficient use of its available time, focusing its energy transfer efforts on critical locations without unnecessary hovering.

It should be noted that, as a specific case of the 2D scenario, the above analysis also applies to the 1D scenario. In this case, the maximum number of hovering points previously indicated in [16] as $K+2$ is actually an overestimation and should be updated to K. Additionally, the proposed SHF structure is applicable in a three-dimensional (3D) scenario where devices are distributed in 3D space. However, in a 3D environment, further constraints and considerations are required to ensure safety, such as maintaining adequate distance between the UAV and GDs and avoiding potential obstacles when the UAV flies near GDs. To simplify the discussion, this section focuses primarily on the 2D topology, with an illustration of the UAV trajectory following the SHF structure in 2D shown in Fig. 2.1. It is important to note that the hovering points do not necessarily coincide exactly with the locations of the GDs. Given the broadcasting nature of WPT, multiple GDs may simultaneously harvest energy from the UAV at any given moment, meaning that the optimal hovering points may be located near, but not directly above, the GDs.

In summary, for both problems (OP) and (P1), an optimal trajectory exists that satisfies two key properties: (i) the trajectory adheres to an SHF structure, and (ii) the number of hovering points, including the initial and final points, does not exceed K.

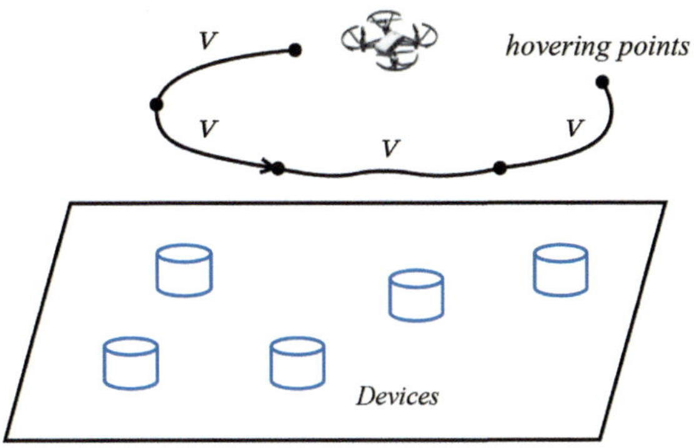

Fig. 2.1 An example of the SHF structure in a 2D topology

2.5 Proposed Trajectory Design

Based on the aforementioned trajectory structure, we reformulate the trajectory design problem and propose an efficient iterative algorithm for solving it. Unlike the 1D scenario discussed in prior work [16], where the trajectory design was relatively straightforward, the introduction of turning points is necessary in the 2D scenario. Here, turning points are distinct from hovering points; they are locations where the UAV changes its direction of flight without the need to hover.

2.5.1 *Problem Reformulation*

We redefine the trajectory following the SHF structure with up to K hovering points. These hovering points include both the starting and ending points, and their locations are represented by (x_i, y_i), where $i \in \mathcal{K}$ at a fixed altitude H. The hovering duration at each point i is denoted as t_i, and the entire set of hovering durations is represented as $\boldsymbol{t} = (t_1, t_2, \ldots, t_K)^T$, where $t_i \geq 0$ for all $i \in \mathcal{K}$ and $(\cdot)^T$ represents the transposition.

To model the UAV flight path between any two hovering points (x_i, y_i) and (x_{i+1}, y_{i+1}), the segment is divided into $N+1$ straight lines by introducing N turning points $\left(x_{i,j}, y_{i,j}\right)$, where $i \in \mathcal{K}$ and $j \in 1, \ldots, N$. These turning points allow us to approximate the curved trajectory between hovering points using a series of straight line segments, as depicted in Fig. 2.2.

The approximation becomes more accurate as the number of turning points increases. Specifically, as the number of turning points becomes sufficiently large, the piecewise linear path between two hovering points converges to the original smooth trajectory. This approach retains the generality of the UAV movement while allowing for a practical representation that can be easily managed computationally. However, there is an inherent trade-off between the number of turning points and the

Fig. 2.2 An example of SHF structure with turning points

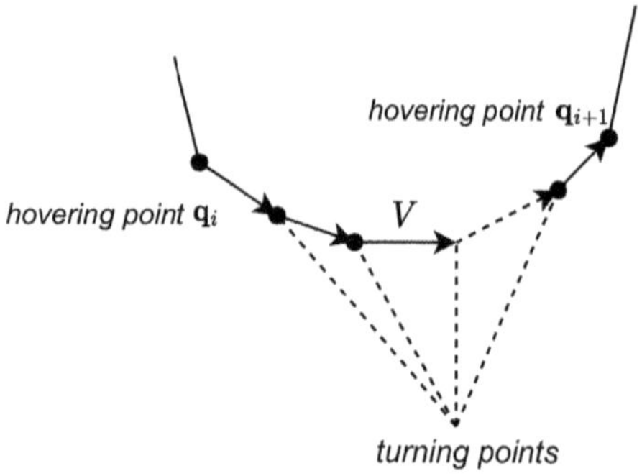

complexity of the solution. While increasing the number of turning points improves the accuracy of the trajectory approximation, it also raises the complexity of solving the optimization problem. Thus, it is crucial to strike a balance that allows for sufficient approximation accuracy without excessively increasing complexity, thereby ensuring that the trajectory solution is both computationally feasible and effective.

To simplify the representation, we define $\left(x_{i,0}, y_{i,0}\right) = (x_i, y_i)$ and $\left(x_{i,N+1}, y_{i,N+1}\right) = (x_{i+1}, y_{i+1})$ for all $i \in \{1, \ldots, K-1\}$. Thus, the set of all hovering points and turning points $(\boldsymbol{x}, \boldsymbol{y})$ can be expressed as $\left[\left(x_{1,0}, \ldots, x_{1,N}, x_{1,N+1}, x_{2,1}, \ldots, x_{K-1,N+1}\right)^T, \left(y_{1,0} \ldots, y_{1,N}, y_{1,N+1}, y_{2,1}, \ldots, y_{K-1,N+1}\right)^T\right]$. In this way, the UAV will visit all the points in $(\boldsymbol{x}, \boldsymbol{y})$ in succession, hovering at the hovering points and changing direction over the turning points. For each line segment from point $\left(x_{i,j}, y_{i,j}\right)$ to $\left(x_{i,j+1}, y_{i,j+1}\right)$, the distance $d_{i,j}$ is calculated as $d_{i,j} = \sqrt{\left(x_{i,j+1} - x_{i,j}\right)^2 + \left(y_{i,j+1} - y_{i,j}\right)^2}$. The corresponding trajectory of the UAV along the straight-line segment is given by

$$\left(\overline{x}_{i,j}(t), \overline{y}_{i,j}(t)\right) = \left(x_{i,j} + \frac{\left(x_{i,j+1} - x_{i,j}\right)Vt}{d_{i,j}}, \quad y_{i,j} + \frac{\left(y_{i,j+1} - y_{i,j}\right)Vt}{d_{i,j}}\right), t \in \left[0, \frac{d_{i,j}}{V}\right]. \tag{2.10}$$

To summarize, the overall trajectory of the UAV during the time period T can be formulated as shown in the following Eq. (2.11). It is expressed with $\delta_{i',j'}(t) = t - \sum_{i=1}^{i'} t_i - \sum_{i=1}^{i'-1} \sum_{j=0}^{N} \frac{d_{i,j}}{V} - \sum_{j=0}^{j'} \frac{d_{i',j}}{V}$.

$$
(x(t), y(t)) = \begin{cases} (x_{i'}, y_{i'}), \quad t \in \left[\sum_{i=1}^{i'-1} t_i + \sum_{i=1}^{i'-1} \sum_{j=0}^{N} \frac{d_{i,j}}{V}, \sum_{i=1}^{i'} t_i + \sum_{i=1}^{i'-1} \sum_{j=0}^{N} \frac{d_{i,j}}{V} \right), \\ \qquad 1 \le i' \le K \\ \left(\bar{x}_{i',j'}\left(\delta_{i',j'}(t)\right), \bar{y}_{i',j'}\left(\delta_{i',j'}(t)\right)\right), \\ \quad t \in \left[\sum_{i=1}^{i'} t_i + \sum_{i=1}^{i'-1} \sum_{j=0}^{N} \frac{d_{i,j}}{V} + \sum_{j=0}^{j'} \frac{d_{i',j}}{V}, \sum_{i=1}^{i'} t_i + \sum_{i=1}^{i'-1} \sum_{j=0}^{N} \frac{d_{i,j}}{V} + \sum_{j=0}^{j'+1} \frac{d_{i',j}}{V} \right) \\ \qquad 1 \le i' < K,\ 0 \le j' < N. \end{cases} \tag{2.11}
$$

It is worth noting that hovering outside the convex hull formed by the GDs does not provide any benefit for the UAV. Therefore, without loss of generality, we consider all hovering and turning points to be confined within a defined rectangular area, defined by $\min_{k \in \mathcal{K}} w_{x,k} \le \boldsymbol{x} \le \max_{k \in \mathcal{K}} w_{x,k}$ and $\min_{k \in \mathcal{K}} w_{y,k} \le \boldsymbol{y} \le \max_{k \in \mathcal{K}} w_{y,k}$.

Consequently, the energy harvested by GD k during the charging period is expressed as:

$$
\begin{aligned} E_k(\boldsymbol{x}, \boldsymbol{y}, \boldsymbol{t}) &= \int_0^T \mathcal{F}_{nl}(Q_k(x(t), y(t)))dt \\ &= \sum_{i=1}^{K} \mathcal{F}_{nl}(Q_k(x_i, y_i))t_i + \sum_{i=1}^{K-1} \sum_{j=0}^{N} \int_0^{\frac{d_{i,j}}{V}} \mathcal{F}_{nl}\left(Q_k\left(\bar{x}_{i,j}(t), \bar{y}_{i,j}(t)\right)\right)dt, \end{aligned} \tag{2.12}
$$

where the term $\int_0^{\frac{d_{i,j}}{V}} \mathcal{F}_{nl}\left(Q_k\left(\bar{x}_{i,j}(t), \bar{y}_{i,j}(t)\right)\right)dt$ represents the energy harvested by GD k as the UAV flies from $\left(x_{i,j}, y_{i,j}\right)$ to $\left(x_{i,j+1}, y_{i,j+1}\right)$ at maximum speed V.

By employing the characterized trajectory structure, we can reformulate problem (P1) into the following problem (P3):

$$
\text{(P3)}: \quad \max_{\boldsymbol{x}, \boldsymbol{y}, \boldsymbol{t}, E} E, \tag{2.13a}
$$

$$
\text{s.t.} \quad E_k(\boldsymbol{x}, \boldsymbol{y}, \boldsymbol{t}) \ge E, \forall k \in \mathcal{K}, \tag{2.13b}
$$

$$
\sum_{i=1}^{K} t_i + \frac{1}{V} \sum_{i=1}^{K-1} \sum_{j=0}^{N} d_{i,j} \le T, \tag{2.13c}
$$

$$
\min\{\boldsymbol{w}_x\} \le \boldsymbol{x} \le \max\{\boldsymbol{w}_x\}, \tag{2.13d}
$$

$$
\min\{\boldsymbol{w}_y\} \le \boldsymbol{y} \le \max\{\boldsymbol{w}_y\}, \tag{2.13e}
$$

$$t \geq 0, \tag{2.13f}$$

where the constraint (2.13c) ensures that the total time for hovering and flying is within the given charging period T.

Problem (P3) is inherently non-convex due to the non-convex constraint described in (2.13b). The harvested energy $E_k(\boldsymbol{x}, \boldsymbol{y}, \boldsymbol{t})$ for the kth GD depends on the UAV trajectory $(\boldsymbol{x}, \boldsymbol{y}, \boldsymbol{t})$, which is influenced by the nonlinear implicit function $\mathcal{F}_{nl}$. This nonlinearity significantly complicates the associated mathematical analysis. To solve problem (P3), we then introduce a convex approximation, leveraging the previously established convexity properties of the nonlinear EH model. Subsequently, an efficient iterative solution is proposed, which jointly optimizes $(\boldsymbol{x}, \boldsymbol{y}, \boldsymbol{t})$.

2.5.2 Convex Approximation

To construct a convex approximation for problem (P3), it is necessary to approximate the non-convex constraint (2.13b) with a convex function. To achieve this, we introduce a concave lower-bound function $E_k^{(r)}$ such that $E_k \geq E_k^{(r)}$, where the equality holds at a given local point $\left(\boldsymbol{x}^{(r)}, \boldsymbol{y}^{(r)}, \boldsymbol{t}^{(r)}\right)$.

For the energy E_k harvested by GD k, we provide convex approximations for the energy harvested during both the hovering period and the flying period. Specifically, the harvested energy at each hovering point i is represented by $p_{i,k}^{(r)}(\boldsymbol{x}, \boldsymbol{y}, \boldsymbol{t})$, and the harvested energy for each flying segment $d_{i,j}$ is denoted by $s_{i,j,k}^{(r)}(\boldsymbol{x}, \boldsymbol{y}, \boldsymbol{t})$, respectively.

(1) **Hovering Period:** During the hovering period, we approximate the harvested energy at each hovering point. Recall that the function $\mathcal{F}_{nl}(Q_k(x_i, y_i))$ is convex with respect to $\left(x_i - w_{x,k}\right)^2 + \left(y_i - w_{y,k}\right)^2$. Therefore, by applying the property of convex functions, we obtain

$$\mathcal{F}_{nl}(Q_k(x_i, y_i))t_i \geq -A_{i,k}^{(r)}\left[\left(x_i - w_{x,k}\right)^2 + \left(y_i - w_{y,k}\right)^2\right]t_i + B_{i,k}^{(r)}t_i. \tag{2.14}$$

where $A_{i,k}^{(r)} = -\left(\mathcal{F}_{nl}\left(Q_k\left(x_i^{(r)}, y_i^{(r)}\right)\right)\right)' > 0$, as the harvested power $\mathcal{F}_{nl}(Q_k)$ is a monotonically increasing function of Q_k, and $B_{i,k}^{(r)} = \mathcal{F}_{nl}\left(Q_k\left(x_i^{(r)}, y_i^{(r)}\right)\right) + \left[\left(x_i^{(r)} - w_{x,k}\right)^2 + \left(y_i^{(r)} - w_{y,k}\right)^2\right]A_{i,k}^{(r)}$.

Based on Eq. (2.14), we can further refine the expression for $\mathcal{F}_{nl}(Q_k(x_i, y_i))t_i$ as follows:

$$\begin{aligned}
\mathcal{F}_{nl}(Q_k(x_i, y_i))t_i &\geq -A_{i,k}^{(r)}\left[\left(x_i - w_{x,k}\right)^2 + \left(y_i - w_{y,k}\right)^2\right]t_i + B_{i,k}^{(r)}t_i \\
&= \underline{-A_{i,k}^{(r)}\left(\left(x_i - w_{x,k}\right)^2 + \left(y_i - w_{y,k}\right)^2 + C_{i,k}^{(r)}\right)\cdot\left(t_i + D_{i,k}^{(r)}t_i\right)} \\
&\quad + A_{i,k}^{(r)}C_{i,k}^{(r)}t_i + B_{i,k}^{(r)}t_i + A_{i,k}^{(r)}D_{i,k}^{(r)}\left[\left(x_i - w_{x,k}\right)^2 + \left(y_i - w_{y,k}\right)^2\right]
\end{aligned} \tag{2.15a}$$

$$\begin{aligned}
&\geq \underline{-A_{i,k}^{(r)}\cdot\frac{1}{2}\left[\left(x_i - w_{x,k}\right)^2 + \left(y_i - w_{y,k}\right)^2 + C_{i,k}^{(r)}\right]^2} \\
&\quad \underline{-A_{i,k}^{(r)}\cdot\frac{1}{2}\left(t_i + D_{i,k}^{(r)}\right)^2} + A_{i,k}^{(r)}C_{i,k}^{(r)}t_i + B_{i,k}^{(r)}t_i \\
&\quad + A_{i,k}^{(r)}D_{i,k}^{(r)}\left[\left(x_i - w_{x,k}\right)^2 + \left(y_i - w_{y,k}\right)^2\right],
\end{aligned} \tag{2.15b}$$

where the approximation from (2.15a) to (2.15b) is derived using the inequality of arithmetic and geometric means, i.e., the inequality $ab \leq \frac{1}{2}a^2 + \frac{1}{2}b^2$ holds for all $a, b \in R^+$, and equality holds when $a = b$. The non-negative constants $C_{i,k}^{(r)}$ and $D_{i,k}^{(r)}$ are defined as $C_{i,k}^{(r)} = \max\{t_i^{(r)} - \left(x_i^{(r)} - w_{x,k}\right)^2 - \left(y_i^{(r)} - w_{y,k}\right)^2, 0\}$ and $D_{i,k}^{(r)} = \left(x_i^{(r)} - w_{x,k}\right)^2 + \left(y_i^{(r)} - w_{y,k}\right)^2 + C_{i,k}^{(r)} - t_i^{(r)}$.

In addition, the function $\left(x_i - w_{x,k}\right)^2 + \left(y_i - w_{y,k}\right)^2$ is convex in x_i and y_i. Therefore, we have:

$$\begin{aligned}
\mathcal{F}_{nl}(Q_k(x_i, y_i))t_i &\geq -\frac{1}{2}A_{i,k}^{(r)}\left(\left(x_i - w_{x,k}\right)^2 + \left(y_i - w_{y,k}\right)^2 + C_{i,k}^{(r)}\right)^2 \\
&\quad - \frac{1}{2}A_{i,k}^{(r)}\left(t_i + D_{i,k}^{(r)}\right)^2 + A_{i,k}^{(r)}C_{i,k}^{(r)}t_i + B_{i,k}^{(r)}t_i \\
&\quad + A_{i,k}^{(r)}D_{i,k}^{(r)}\left[\left(x_i^{(r)} - w_{x,k}\right)\left(2x_i - x_i^{(r)} - w_{x,k}\right)\right. \\
&\qquad \left. + \left(y_i^{(r)} - w_{y,k}\right)\left(2y_i - y_i^{(r)} - w_{y,k}\right)\right] \\
&\triangleq p_{i,k}^{(r)}(\boldsymbol{x}, \boldsymbol{y}, \boldsymbol{t}),
\end{aligned} \tag{2.16}$$

where the equality holds at local points. As a polynomial, the concavity of $p_{i,k}^{(r)}(\boldsymbol{x}, \boldsymbol{y}, \boldsymbol{t})$ in $(\boldsymbol{x}, \boldsymbol{y}, \boldsymbol{t})$ can be easily shown.

(2) **Maximum-Speed Flying Period**: To approximate the energy harvested during each flying segment, we need to address the integral over the segment, which encourages a differentiation-based convex approximation approach.

We divide each flying segment into L subsegments, where L is chosen to be sufficiently large to make each subsegment small enough, allowing the UAV to be assumed static over each. Effectively, each subsegment represents a static UAV position with an associated virtual hovering duration. Considering each linear segment between points $(x_{i,j}, y_{i,j})$ and $(x_{i,j+1}, y_{i,j+1})$, the trajectory $(\overline{x}_{i,j}(t), \overline{y}_{i,j}(t))$ is represented by subdividing it into L static positions when L is sufficiently large. Specifically, we have $\overline{x}_{i,j,l} = x_{i,j} + \frac{l}{L}(x_{i,j+1} - x_{i,j})$, $\overline{y}_{i,j,l} = y_{i,j} + \frac{l}{L}(y_{i,j+1} - y_{i,j})$, $\forall l \in \{0, 1, \ldots, L-1\}$. As $d_{i,j}$ represents the segment length, each subsegment is assigned a virtual hovering duration of $\frac{d_{i,j}}{LV}$. Thus, the energy harvested by GD k during the UAV's flight over segment $(x_{i,j}, y_{i,j}) \rightarrow (x_{i,j+1}, y_{i,j+1})$ can be given by:

$$\int_0^{\frac{d_{i,j}}{V}} \mathcal{F}_{nl}\left(Q_k\left(\overline{x}_{i,j}(t), \overline{y}_{i,j}(t)\right)\right)dt = \lim_{L\to\infty} \sum_{l=0}^{L-1} \mathcal{F}_{n\ell}\left(Q_k\left(\overline{x}_{i,j,l}, \overline{y}_{i,j,l}\right)\right) \cdot \frac{d_{i,j}}{LV}, \tag{2.17}$$

where $\overline{x}_{i,j,l}$, $\overline{y}_{i,j,l}$, and $d_{i,j}$ are all functions of the variables $x_{i,j}$, $y_{i,j}$, $x_{i,j+1}$, and $y_{i,j+1}$. Next, we approximate each subsegment individually. By doing this, we can ultimately derive a concave approximation for the integrated harvested energy on each straight segment, as the sufficiently large value L approaches infinity.

Adopting the convex property of the linear process, which is used in the approximation during the hovering period, we obtain:

$$\begin{aligned} &\int_0^{\frac{d_{i,j}}{V}} \mathcal{F}_{nl}\left(Q_k\left(\overline{x}_{i,j}(t), \overline{y}_{i,j}(t)\right)\right)dt \\ &\geq \lim_{L\to\infty} \sum_{l=0}^{L-1}\left[-A_{i,j,k,l}^{(r)}\left(\left(\overline{x}_{i,j,l} - w_{x,k}\right)^2 + \left(\overline{y}_{i,j,l} - w_{y,k}\right)^2\right) + B_{i,j,k,l}^{(r)}\right] \cdot \frac{d_{i,j}}{LV} \\ &= \left(G_{i,j,k}^{(r)} - h_{i,j,k}^{(r)}(x, y, t)\right) \cdot d_{i,j}, \end{aligned} \tag{2.18}$$

where $A_{i,j,k,l}^{(r)} = -\left(\mathcal{F}_{n\ell}\left(Q_k\left(\overline{x}_{i,j,l}^{(r)}, \overline{y}_{i,j,l}^{(r)}\right)\right)\right)' > 0$, $B_{i,j,k,l}^{(r)} = \mathcal{F}_{n\ell}\left(Q_k\left(\overline{x}_{i,j,l}^{(r)}, \overline{y}_{i,j,l}^{(r)}\right)\right) + \left(\left(\overline{x}_{i,j,l}^{(r)} - w_{x,k}\right)^2 + \left(\overline{y}_{i,j,l}^{(r)} - w_{y,k}\right)^2\right) \times A_{i,j,k,l}^{(r)}$, $h_{i,j,k}^{(r)}(\boldsymbol{x}, \boldsymbol{y}, \boldsymbol{t}) = \lim_{L\to\infty} \sum_{l=0}^{L-1} \frac{1}{LV} A_{i,j,k,l}^{(r)} \cdot \left(\left(\overline{x}_{i,j,l} - w_{x,k}\right)^2 + \left(\overline{y}_{i,j,l} - w_{y,k}\right)^2 + H^2\right)$, $G_{i,j,k}^{(r)} = \lim_{L\to\infty} \sum_{l=0}^{L-1} \frac{1}{LV} B_{i,j,k,l}^{(r)}$.

Note that $h_{i,j,k}^{(r)}(\boldsymbol{x}, \boldsymbol{y}, \boldsymbol{t})$ is a positive convex function and $G_{i,j,k}^{(r)}$ is a positive constant. Although the function $h_{i,j,k}^{(r)}(\boldsymbol{x}, \boldsymbol{y}, \boldsymbol{t})$ is defined as an accumulation of infinitely many convex functions, as L approaches infinity, $h_{i,j,k}^{(r)}(\boldsymbol{x}, \boldsymbol{y}, \boldsymbol{t})$ can be equivalently expressed as a polynomial function $h_{i,j,k}^{(r)}(x_{i,j}, x_{i,j+1}, y_{i,j}, y_{i,j+1})$, where the coefficients are obtained through integration. Specifically, both $A_{i,j,k,l}^{(r)}$ and $B_{i,j,k,l}^{(r)}$ are obtained based on l/L. By introducing a variable $z = l/L$, where $z \in [0, 1]$, we can define $A_{i,j,k}^{(r)}(z) \triangleq A_{i,j,k,\,l|_{L}^{l}=z}^{(r)}$, $B_{i,j,k}^{(r)}(z) \triangleq B_{i,j,k,\,l|_{L}^{l}=z}^{(r)}$, $\overline{x}_{i,j}{}'(z) \triangleq x_{i,j,l}\Big|_{\frac{l}{L}=z} = x_{i,j} + z\left(x_{i,j+1} - x_{i,j}\right)$ and $\overline{y}_{i,j}{}'(z) \triangleq y_{i,j,l}\Big|_{\frac{l}{L}=z} = y_{i,j} + z\left(y_{i,j+1} - y_{i,j}\right)$. The factor $1/L$ can be directly considered as the differentiation of variable z. Thus, we have $h_{i,j,k}^{(r)}(x, y, t) = \int_0^1 \frac{1}{V} A_{i,j,k}^{(r)}(z)\left[\left(\overline{x}_{i,j}{}'(z) - w_{x,k}\right)^2 + \left(\overline{y}_{i,j}{}'(z) - w_{y,k}\right)^2 + H^2\right]dz$ and $G_{i,j,k}^{(r)} = \int_0^1 \frac{1}{V} B_{i,j,k}^{(r)}(z)dz$.

Since the segment length $d_{i,j}$ is a function of $(\boldsymbol{x}, \boldsymbol{y})$, i.e., $d_{i,j} = d_{i,j}(\boldsymbol{x}, \boldsymbol{y})$, a further approximation is necessary. According to the inequality of arithmetic and geometric means, for all $a, b \in R^+$ and $F > 0$, we have $ab \le \frac{1}{F}\left(\frac{1}{2}F^2a^2 + \frac{1}{2}b^2\right)$, where equality holds when $Fa = b$. Using this property, for $F_{i,j,k}^{(r)} = \frac{d_{i,j}}{h_{i,j,k}^{(r)}\left(\boldsymbol{x}^{(r)}, \boldsymbol{y}^{(r)}, \boldsymbol{t}^{(r)}\right)}$, we can obtain

$$\int_0^{\frac{d_{i,j}}{V}} \mathcal{F}_{nl} Q_k\left(\left(\overline{x}_{i,j}(t), \overline{y}_{i,j}(t)\right)\right)dt \ge -\frac{F_{i,j,k}^{(r)}}{2}\left(h_{i,j,k}^{(r)}(x, y, \mathrm{t})\right)^2 - \frac{1}{2F_{i,j,k}^{(r)}} d_{i,j}^2 + G_{i,j,k}^{(r)} d_{i,j}. \tag{2.19}$$

Note that $d_{i,j} = \sqrt{\left(x_{i,j+1}^{(r)} - x_{i,j}^{(r)}\right)^2 + \left(y_{i,j+1}^{(r)} - y_{i,j}^{(r)}\right)^2} > 0$, which ensures $F_{i,j,k}^{(r)} > 0$. Otherwise, $\left(x_{i,j}^{(r)}, y_{i,j}^{(r)}\right)$ and $\left(x_{i,j+1}^{(r)}, y_{i,j+1}^{(r)}\right)$ are the same point, and thus can be treated as a single point. Since $d_{i,j} = d_{i,j}(\boldsymbol{x}, \boldsymbol{y})$ is a convex function, we have:

$$\begin{aligned}
&\int_0^{\frac{d_{i,j}}{V}} \mathcal{F}_{nl}\left(Q_k\left(\overline{x}_{i,j}(t), \overline{y}_{i,j}(t)\right)\right)dt \\
&\ge -\frac{F_{i,j,k}^{(r)}}{2}\left(h_{i,j,k}^{(r)}(\boldsymbol{x}, \boldsymbol{y}, \boldsymbol{t})\right)^2 - \frac{d_{i,j}^2}{2F_{i,j,k}^{(r)}} \\
&+\frac{G_{i,j,k}^{(r)}}{d_{i,j}^{(r)}}\left[\left(x_{i,j+1}^{(r)} - x_{i,j}^{(r)}\right)\left(x_{i,j+1} - x_{i,j}\right) + \left(y_{i,j+1}^{(r)} - y_{i,j}^{(r)}\right)\left(y_{i,j+1} - y_{i,j}\right)\right] \\
&\triangleq s_{i,j,k}^{(r)}(\boldsymbol{x}, \boldsymbol{y}, \boldsymbol{t}).
\end{aligned} \tag{2.20}$$

Since both $\left(h_{i,j,k}^{(r)}(\boldsymbol{x}, \boldsymbol{y}, \boldsymbol{t})\right)^2$ and $d_{i,j}^2$ are convex in $(\boldsymbol{x}, \boldsymbol{y}, \boldsymbol{t})$, the function $s_{i,j,k}^{(r)}(\boldsymbol{x}, \boldsymbol{y}, \boldsymbol{t})$ is concave in $(\boldsymbol{x}, \boldsymbol{y}, \boldsymbol{t})$. Based on the approximations for both the hovering and flying periods, at a given point $\left(x^{(r)}, y^{(r)}, t^{(r)}\right)$, for any feasible point (x, y, t) of problem (P3), we have:

$$E_k(\boldsymbol{x}, \boldsymbol{y}, \boldsymbol{t}) \geq \sum_{i=1}^{K} p_{i,k}^{(r)}(\boldsymbol{x}, \boldsymbol{y}, \boldsymbol{t}) + \sum_{i=1}^{K-1} \sum_{j=0}^{N} s_{i,j,k}^{(r)}(\boldsymbol{x}, \boldsymbol{y}, \boldsymbol{t}) = E_k^{(r)}(\boldsymbol{x}, \boldsymbol{y}, \boldsymbol{t}). \tag{2.21}$$

Note that $E_k^{(r)}(\boldsymbol{x}, \boldsymbol{y}, \boldsymbol{t})$ is a concave function, and equality in Eq. (2.21) holds at the local point $\left(\boldsymbol{x}^{(r)}, \boldsymbol{y}^{(r)}, \boldsymbol{t}^{(r)}\right)$.

2.5.3 Proposed Iterative Solution

For the iterative process, we start with an initial feasible local point $\left(\boldsymbol{x}^{(0)}, \boldsymbol{y}^{(0)}, \boldsymbol{t}^{(0)}\right)$ and set the iteration counter $r = 0$. During each iteration r, we create a concave approximation of the harvested energy at each device using the current local point $\left(\boldsymbol{x}^{(r)}, \boldsymbol{y}^{(r)}, \boldsymbol{t}^{(r)}\right)$, as defined in Eq. (2.21). By replacing the harvested energy in problem (P3) with this approximation, we have the following convex problem:

$$(\text{P4}): \max_{\boldsymbol{x},\boldsymbol{y},\boldsymbol{t},E} E, \tag{2.22a}$$

$$\begin{aligned} \text{s.t. } & E_k^{(r)}(\boldsymbol{x}, \boldsymbol{y}, \boldsymbol{t}) \geq E, \forall k \in \mathcal{K}, \\ & (2.13\text{c}),\ (2.13\text{d}),\ (2.13\text{e}),\ (2.13f) \end{aligned} \tag{2.22b}$$

The introduction of the concave approximation $E_k^{(r)}(\boldsymbol{x}, \boldsymbol{y}, \boldsymbol{t})$ converts the original non-convex constraint (2.13b) into a convex constraint (2.22b). As a result, problem (P4) becomes convex and can be effectively solved using various optimization techniques. In this context, we utilize a sub-gradient method, specifically the ellipsoid method with constraints [28], to solve problem (P4).

The solution obtained from problem (P4) is then used as the updated local point for the next iteration. The relationship between the objective values of two consecutive iterations can be expressed as follows:

$$\begin{aligned} \min_k\left\{E_k\left(\boldsymbol{x}^{(r)}, \boldsymbol{y}^{(r)}, \boldsymbol{t}^{(r)}\right)\right\} &= \min_k\left\{E_k^{(r)}\left(\boldsymbol{x}^{(r)}, \boldsymbol{y}^{(r)}, \boldsymbol{t}^{(r)}\right)\right\} \\ &\leq \min_k\left\{E_k^{(r)}\left(\boldsymbol{x}^{(r+1)}, \boldsymbol{y}^{(r+1)}, \boldsymbol{t}^{(r+1)}\right)\right\} \\ &\leq \min_k\left\{E_k\left(\boldsymbol{x}^{(r+1)}, \boldsymbol{y}^{(r+1)}, \boldsymbol{t}^{(r+1)}\right)\right\}. \end{aligned} \tag{2.23}$$

In other words, the objective function value improves after each iteration, leading to convergence of the algorithm.

2.6 Numerical Simulation

In this section, we numerically evaluate the performance of the proposed iterative algorithm in terms of the minimum average harvested power among the ground devices (GDs). The evaluation is conducted within a square charging area with a width of 50 m, where the GDs are randomly distributed. Unless otherwise specified, the following simulation parameters are used: $K = 5$, $N = 1$, $\beta_0 = -30$ dB, $H = 10$ m, $V = 10$ m/s, $P = 40$ dBm, and $T = 50$ s. The results are averaged over 40 different random topologies. To assess the performance of the proposed algorithm, we compare it with the following benchmark methods:

- Upper bound: the upper bound for each topology is obtained by ignoring the UAV maximum speed constraint, representing the optimal performance achievable without speed limits. A small performance gap between the proposed trajectory design (with speed constraints) and this upper bound, especially as UAV speed increases, indicates high design efficiency.
- Linear SHF with turning points: this scheme optimizes the minimum harvested power among devices while assuming a linear approximation of the charging process. The optimization procedure is similar to our proposed strategy, and the average harvested power is then calculated based on the derived trajectory using the accurate nonlinear EH process. The performance difference between this scheme and the proposed design highlights the benefits of incorporating the nonlinear EH model.
- Nonlinear successive convex programming (SCP) based on SHF: to demonstrate the benefit of incorporating turning points into the SHF trajectory design, we also include a benchmark SCP-based solution. SCP, a popular method in trajectory optimization, approximates the trajectory by dividing it into numerous points based on time intervals and then optimizes their positions. In this scheme, we apply SCP using the nonlinear EH model, with an initialized SHF trajectory, and evaluate the resulting trajectory by assuming constant speed between consecutive points.
- Naive SHF: this scheme represents the initial trajectory, providing a simple and intuitive design without further optimization.

We start with Fig. 2.3 to illustrates the minimum average harvested power of the proposed design in comparison with various benchmarks over a wide range of charging durations. The proposed SHF consistently outperforms Linear SHF, Naive SHF, and Nonlinear SCP for all charging durations and UAV speed settings. A noticeable performance improvement is achieved by considering the nonlinear EH process, as indicated by the difference between the proposed SHF and Linear SHF results. This validates the necessity of including nonlinear EH in the design. As the charging duration increases, the proposed SHF gradually converges to the upper bound, whereas Linear SHF and Naive SHF converge to significantly lower levels. Interestingly, Naive SHF surpasses Linear SHF under higher speed limits and longer durations,

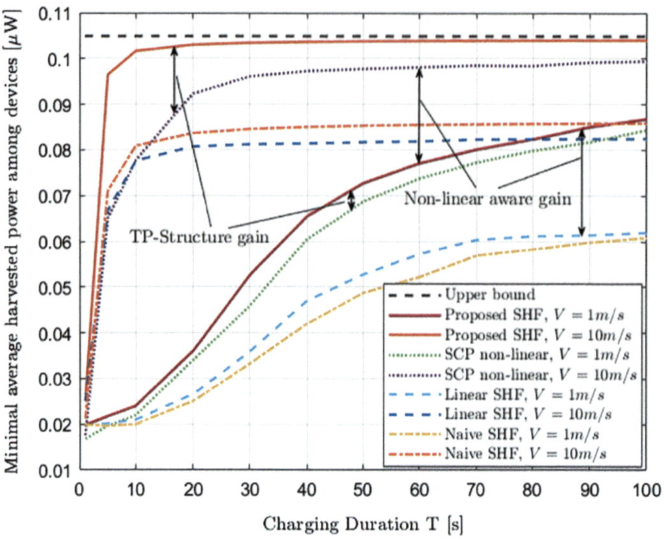

Fig. 2.3 Average performance comparison with varying charging duration

highlighting the shortcomings of the linear EH assumption. Moreover, the performance gap observed between Nonlinear SCP and the proposed SHF demonstrates the advantage of incorporating turning points in the trajectory design, which helps achieve a better balance between trajectory resolution and computational efficiency. While both the proposed SHF with turning points and the SCP approach are promising ways to approximate the optimal solution, the proposed SHF offers a significantly reduced computational cost. Indeed, simulations have shown that the proposed SHF achieves approximately 30 times greater efficiency than the Nonlinear SCP. Detailed discussions of this computational complexity can be found in our previous work [16], highlighting the significant drawbacks of SCP in terms of computational demands.

To provide further insights into the behaviour of the proposed design, Fig. 2.4 compares the trajectory solutions of the proposed design and other benchmarks under four scenarios with the same ground device (GD) topology but different maximum UAV speeds. In Fig. 2.4a and b, representing low UAV speed limits of $V = 1$ m/s and $V = 2$ m/s respectively, the Naive SHF trajectory appears as a scaled path connecting the device locations, which leads to poor performance under strict speed and charging duration constraints, as also indicated in Fig. 2.3. With higher speed limits, as shown in Fig. 2.4c and d, the proposed SHF design and the upper bound case both exhibit hovering points positioned closer to the GD locations, which helps maximize energy transfer efficiency. In contrast, Linear SHF trajectories do not follow this strategy, resulting in relatively lower performance. The results indicate that considering the non-linear EH model generally leads to UAV trajectories that hover closer to the GD locations, maximizing harvested energy. Although the trajectories of different schemes might look similar when the UAV speed limit is high, subtle differences in

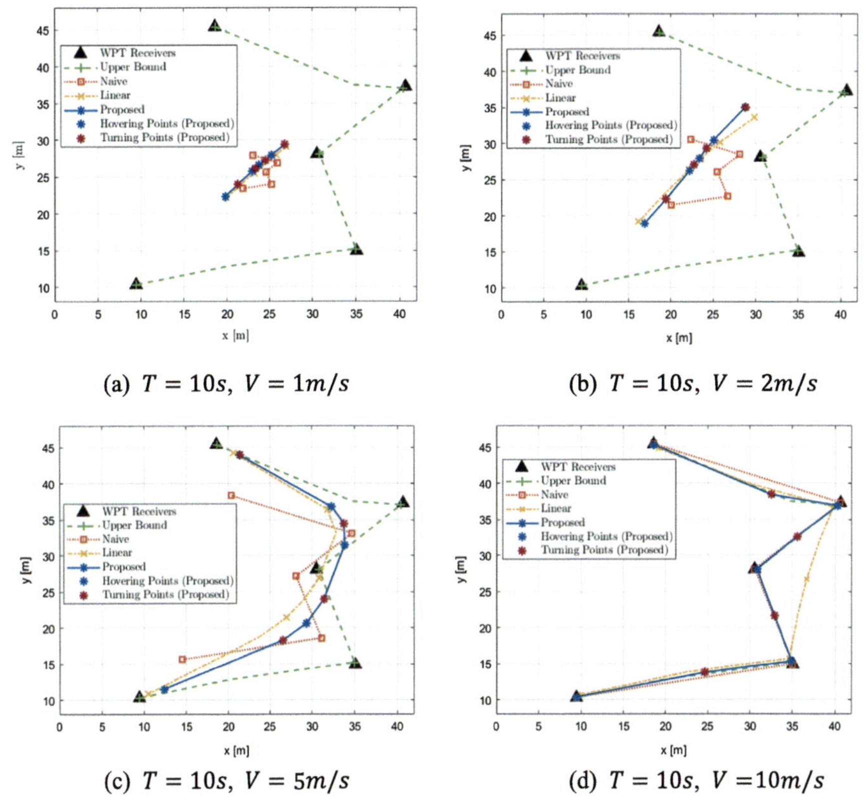

(a) $T = 10s,\ V = 1m/s$

(b) $T = 10s,\ V = 2m/s$

(c) $T = 10s,\ V = 5m/s$

(d) $T = 10s,\ V = 10m/s$

Fig. 2.4 Trajectory examples of the proposed design in comparison to other benchmarks

placement of hovering points and allocation of hovering durations lead to notable performance gaps in terms of harvested energy, as illustrated in Fig. 2.3.

Moreover, we study the average WPT performance of the proposed design over 40 different network topologies. The influence of UAV's transmit power on the minimal average harvested power among GDs is shown in Fig. 2.5. It is clear that the performance gap between the proposed SHF design and the upper bound is not consistent. This variability arises because the upper bound ignores UAV speed constraints, allowing it to instantaneously move from one point to another. This is the reason why it can be only treated as a benchmark performance bound but not the objective of the design. In Fig. 2.5a with a maximum speed of 1 m/s, this performance gap expands as the transmit power increases, reflecting the effect of speed limitations. Conversely, in Fig. 2.5b, with a higher speed of 10 m/s, the gap becomes negligible, indicating that at higher speeds, the proposed design closely approaches the upper bound. Furthermore, the proposed SHF consistently outperforms the Linear SHF and Naive SHF across all ranges, particularly when transmit power is high, demonstrating the effectiveness of considering nonlinear EH model in the design.

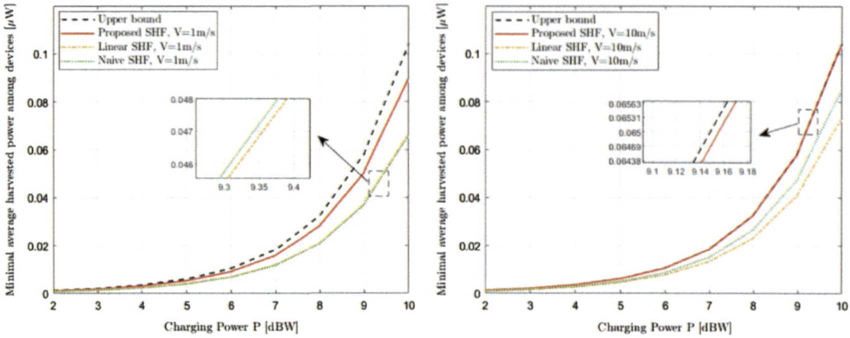

(a) $T = 40s,\ V = 1m/s$ (b) $T = 40s,\ V = 10m/s$

Fig. 2.5 Average performance comparison with varying charging power P

Next, the influence of the number of GDs on the minimal average harvested power is investigated. The results are provided in Fig. 2.6. The results at each number of GDs are also averaged over 40 different devices topologies in the following way. Starting from $K = 3$, we randomly generate the positions of three devices for each topology, and progressively add one more device at each subsequent step while retaining the earlier topologies. This ensures that we eliminate the randomness of device density and focus on evaluating performance as the number of devices increases. It is observed that the WPT performance consistently decreases as more devices are added, due to the increased challenge of maintaining efficient energy transfer across all receivers. The proposed SHF design remains close to the upper bound across all tested scenarios, indicating the robustness of our approach regardless of the number of GDs.

Furthermore, we compare the proposed design against various benchmarks regarding the harvested power at each device, as illustrated in Fig. 2.7. The results are based on a randomly generated topology with five devices. It can be observed that the harvested power across all devices is nearly equal for both the upper bound and proposed SHF solutions, which validates the efficiency of the proposed method in achieving a balanced energy distribution. The Naive SHF, on the other hand, does not prioritize fairness, resulting in poor minimal harvested power. While the Linear SHF takes fairness into account, the oversimplification of using a linear EH model causes even worse performance compared to the Naive SHF. Additionally, we also examined the corresponding UAV trajectories (not depicted here). Consistent with the observations from Fig. 2.4d, for larger values of T and V, the flight paths of different schemes appear similar. However, the hovering durations vary, which is a critical factor contributing to the performance difference. As a result, both the upper bound and proposed SHF outperform the Linear and Naive SHF schemes.

Finally, to evaluate the effect of the number of turning points, we first visualize the trajectory changes with varying turning points in Fig. 2.8. As seen in Fig. 2.8b, adding more turning points smooths the path, particularly at sharp corners. However,

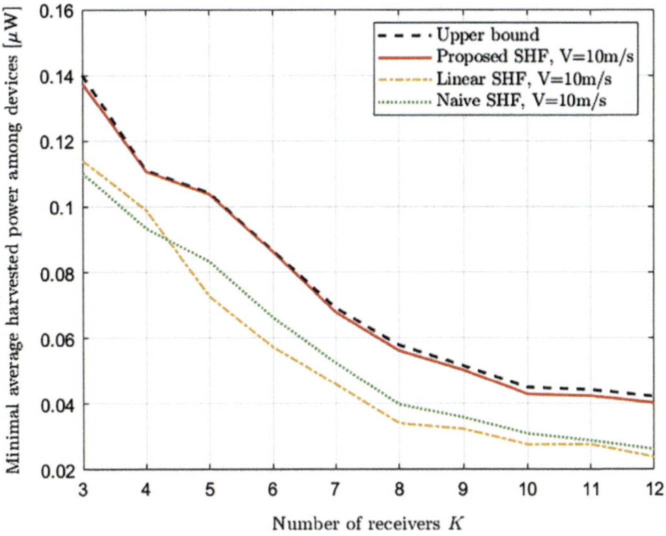

Fig. 2.6 Average performance comparison with varying number of GDs K

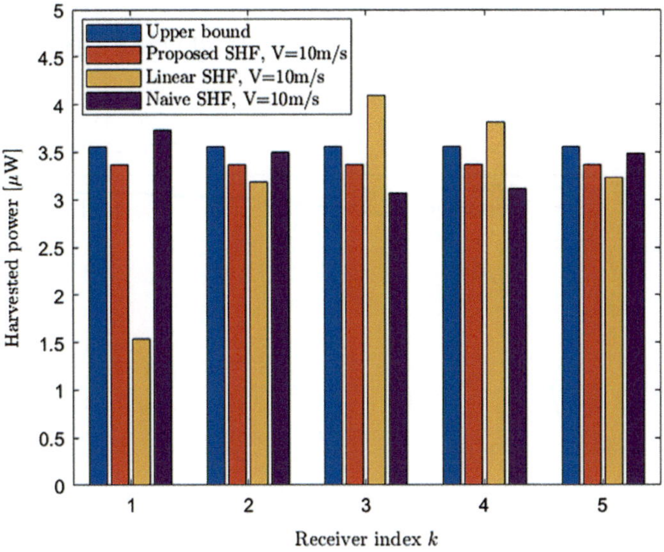

Fig. 2.7 An example on harvested power of each GD

after reaching more than one turning point, further changes become less noticeable. Figure 2.9 presents the impact of turning points on both harvested power and computing time. As turning points increase, performance improves but with diminishing returns, eventually stabilizing. Meanwhile, the computation time rises significantly, highlighting the trade-off between trajectory smoothness and computational complexity. Considering both figures, using a moderate number of turning points (e.g., 3–5) seems optimal, balancing performance and computational costs effectively.

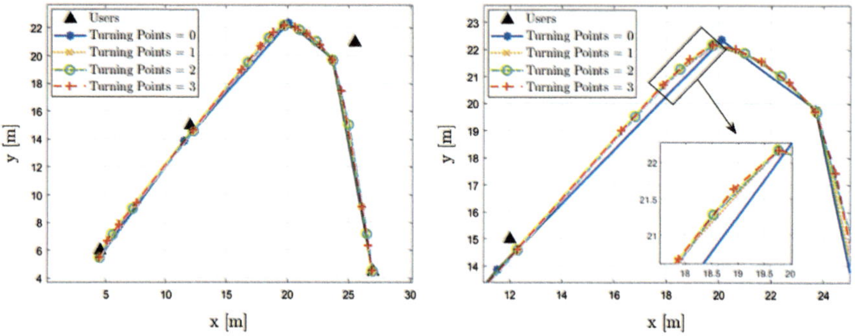

(a) Full trajectories (b) Local enlargement

Fig. 2.8 Trajectories with various number of turning points

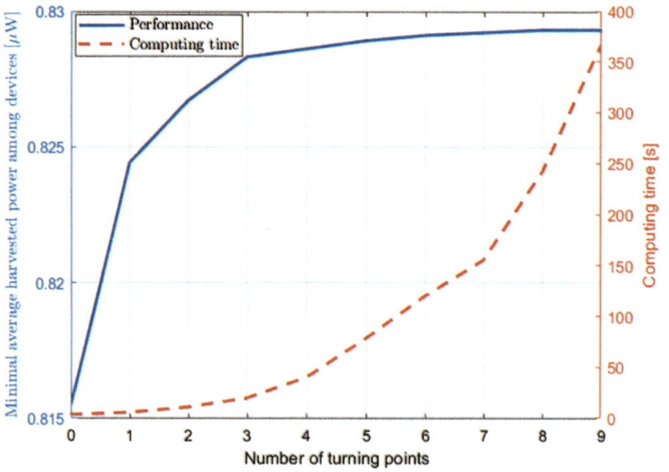

Fig. 2.9 Impact of turning points on the performance and computing time

2.7 Conclusion

In this chapter, we studied a UAV-enabled WPT network, taking into account the nonlinear characteristics of the EH process. We focused on the design of the UAV trajectory that maximize the minimum energy harvested by GDs, under the constraint of the maximum flight speed. We began by characterizing the optimal UAV trajectory structure in a 2D environment where GDs are distributed within a specified area. This led to the establishment of the SHF structure as the optimal solution. To further refine this approach, we introduced turning points to approximate potential curved paths between hovering locations. This modification transformed the original problem, which inherently involved infinite variables, into one with limited variables, making it more computationally tractable. The introduction of turning points also enhanced the flexibility of the SHF structure, enabling its application in a wider range of optimization problems and scenarios. Building on this, we utilized the convex properties of the nonlinear EH model to formulate a convex approximation of the problem, following the SHF trajectory with turning points. Then, we proposed an iterative algorithm to solve the trajectory optimization efficiently. Through simulations, we demonstrated that the proposed design achieves near-optimal performance, especially as the charging duration or maximum UAV speed increases, which highlights the efficiency of the trajectory design. Additionally, our findings showed that increasing the transmit power and reducing the number of energy receivers can significantly improve the overall harvesting performance. Moreover, when the nonlinear EH process is ignored, the linear SHF design fails to maximize the harvested energy effectively, underscoring the necessity of considering nonlinearity in the EH process.

Notably, the concept of turning points, along with the established optimal SHF structure, can potentially be generalized to other UAV trajectory design challenges, such as those in UAV-assisted wireless communications, mobile edge computing, and sensing, to better support the modelling of curved trajectories. Furthermore, during the approximation of harvested energy in the flying period, we employed a differentiation approach to approximate an integral function, as described in the derivations from Eqs. (2.17) to (2.20). This approach proved to be highly effective in our design and holds considerable potential for addressing other trajectory design challenges. Additionally, the methodology we developed for managing the nonlinear EH process can be adapted to create new designs for WPT networks, further expanding its applicability.

References

1. Mozaffari, M., Saad, W., Bennis, M., Nam, Y.-H., & Debbah, M. (2019). A tutorial on UAVs for wireless networks: Applications, challenges, and open problems. *IEEE Communications Surveys & Tutorials, 21*, 2334–2360.
2. Zeng, Y., Zhang, R., & Lim, T. J. (2016). Wireless communications with unmanned aerial vehicles: Opportunities and challenges. *IEEE Communications Magazine, 54*, 36–42.

3. Gupta, L., Jain, R., & Vaszkun, G. (2016). Survey of important issues in UAV communication networks. *IEEE Communications Surveys & Tutorials, 18*, 1123–1152.
4. Al-Hourani, A., Kandeepan, S., & Lardner, S. (2014). Optimal LAP altitude for maximum coverage. *IEEE Wireless Communications Letters, 3*, 569–572.
5. Lyu, J., Zeng, Y., Zhang, R., & Lim, T. J. (2017). Placement optimization of UAV-mounted mobile base stations. *IEEE Communications Letters, 21*, 604–607.
6. Fan, R., Cui, J., Jin, S., Yang, K., & An, J. (2018). Optimal node placement and resource allocation for UAV relaying network. *IEEE Wireless Communications Letters, 22*, 808–811.
7. Li, P., & Xu, J. (2020). Fundamental rate limits of UAV-enabled multiple access channel with trajectory optimization. *IEEE Transactions on Wireless Communications, 19*, 458–474.
8. Zeng, Y., Xu, J., & Zhang, R. (2019). Energy minimization for wireless communication with rotary-wing UAV. *IEEE Transactions on Wireless Communications, 18*, 2329–2345.
9. Hu, Y., Zhu, Y., Gursoy, M. C., & Schmeink, A. (2019). SWIPT-enabled relaying in IoT networks operating with finite blocklength codes. *IEEE Journal on Selected Areas in Communications, 37*, 1–16.
10. Bi, S., Ho, C. K., & Zhang, R. (2015). Wireless powered communication: Opportunities and challenges. *IEEE Communications Magazine, 53*, 117–125.
11. Lu, X., Wang, P., Niyato, D., Kim, D. I., & Han, Z. (2015). Wireless networks with RF energy harvesting: A contemporary survey. *IEEE Communications Surveys & Tutorials, 17*, 757–789.
12. Dong, Y., Hossain, M. J., & Cheng, J. (2016). Performance of wireless powered amplify and forward relaying over Nakagami-*m* fading channels with nonlinear energy harvester. *IEEE Communications Letters, 20*, 672–675.
13. Hua, M., Li, C., Huang, Y., & Yang, L. (2017). Throughput maximization for UAV-enabled wireless power transfer in relaying system. In *2017 9th International Conference on Wireless Communications and Signal Processing (WCSP)*, Nanjing (pp. 1–5). https://doi.org/ https://doi.org/10.1109/WCSP.2017.8170970
14. Liu, Y., Xiong, K., Ni, Q., Fan, P., & Letaief, K. B. (2020). UAV-assisted wireless powered cooperative mobile edge computing: Joint offloading, CPU control, and trajectory optimization. *IEEE Internet of Things Journal, 7*, 2777–2790.
15. Xu, J., Zeng, Y., & Zhang, R. (2018). UAV-enabled wireless power transfer: Trajectory design and energy optimization. *IEEE Transactions on Wireless Communications, 17*, 5092–5106.
16. Hu, Y., Yuan, X., Xu, J., & Schmeink, A. (2019). Optimal 1D trajectory design for UAV-enabled multiuser wireless power transfer. *IEEE Transactions on Communications, 67*, 5674–5688.
17. Mo, X., Huang, Y., & Xu, J. (2020). Radio-map-based robust positioning optimization for UAV-enabled wireless power transfer. *IEEE Wireless Communications Letters, 9*, 179–183.
18. Xie, L., Xu, J., & Zhang, R. (2019). Throughput maximization for UAV-enabled wireless powered communication networks. *IEEE Internet of Things Journal, 6*, 1690–1703.
19. Xie, L., Xu, J., & Zeng, Y. (2020). Common throughput maximization for UAV-enabled interference channel with wireless powered communications. *IEEE Transactions on Communications, 68*, 3197–3212.
20. Park, J., Lee, H., Eom, S., & Lee, I. (2019). UAV-aided wireless powered communication networks: Trajectory optimization and resource allocation for minimum throughput maximization. *IEEE Access, 7*, 134978–134991.
21. Wang, Y., Hua, M., Liu, Z., Zhang, D., Dai, H., & Hu, Y. (2019). Joint scheduling and trajectory design for UAV-aided wireless power transfer system. In *International Conference on 5G for Future Wireless Networks* (pp. 3–17). Springer.
22. Valenta, C. R., & Durgin, G. D. (2014). Harvesting wireless power: Survey of energy-harvester conversion efficiency in far-field, wireless power transfer systems. *IEEE Microwave Magazine, 15*, 108–120.
23. Clerckx, B., Zhang, R., Schober, R., Ng, D. W. K., Kim, D. I., & Poor, H. V. (2019). Fundamentals of wireless information and power transfer: From RF energy harvester models to signal and system designs. *IEEE Journal on Selected Areas in Communications, 37*, 4–33.
24. Boshkovska, E., Ng, D. W. K., Zlatanov, N., & Schober, R. (2015). Practical non-linear energy harvesting model and resource allocation for SWIPT systems. *IEEE Communications Letters, 19*, 2082–2085.

25. Xiong, K., Wang, B., & Liu, K. J. R. (2017). Rate-energy region of SWIPT for MIMO broadcasting under nonlinear energy harvesting model. *IEEE Transactions on Wireless Communications, 16*, 5147–5161.
26. Zhang, P., Wang, Z., Zhang, Q., Liu, Y., Wan, X., & Fan, Z. (2020). Max-min placement optimization for UAV enabled wireless powered networks with non-linear energy harvesting model. In: *2020 IEEE 5th International Conference on Cloud Computing and Big Data Analytics (ICCCBDA)*, Chengdu (pp. 436–439).
27. Hu, Y., Yuan, X., Yang, T., Clerckx, B., Schmeink, A. (2020). On the convex properties of wireless power transfer with nonlinear energy harvesting. *IEEE Transactions on Vehicular Technology, 69*, 5672–5676. http://arxiv.org/abs/1912.04785
28. Boyd, S. (2017, June 29) *Convex Optimization II, Course Notes*. http://www.stanford.edu/class/ee364b/

Chapter 3
Sustainable Wireless Sensor Networks with UAV-Enabled Wireless Power Transfer

Yulin Hu, Yuxi Huang, Qinwei He, and Xiaopeng Yuan

Abstract The development of unmanned aerial vehicle (UAV)-enabled wireless power transfer (WPT) networks has opened up new possibilities for enabling sustained power delivery to all ground devices (GDs). By leveraging the mobility and controllability of UAVs, these networks can provide continuous charging to devices in remote or difficult-to-access areas. This chapter explores a novel design for UAV-assisted WPT networks, where the UAV periodically recharges devices while returning to a fixed landing position between working cycles. We begin with the assumption that only the initial remaining network lifetime (RNL) is known, without additional updates on the network's lifespan. We then focus on a sustainable design aimed at minimizing the average UAV consumed power while ensuring the network's sustainability. Moreover, to mitigate the impact of the initial RNL, we further introduce the transition-based sustainable design. This approach incorporates a transitional phase aimed at enhancing the RNL, during which the UAV adjusts its operations in each task period. After transitions, the network shifts to a steady state and operates in a periodic manner, similar to the first design. Finally, we verify the sustainable performance and effectiveness of both designs in simulations.

Keywords Unmanned aerial vehicle · Multiple users · Network lifetime · Trajectory design · Nonlinear energy harvesting · Sustainable network

Y. Hu · Y. Huang
School of Electronic Information, Wuhan University, Wuhan, China
e-mail: yulin.hu@whu.edu.cn

Y. Huang
e-mail: hyxxii@whu.edu.cn

Q. He
Global Energy Interconnection Research Institute Europe GmbH, Berlin, Germany
e-mail: qinwei.he@geiri.eu

X. Yuan (✉)
Chair of Information Theory and Data Analytics, RWTH Aachen University, Aachen, Germany
e-mail: yuan@inda.rwth-aachen.de

Y. Hu et al. (eds.), *Reliable and Efficient Trajectory Design in UAV-Aided Wireless Networks*, https://doi.org/10.1007/978-981-96-5988-3_3

Abbreviations

DC	Direct Current
ET	Energy Transmitter
GD	Ground Device
IoT	Internet of Things
LoS	Line of Sight
PLoS	Probabilistic Line of Sight
RF	Radio-Frequency
RNL	Remaining Network Lifetime
SHF	Successive Hover and Fly
SHF-T	Successive Hover and Fly with Turning Points
UAV	Unmanned Aerial Vehicle
WPT	Wireless Power Transfer

3.1 Introduction

3.1.1 Background

The progress in the Internet of Things (IoT) is set to pave the way for numerous innovative applications across a broad spectrum of domains, from smart cities and automated factories to sophisticated transportation systems [1]. Towards this end, low-power wireless IoT networks are comprised of a set of battery-powered sensor devices, which are deployed for the purpose of monitoring and detecting specific events of significance [2–4]. Due to the limited battery capacity, the lifetimes of the sensor nodes are normally short [4, 5]. Ensuring the sustainability of IoT network powered by limited batteries is a primary design challenge. Thus, there is a pressing need for an effective energy management strategy to extend device lifetime and maintain system sustainability. Besides, as these devices are occasionally placed in risky areas for humans, e.g., tropical rainforests, steep mountains, bridge dam [6, 7], recharging their batteries manually or via wires can be difficult and inconvenient. To address above challenge, the majority of existing technologies rely on reducing the power consumption and improving the energy efficiency to improve the lifetime of battery-limited networks [7–10]. However, they cannot guarantee a sustainable network operation. Therefore, the radio-frequency (RF) transmission-based wireless power transfer (WPT) technologies [11] has emerged as a promising new solution to provide sustainable energy supply for such networks [12–14]. The main issue comes from significant RF signal loss over distance, which causes the energy transfer efficiency to decrease sharply when devices are far from the ETs. Additionally, when multiple devices are spread out, those closer to an ET tend to receive much more energy than those further away, creating a near-far fairness problem [15–17].

Unmanned Aerial Vehicle (UAV)-enabled WPT has arisen public attention for addressing key technical challenges in WPT from a fresh perspective. Thanks to its flexibility and mobility, UAVs can serve as an aerial ET that can fly to charge ground devices (GDs) of network efficiently. In particular, UAVs can not only take advantage of direct line-of-sight (LoS) connections with ground devices [18, 19] but also dynamically adjust their flight paths in real-time to minimize the transmission distance to GDs, significantly improving energy transfer efficiency while also ensuring fair energy distribution among devices. As such, deploying UAV in WPT system will bring significant performance improvements.

3.1.2 Motivation

While the majority of existing researches on UAV-enabled WPT networks assume a static working phase and concentrate on short-term planning, they do not ensure the sustainable operation of the entire WPT network over an extended period. Moreover, the battery capacity limits at all ground devices (GDs) are often overlooked, which could potentially restrict the actual energy harvesting capabilities, as UAVs might attempt to charge GDs when their batteries are already saturate. It is essential to design a sustainable UAV-enabled WPT network in long-term planning, particularly when accounting for the effects of nonlinear energy harvesting processes and battery limitations. Therefore, we propose two approaches for a systematic sustainable design, i.e., a Sustainable Design and a Transition-Based Sustainable Design.

3.1.3 Main Contributions

Main contributions to this chapter are as follows:

(1) Characterization on the Systematic Sustainable Design Problem: In this chapter, we for the first time present a systematic sustainable design for a UAV-assisted WPT network. We adopt a nonlinear energy harvesting model and a practical assumption: instead of knowing the instantaneous battery states of all GDs, we consider only the network lifetime of all GDs.
(2) Sustainable Design: In all task periods, the UAV follows the same trajectory, allowing it to sustain the battery-limited network over an extended time. We consider a joint optimization problem aiming at minimizing the average UAV consumed power by optimizing the task period length, the working phase length and the loop trajectory. We then propose an iterative algorithm to solve this problem successfully.
(3) Transition-Based Sustainable Design: To overcome the constraints of the initial RNL, we enhance the proposed Sustainable Design by incorporating a transitional strategy prior to the commencement of recurring operational phases.

Specifically, the network's lifespan is significantly bolstered during these transitional periods through the employment of an adaptive UAV trajectory. Once the network's lifespan surpasses a predefined threshold, the UAV proceeds to operate in a stable state. The effectiveness of this approach is confirmed through simulated evaluations.

3.1.4 Structural Organization

Relevant works are introduced in Sect. 3.2, including UAV-enabled WPT technology and UAV trajectory design methods. Section 3.3 shows the system model and displays the UAV trajectory structure we adopted. Section 3.4 presents the original problem statement in sustainable network, while Sects. 3.5 and 3.6 respectively shows the process of sustainable design and transitioned-based design. In addition, the numerical results of the simulation and the conclusions are provided and discussed in Sects. 3.7 and 3.8.

3.2 Related Works

3.2.1 UAV-Enabled WPT Technology

Among all WPT technologies, UAVs are seen as a promising addition to WPT systems, serving as energy providers to overcome the constraints of conventional energy sources, namely UAV-enabled WPT [20–23]. Since UAVs consistently operate at high altitudes, they are more likely to establish LoS links with ground devices and cover a wider area compared to cellular networks, resulting in a significant improvement in WPT. For instance, the authors in [1] consider a UAV-enabled WPT system and investigate the joint design aimed at maximizing the GDs' received energy by optimizing the UAV's hover position and charging duration. By combining the UAV trajectory design and resource allocation, the authors in [20] aim to maximize the uplink throughput among all ground users in WPT networks. Furthermore, the authors in [24] consider a different optimization objective in a UAV-enabled WPT network, aiming to minimize the total energy consumption of the UAV while meeting the minimum data transmission requests of the users. The authors in [25] consider a probabilistic line-of-sight (PLoS) channel model and aim to minimize the completion time of a WPT task by designing the UAV trajectory. Besides, the author in [26] study a UAV-enabled WPT and relay network, where the WPT process is divided into two stages. They propose an efficient algorithm to successfully solve the design with respect to the time durations of the two stages and the UAV's transmit powers and trajectory, respectively.

The studies mentioned above have enhanced the energy harvesting performance in IoT networks with limited battery lifetimes, demonstrating that UAV-enabled WPT technology can emerge as a viable solution to address the key technical challenges in WPT from a fundamentally new perspective.

3.2.2 UAV Trajectory Design

To leverage the mobility of UAVs, it is essential to design their trajectories effectively to achieve significant performance improvements. However, since UAV trajectories involve an infinite number of variables across infinite time slots, optimizing them presents a considerable challenge [27]. Therefore, many studies focus on designing UAV trajectories, particularly in UAV-enabled WPT networks. For instance, the authors in [21] investigate a max-min problem related to the received energy of all GDs by optimizing the UAV's hover positions and hover durations. The researchers in [22] introduced a suboptimal trajectory strategy known as successive-hover-and-fly (SHF). This method divides UAV trajectory into two parts, including the UAV hovering at various locations sequentially and then moving at top speed along the straight line between these locations. Building on our previous research, as documented in Refs. [23, 28], we pioneered the analysis of UAV flight trajectory in both one-dimensional and two-dimensional settings, revealing that it adheres to the SHF pattern. Significantly, by discretizing the path or arc between hover points within the SHF approach using turning points—where the UAV changes direction in flight without stopping—it was demonstrated in [28] that the SHF method, when enhanced with these turning points (referred to as SHF-T), closely approximates the optimal trajectory as the count of turning points grows.

Although these efforts have contributed a lot to obtain the efficient UAV trajectory, there is still room for improvement in practical applications focusing on long-term design. This long-term design includes a sufficiently large number of task periods and in each period, the UAV performs the same trajectory for charging and switches to sleep at the certain position. Hence, we explore a sustaining design in UAV-assisted WPT network, which is optimized over an extended operational period. We employ the optimal SHF-T trajectory to bolster this design, aiming to enhance its efficiency and effectiveness.

3.3 System Model

3.3.1 IoT Network Description

In the considered IoT network, a set of K IoT GDs are randomly fixed in ground and the corresponding locations are denoted by $(s_{k,x}, s_{k,y})$, $k \in \mathcal{K} \triangleq \{1, ..., K\}$, while the UAV flies at a fixed height $H > 0$ and charge the GDs. The UAV's horizontal position at time $t \geq 0$ is represented by $(x(t), y(t))$. Additionally, the speed of the UAV at any given time t is restricted by a maximum speed $V_{\max}$, i.e., $||(\dot{x}(t), \dot{y}(t))||_2 \leq V_{\max}, \forall t \geq 0$.

To maintain the entire IoT network, a UAV-assisted WPT is investigated for charging the GDs. It is presumed that the UAV is equipped with an omni-directional antenna, meaning that all GDs can simultaneously harvest energy. The distance between the UAV and GD k at time point t is given by $\sqrt{(x(t) - s_{k,x})^2 + (y(t) - s_{k,y})^2 + H^2}$. The landing position of UAV is denoted as (x_0, y_0).

Specifically, to sustain the network, we assume the UAV operates in two modes: a working mode and a sleeping mode. Therefore, the UAV executes the same task in each task period of sufficiently long time. Each task period T_{total} includes two parts: during the working duration T_{work}, the UAV flies to charge all GDs, and during the sleeping duration $T_{total} - T_{work}$, it switches to sleeping mode at the landing position to rest, as shown in Fig. 3.1. Once the UAV enters sleeping mode, its stored energy is sufficient to maintain the operation of the network.

3.3.2 Channel Model

We adopt a free-space path loss model, as the links between the UAV and GDs have a high probability of LoS, as typically adopted in literature [21, 23]. Therefore, the RF power received by GD k at time point t is given by

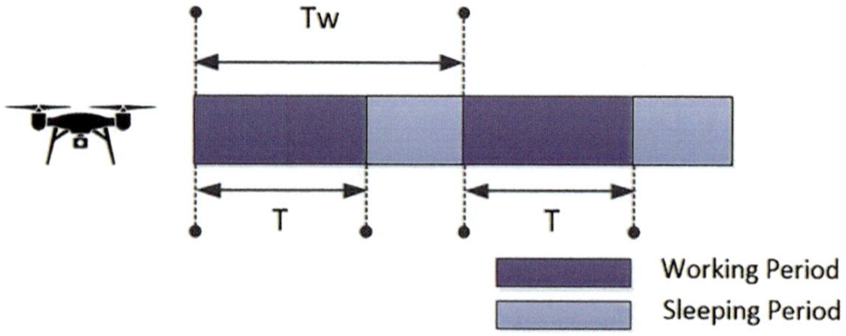

Fig. 3.1 An example of task period in UAV-assisted sustainable design

$$\Phi_{k,RF}(x(t), y(t)) = \frac{\beta_0 P}{(x(t) - s_{k,x})^2 + (y(t) - s_{k,y})^2 + H^2}, \tag{3.1}$$

where β_0 is the unit channel gain, and P represents the UAV transmission power.

3.3.3 Nonlinear Energy Harvesting

Moreover, in the course of wireless power transfer (WPT), each GD is tasked with transforming the received RF signal into a DC signal. The conversion from RF to DC is inherently a non-linear process, as detailed in Ref. [29]. To manage the non-linearity of the rectification process at the GDs, we employ a small-signal model, as outlined in [16]. This model uses a Taylor series expansion to approximate the non-linear behavior of the diode, allowing us to express the non-linear relationship between the output current I_{out} and the received RF power Φ_{RF} through an implicit function $I_{\text{out}}(\Phi_{RF})$. Drawing from Eq. (6) in [30], we have:

$$e^{\frac{R_L I_{\text{out}}(\Phi_{RF})}{n v_t}} (I_{\text{out}}(\Phi_{RF}) + I_s) \approx \sum_{j=0}^{n_p} \alpha_j \Phi_{RF}^j, \tag{3.2}$$

where R_L, n, v_t and I_s represent the load resistance, ideality factor, thermal voltage and the reverse bias saturation current of the diode, respectively. Note that the converted charging power P_{ch} satisfies $P_{\text{ch}} = [I_{\text{out}}(\Phi_{RF})]^2 R_L$. Therefore, for a given GD k, the relationship between the converted DC power $P_{k,ch}$ and the received RF power $\Phi_{k,RF}$ is expressed as

$$P_{k,ch} = \mathcal{F}(\Phi_{k,RF}) \triangleq [I_{\text{out}}(\Phi_{k,RF})]^2 R_L. \tag{3.3}$$

3.3.4 Remaining Network Lifetime (RNL)

In this chapter, we introduce the concept of network lifetime to assess sustainability, considering varying energy states across GDs due to differences in battery capacity, stored energy, and power consumption rates. Specifically, the power consumption rate for GD k is given by $\rho_k > 0$,while the battery capacity for GD k is given by $\chi_k > 0$. Moreover, the remaining stored energy for GD k is given by $\alpha_k(t)$, $k \in \mathcal{K}$, $t \geq 0$, which varies over time. Apparently, we have $0 \leq a_k(t) \leq \chi_k$.

We define the lifetime of GD k as the duration $\frac{\alpha_k(t)}{\rho_k}$ until it runs out of remaining energy without being charged. Because of the variations in the energy states of the GDs, they are likely to have different lifetimes. The RNL represents how long the entire network can operate without further charging, which is limited by the GD with the shortest lifetime. Specifically, the RNL after t is given by

$$\kappa(t) \triangleq \min_{k \in \mathcal{K}} \left\{ \frac{\alpha_k(t)}{\rho_k} \right\} \tag{3.4}$$

Clearly, a positive RNL indicates that all GDs have a positive remaining energy, and the network remains operational only if the RNL stays positive. Thus, to design the system for network sustainability, we adopt a more practical assumption that we use only the initial RNL $\kappa_0 = \kappa(0)$ or an estimated lower bound $\kappa_0 \leq \kappa(0)$, rather than the lifetime information from all the GDs.

3.3.5 UAV Trajectory Structure

As previously stated, the UAV performs the same looped trajectory behavior during each task period. A series of repeated task periods over an extended duration will form a long-term design, maintaining the normal operation of the network. In this subsection, we adopt the SHF-T structure for our UAV trajectory design, as it has been demonstrated to be the optimal trajectory structure in WPT-enabled networks [28], as illustrated in Fig. 3.2.

According to the SHF-T structure, the UAV trajectory behavior divides into two parts: hovering behavior at maximum hovering points K for specific positive hovering durations and flying behavior with maximum speed $V_{\max}$ between adjacent hovering points. Additionally, the curved path between each pair of neighboring hovering points consists of M turning points and the straight lines between them. We denote the horizontal positions of K hovering points by $(x_{i,0}, y_{i,0})$, $i \in \mathcal{K}$ and the corresponding hovering durations by $\boldsymbol{t} = (t_1, t_2, ..., t_K)^T$. Since the UAV trajectory $\{x(t), y(t)\}$ starts and ends at the landing position (x_0, y_0) to create a loop, we define (x_0, y_0) as an

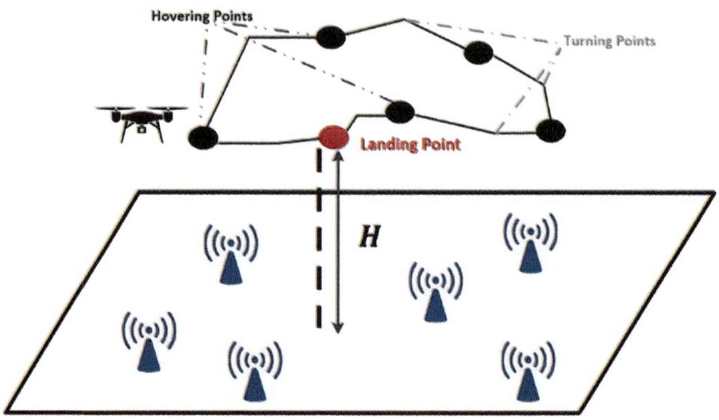

Fig. 3.2 UAV-enabled WPT network with SHF-T structure.

additional hovering point with zero hovering time duration. Besides, the starting point and ending point are denoted by $(x_{0,0}, y_{0,0}) = (x_0, y_0)$ and $(x_{K+1,0}, y_{K+1,0}) = (x_0, y_0)$, respectively. The M turning points between hovering points $(x_{i,0}, y_{i,0})$ and $(x_{i+1,0}, y_{i+1,0})(i \in \mathcal{K} \cup \{0\})$ are denoted by $(x_{i,j}, y_{i,j}), j = 1, ..., M$. For simplification, we define $(x_{i,0}, y_{i,0}) = (x_i, y_i)$ and $(x_{i,M+1}, y_{i,M+1}) = (x_{i+1}, y_{i+1}), \forall i \in \mathcal{K}$. Therefore, the set of hovering points and turning points $(\boldsymbol{x}, \boldsymbol{y})$ is expressed as

$$\Big[\big(x_{0,0}, x_{0,1}, ..., x_{0,M}, x_{0,N+1}, x_{1,1}, ..., x_{K,M+1}\big)^T, \\ \big(y_{0,0}, y_{0,1}, ..., y_{0,M}, y_{0,N+1}, y_{1,1}, ..., y_{K,M+1}\big)\Big]^T \tag{3.5}$$

Note that the landing position (x_0, y_0) has been included in $(\boldsymbol{x}, \boldsymbol{y})$ as both the starting and ending points to ensure that the sleeping mode operates normally. Then, the UAV will successively visit all the points during $(\boldsymbol{x}, \boldsymbol{y})$ while hovering at the hovering points and flying directly over the turning points. For each line segment from point $(x_{i,j}, y_{i,j})$ to $(x_{i,j+1}, y_{i,j+1})$, the distance $d_{i,j}$ is given by

$$d_{i,j} = \sqrt{(x_{i,j+1} - x_{i,j})^2 + (y_{i,j+1} - y_{i,j})^2}. \tag{3.6}$$

The corresponding trajectory on the straight-line segment is denoted by:

$$(\overline{x}_{i,j}(t), \overline{y}_{i,j}(t)) = \left(x_{i,j} + \frac{(x_{i,j+1} - x_{i,j})V_{\max}t}{d_{i,j}}, y_{i,j} + \frac{(y_{i,j+1} - y_{i,j})V_{\max}t}{d_{i,j}}\right), \\ t \in \left[0, \frac{d_{i,j}}{V_{\max}}\right]. \tag{3.7}$$

3.4 Problem Statement

3.4.1 Problem Constraints About RNL

To achieve a self-sustaining network operation, it is imperative that all GDs remain operational, meaning the RNL must remain above zero. Utilizing the trajectory formulation established for a task period, the converted energy at GD k during working phase $[0, T_{work}]$ is expressed as

$$E_k(\boldsymbol{x}, \boldsymbol{y}, \boldsymbol{t}) = \int_0^{T_{work}} P_k(\tilde{x}(t), \tilde{y}(t))dt$$
$$= \sum_{i=1}^{K} \mathcal{F}\big(\Phi_{k,RF}(x_i, y_i)\big)t_i + \sum_{i=0}^{K}\sum_{j=0}^{M} \int_0^{\frac{d_{i,j}}{V_{\max}}} \mathcal{F}\big(\Phi_{k,RF}(\overline{x}_{i,j}(t), \overline{y}_{i,j}(t))\big)dt. \tag{3.8}$$

Additionally, the integral $\int_0^{\frac{d_{i,j}}{V_{\max}}} \mathcal{F}\big(\Phi_{k,RF}(\overline{x}_{i,j}(t), \overline{y}_{i,j}(t))\big)dt$ represents the energy converted by GD k during the time the UAV flies from $(x_{i,j}, y_{i,j})$ to $(x_{i,j+1}, y_{i,j+1})$ along the straight line at the maximum speed $V_{\max}$.

Since the UAV trajectory in each task period is identical, the converted energy at GD k in any time slot with length T_{total}, i.e., the converted energy at GD k in $[t_0, t_0 + T_{total}]$, $\forall t_0 \geq 0$, is exactly equal to the converted energy $E_k(\boldsymbol{x}, \boldsymbol{y}, \boldsymbol{t})$. To ensure a long-term sustainable design, the consumed energy by all GDs should not exceed the converted energy, namely

$$E_k(\boldsymbol{x}, \boldsymbol{y}, \boldsymbol{t}) \geq \rho_k T_{total}, \forall k \in \mathcal{K}. \tag{3.9}$$

Nevertheless, while the constraint (3.9) alone cannot guarantee a positive remaining energy at all GDs, the actual remaining energy is also influenced by the battery capacity χ_k at the GDs during a task period T_{total}. Hence, we introduce an additional constraint to account for potential harvesting losses, which can be expressed as:

$$T_{total} \leq \min_{k \in \mathcal{K}} \left\{ \frac{\chi_k}{\rho_k} \right\} \triangleq \kappa_{\max}. \tag{3.10}$$

If the designed task period satisfies the conditions (3.9) and (3.10), we obtain that $\alpha_k(t_0 + T_{total}) > 0$ must hold when $\alpha_k(t_0) > 0$, $t_0 \geq 0$. By combining constraints (3.9) and (3.10), the designed task period ensures that the periodic repetitions of task periods can continuously sustain the entire network, as long as the network remains operational during the first task period, i.e., $[0, T_{total}]$.

3.4.2 Optimization Problem Objective

Next, we calculate the power consumption at the UAV based on the trajectory formulation in SHF-T structure. Specifically, we denote the power consumed during hovering, flying at maximum speed, and sleeping as P_{hov}, P_{fly} and P_{slp}, respectively. The average consumed UAV power P_{avg} during the task period is expressed as

$$P_{\text{avg}}(\boldsymbol{x}, \boldsymbol{y}, T_{work}, T_{total}) = \frac{1}{T_{total}}(P_{\text{slp}}(T_{total} - T_{work}) + (P + P_{\text{hov}})T_{work} + (P_{\text{fly}} - P_{\text{hov}})\frac{1}{V}\sum_{i=0}^{K}\sum_{j=0}^{N} d_{i,j}). \tag{3.11}$$

For simplification, we define $P_1 = P_{\text{slp}}$, $P_2 = P + P_{\text{hov}} - P_{\text{slp}} > 0$, $P_3 = P_{\text{fly}} - P_{\text{hov}} > 0$, and P_1, P_2 and P_3 are constants. Then, P_{av} can be further expressed as

$$P_{\text{avg}}(\boldsymbol{x}, \boldsymbol{y}, T_{work}, T_{total}) = P_1 + \frac{P_2 T_{work}}{T_{total}} + \frac{P_3}{VT_{total}}\sum_{i=0}^{K}\sum_{j=0}^{M} d_{i,j}. \tag{3.12}$$

3.4.3 Original Problem Construction

Our focus is on designing a system that minimizes the UAV's average power consumption by optimizing UAV trajectory which involves the decisions regarding when the UAV switches between charging and sleeping modes. Consequently, the original problem is formulated as

$$\begin{aligned}
(\text{P1}):\quad & \min_{\boldsymbol{x},\boldsymbol{y},\boldsymbol{t},T_{work},T_{total}} \quad P_{\text{avg}}(\boldsymbol{x}, \boldsymbol{y}, T_{work}, T_{total}) \\
\text{s.t.}\quad & \text{C1: } T_{work} \le \kappa_0; \\
& \text{C2: } E_k(\boldsymbol{x}, \boldsymbol{y}, \boldsymbol{t}) \ge \rho_k T_{total}, \forall k \in \mathcal{K}; \\
& \text{C3: } T_{total} \le \kappa_{\max}; \\
& \text{C4: } 0 < T_{work} \le T_{total}; \\
& \text{C5: } \sum_{i=1}^{K} t_i + \frac{1}{V_{\max}}\sum_{i=0}^{K}\sum_{j=0}^{M} d_{i,j} \le T_{work}; \\
& \text{C6: } \boldsymbol{t} \ge 0,
\end{aligned} \tag{3.13}$$

where the constraint (3.13)-C1 ensures that the working phase is short enough to allow the initial RNL to sustain the network during the working phase, therefore a positive RNL in the first task period $[0, T_{total}]$ can be guaranteed. Moreover, as discussed, constraints (3.13)-C2 and C3 ensure that the RNL remains positive after the first task period. Hence, we can successfully guarantee the network's sustainability by repeating the solution to problem (P1).

3.5 Sustainable Design

In this section, we concentrate on addressing problem (P1) by devising an iterative solution. It is important to note that (P1) is inherently non-convex due to the non-convexity of the objective function and constraint C2. To address this, we propose convex approximations for both the converted energy at GDs and the UAV average consumed power, offering an iterative approach to (P1).

Specifically, we carefully design a concave function $E_k^{(r)}(\boldsymbol{x}, \boldsymbol{y}, \boldsymbol{t})$ and a convex function $P_{\text{av}}^{(r)}(\boldsymbol{x}, \boldsymbol{y}, T, T_{\text{w}})$, ensuring the tight approximations at the local point $(\boldsymbol{x}^{(r)}, \boldsymbol{y}^{(r)}, \boldsymbol{t}^{(r)}, T_{work}^{(r)}, T_{total}^{(r)})$, where the equalities hold.

To have $E_k^{(r)}(\boldsymbol{x}, \boldsymbol{y}, \boldsymbol{t})$, we first examine the energy E_k converted at GD k, offering two convex approximations: one for the hovering phase and another for the flying phase. Hence, we introduce two concave functions $\xi_{i,k}^{(r)}(\boldsymbol{x}, \boldsymbol{y}, \boldsymbol{t})$ and $\varsigma_{i,j,k}^{(r)}(\boldsymbol{x}, \boldsymbol{y}, \boldsymbol{t})$, ensuring that:

$$\begin{cases} \mathcal{F}(\Phi_{k,RF}(x_i, y_i))t_i & \geq \xi_{i,k}^{(r)}(\boldsymbol{x}, \boldsymbol{y}, \boldsymbol{t}), \\ \int_0^{\frac{d_{i,j}}{V_{\max}}} \mathcal{F}\big(\Phi_{k,RF}(\overline{x}_{i,j}(t), \overline{y}_{i,j}(t))\big)dt & \geq \varsigma_{i,j,k}^{(r)}(\boldsymbol{x}, \boldsymbol{y}, \boldsymbol{t}), \end{cases} \tag{3.14}$$

while the equality of the above inequalities holds at the local point $(\boldsymbol{x}^{(r)}, \boldsymbol{y}^{(r)}, \boldsymbol{t}^{(r)})$.

According to [28], these two approximated functions are constructed as

$$\begin{aligned} \xi_{i,k}^{(r)}(\boldsymbol{x}, \boldsymbol{y}, \boldsymbol{t}) = & -\frac{1}{2}U_{i,k}^{(r)}((x_i - w_{x,k})^2 + (y_i - w_{y,k})^2 + W_{i,k}^{(r)})^2 \\ & - \frac{1}{2}U_{i,k}^{(r)}(t_i + X_{i,k}^{(r)})^2 + U_{i,k}^{(r)}W_{i,k}^{(r)}t_i + V_{i,k}^{(r)}t_i \\ & + U_{i,k}^{(r)}X_{i,k}^{(r)}[(x_i^{(r)} - w_{k,x})(2x_i - x_i^{(r)} - w_{k,x}) \\ & + (y_i^{(r)} - w_{k,y})(2y_i - y_i^{(r)} - w_{k,y})], \end{aligned} \tag{3.15}$$

$$\begin{aligned} \varsigma_{i,j,k}^{(r)}(\boldsymbol{x}, \boldsymbol{y}, \boldsymbol{t}) = & -\frac{A_{i,j,k}^{(r)}}{2}\left(\eta_{i,j,k}^{(r)}(\boldsymbol{x}, \boldsymbol{y}, \boldsymbol{t})\right)^2 - \frac{1}{2A_{i,j,k}^{(r)}}d_{i,j}^2 \\ & + \frac{B_{i,j,k}^{(r)}}{d_{i,j}^{(r)}}\Big[(x_{i,j+1}^{(r)} - x_{i,j}^{(r)})(x_{i,j+1} - x_{i,j}) \\ & + (y_{i,j+1}^{(r)} - y_{i,j}^{(r)})(y_{i,j+1} - y_{i,j})\Big], \end{aligned} \tag{3.16}$$

where the non-negative constants $U_{i,k}^{(r)}, V_{i,k}^{(r)}, W_{i,k}^{(r)}, X_{i,k}^{(r)}$ are defined as

$$
\begin{cases}
U_{i,k}^{(r)} = -\left(\mathcal{F}(\Phi_{k,RF}(x_i^{(r)}, y_i^{(r)}))\right)' > 0, \\
V_{i,k}^{(r)} = \mathcal{F}(\Phi_{k,RF}(x_i^{(r)}, y_i^{(r)})) + [(x_i^{(r)} - w_{k,x})^2 + (y_i^{(r)} - w_{k,y})^2]\, U_{i,k}^{(r)} > 0, \\
W_{i,k}^{(r)} = \max\{t_i^{(r)} - (x_i^{(r)} - w_{k,x})^2 - (y_i^{(r)} - w_{k,y})^2, 0\} \geq 0, \\
X_{i,k}^{(r)} = (x_i^{(r)} - w_{k,x})^2 + (y_i^{(r)} - w_{k,y})^2 + W_{i,k}^{(r)} - t_i^{(r)} \geq 0.
\end{cases}
\tag{3.17}
$$

The positive value $B_{i,j,k}^{(r)}$ and positive convex function $\eta_{i,j,k}^{(r)}(\boldsymbol{x}, \boldsymbol{y}, \boldsymbol{t})$ are constructed through some integrals, as shown in Eqs. (31), (32) in [28]. The positive constant $F_{i,j,k}^{(r)}$ is given by $A_{i,j,k}^{(r)} = \frac{d_{i,j}^{(r)}}{\eta_{i,j,k}^{(r)}(\boldsymbol{x}^{(r)}, \boldsymbol{y}^{(r)}, \boldsymbol{t}^{(r)})}$.

Therefore, we obtain the concave function $E_k^{(r)}(\boldsymbol{x}, \boldsymbol{y}, \boldsymbol{t})$ as

$$
E_k^{(r)}(\boldsymbol{x}, \boldsymbol{y}, \boldsymbol{t}) \triangleq \sum_{i=1}^{K} \xi_{i,k}^{(r)}(\boldsymbol{x}, \boldsymbol{y}, \boldsymbol{t}) + \sum_{i=0}^{K} \sum_{j=0}^{M} \varsigma_{i,j,k}^{(r)}(\boldsymbol{x}, \boldsymbol{y}, \boldsymbol{t}) \leq E_k(\boldsymbol{x}, \boldsymbol{y}, \boldsymbol{t}), \tag{3.18}
$$

where $E_k^{(r)}(\boldsymbol{x}, \boldsymbol{y}, \boldsymbol{t}) \leq E_k(\boldsymbol{x}, \boldsymbol{y}, \boldsymbol{t})$ holds for any random feasible point $(\boldsymbol{x}, \boldsymbol{y}, \boldsymbol{t}, T_{work}, T_{total})$ to problem (P1) while the equality holds at the local point $(\boldsymbol{x}^{(r)}, \boldsymbol{y}^{(r)}, \boldsymbol{t}^{(r)})$. Next, we approximate the objective of the problem (P1) at a local point $(\boldsymbol{x}^{(r)}, \boldsymbol{y}^{(r)}, \boldsymbol{t}^{(r)}, T_{work}^{(r)}, T_{total}^{(r)})$. To facilitate the derivation, we simplify two constant terms at the local point $A_1^{(r)} = \frac{1}{T_{work}^{(r)} T_{total}^{(r)}}$ and $A_2^{(r)} = \frac{1}{T_{total}^{(r)} \sum_{i=0}^{K} \sum_{j=0}^{M} d_{i,j}^{(r)}}$. Then, according to the inequality of arithmetic and geometric means, we have

$$
\begin{aligned}
P_{\mathrm{avg}}(\boldsymbol{x}, \boldsymbol{y}, T_{work}, T_{total}) &= P_1 + \frac{P_2 T_{work}}{T_{total}} + \frac{P_3}{V_{\max} T_{total}} \sum_{i=0}^{K} \sum_{j=0}^{M} d_{i,j} \\
&= P_1 + \frac{P_2}{A_1^{(r)}} \cdot \underline{\underline{A_1^{(r)} T_{work}}} \cdot \underline{\underline{\frac{1}{T_{total}}}} \\
&\quad + \frac{P_3}{A_2^{(r)} V_{\max}} \cdot \underline{\frac{1}{T_{total}}} \cdot \overline{F_2^{(r)} \sum_{i=0}^{K} \sum_{j=0}^{M} d_{i,j}} \\
&\leq P_1 + \frac{P_2}{A_1^{(r)}} \underline{\underline{\frac{1}{2} (A_1^{(r)} T_{work})^2}} + \frac{P_2}{A_1^{(r)}} \underline{\underline{\frac{1}{2} \frac{1}{(T_{total})^2}}} \\
&\quad + \frac{P_3}{A_2^{(r)} V_{\max}} \underline{\frac{1}{2} \frac{1}{(T_{total})^2}} + \frac{P_3}{A_2^{(r)} V_{\max}} \underline{\frac{1}{2} \left(A_2^{(r)} \sum_{i=0}^{K} \sum_{j=0}^{M} d_{i,j} \right)^2} \\
&\triangleq P_{\mathrm{avg}}^{(r)}(\boldsymbol{x}, \boldsymbol{y}, T_{work}, T_{total}),
\end{aligned}
\tag{3.19}
$$

where $P_{\text{avg}}^{(r)}(\boldsymbol{x}, \boldsymbol{y}, T_{work}, T_{total})$ is convex and $P_{\text{avg}}^{(r)}(\boldsymbol{x}, \boldsymbol{y}, T_{work}, T_{total}) = P_{\text{avg}}(\boldsymbol{x}, \boldsymbol{y}, T_{work}, T_{total})$ holds at the local point $(\boldsymbol{x}^{(r)}, \boldsymbol{y}^{(r)}, \boldsymbol{t}^{(r)}, T_{work}^{(r)}, T_{total}^{(r)})$.

So far, we have obtained tight convex approximations $E_k^{(r)}(\boldsymbol{x}, \boldsymbol{y}, \boldsymbol{t})$ and $P_{\text{avg}}^{(r)}(\boldsymbol{x}, \boldsymbol{y}, T_{work}, T_{total})$ to $E_k(\boldsymbol{x}, \boldsymbol{y}, \boldsymbol{t})$ and $P_{\text{avg}}(\boldsymbol{x}, \boldsymbol{y}, T_{work}, T_{total})$. Finally, problem (P1) is approximated to a convex problem

$$
\begin{aligned}
&(\text{P2}): \min_{\boldsymbol{x},\boldsymbol{y},\boldsymbol{t},T_{work},T_{total}} \quad P_{\text{avg}}^{(r)}(\boldsymbol{x}, \boldsymbol{y}, T_{work}, T_{total}) \\
&\text{s.t.} \;\; \text{C1}{:}E_k^{(r)}(\boldsymbol{x}, \boldsymbol{y}, \boldsymbol{t}) \geq \rho_k T_{total}, \forall k \in \mathcal{K}; \\
&(\text{P1})\text{-C1,C3,C4,C5,C6.}
\end{aligned}
\tag{3.20}
$$

Using the convex approximations at each local point provided in problem (P2), we can efficiently solve problem (P1) in an iterative manner.

3.6 Transition-Based Sustainable Design

In our sustainable design, the UAV's flight path, task duration, and operational phases are optimized holistically, with a consistent solution for all task periods. However, as problem (P1) indicates, the operational phase in this design is significantly limited by the initial RNL κ_0 at the onset of the first task period. For example, when the initial RNL κ_0 is low, the working phase will be significantly restricted by constraint C1 in (P1). In this case, the UAV will have short hovering durations and spend most of the working phase flying, which is likely to result in poor charging performance. Over the long term, with extended operation time and an increasing number of task periods, the proposed sustainable design could lead to lower energy efficiency for the UAV.

For long-term design purposes, to overcome the limitations imposed by the initial RNL, we introduce a transition approach in this section, resulting in an enhanced sustainable design called the Transition-Based Sustainable Design. In this design, the UAV operates in following two working states:

- **Transition State**: The state focuses on quickly improve the RNL, with the UAV using adaptive loop trajectories and no sleeping phase.
- **Steady State**: The state where UAV periodically repeats a designed task period.

According to the definition, the transition-based sustainable design allows the UAV to adjust its flight path between task periods in the transition phase, as illustrated in Fig. 3.3. This differs from the original sustainable design, where the UAV follows the same behavior across all task periods. In the transition state, the RNL improves significantly after enough task periods. Once the RNL surpasses a predefined threshold, the UAV switches to the steady state, with a periodic task period design based on the improved RNL, repeating the same behavior across task periods as indicated in Fig. 3.3. In fact, the original sustainable design can be viewed as a

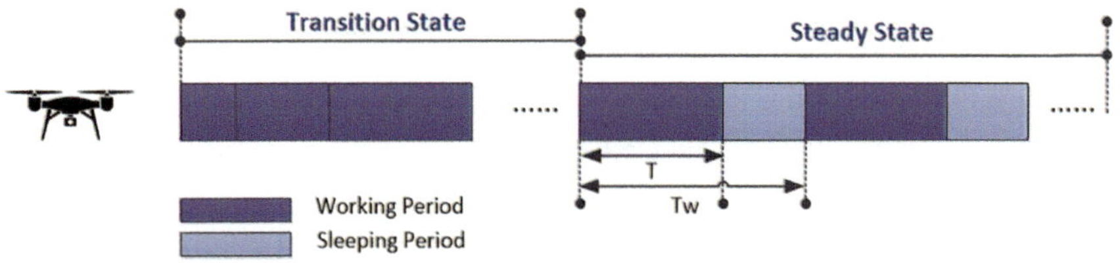

Fig. 3.3 An example for transition-based sustainable design

special case of the enhanced design without a transition state. Both designs yield the same solution when the initial RNL is sufficiently high, exceeding the designed threshold.

In the subsequent part of this section, we first determine the conditions for an energy-efficient UAV trajectory in the steady state, focusing on the RNL threshold that triggers the transition from the transition state to the steady state. Then, we present a transition design that incrementally raises the RNL lower bound through each task period until the threshold is reached. Lastly, we provide a comprehensive algorithm detailing the entire design process.

3.6.1 Steady Condition: Determining the RNL Threshold T^*

With a sufficient initial RNL, constraint C1 in (P1) can be ignored. The optimal steady state can be found by solving the following problem (P3).

$$\begin{aligned} \text{(P3)}: \quad & \min_{\boldsymbol{x},\boldsymbol{y},\boldsymbol{t},T_{work},T_{total}} \quad P_{\text{avg}}(\boldsymbol{x},\boldsymbol{y},T_{work},T_{total}) \\ \text{s.t.} \quad & \text{C1}: E_k(\boldsymbol{x},\boldsymbol{y},\boldsymbol{t}) \geq \rho_k T_{total}, \forall k \in \mathcal{K}; \\ & \text{(P1)-C3,C4,C5,C6.} \end{aligned} \tag{3.21}$$

We denote by $T^{\star}$ the charging period T_{work} in the optimal steady state from problem (P3). When the initial RNL is larger than $T^{\star}$, UAV can directly switch to the optimal steady state and repeat the designed task period to sustain the entire network. Conversely, if the RNL is lower than $T^{\star}$, a successful switch cannot be guaranteed, which may lead to high average power consumption or even failure to sustain the network. In other words, $T^{\star}$ is the RNL threshold for switching to the optimal steady state.

Furthermore, with the assistance of aforementioned convex approximation (3.18) and (3.19), we can further build a convex problem (P4).

$$
\begin{aligned}
&(\text{P4}): \min_{\boldsymbol{x},\boldsymbol{y},\boldsymbol{t},T_{work},T_{total}} P_{\text{avg}}^{(r)}(\boldsymbol{x},\boldsymbol{y},T_{work},T_{total}) \\
&\text{s.t. C1:} E_k^{(r)}(\boldsymbol{x},\boldsymbol{y},\boldsymbol{t}) \geq \rho_k T_{total}, \forall k \in \mathcal{K}; \\
&\text{(P1)-C3,C4,C5,C6.}
\end{aligned} \tag{3.22}
$$

The average consumed power in problem (P3) will be sminimized by solving the convex problem (P4). By combining the feasible points from problem (P3) with the convex approximations in (P4), we can iteratively solve problem (P3) and ultimately find a suboptimal solution for problem (P4).

3.6.2 Transition Design Approaching T^*

To achieve $T^\star$, we aim to maximize the improvement of RNL by designing the loop trajectory for each transition. By establishing a lower bound for RNL after each transition and working to improve this lower bound, we can gradually approach the RNL threshold $T^\star$.

We denote by $\alpha_{k,b}$ and κ_b the remaining stored energy at GD k and the lower bound of RNL after bth transition. Clearly, the lower bound of RNL before transition is known as κ_0 and we have $\kappa_b \leq \min_{k\in\mathcal{K}}\{\frac{\alpha_{k,b}}{\rho_k}\}, \forall b \geq 1$. Recall that the sleeping phase is set to zero in the transition state, meaning $T_{\text{work}} = T_{total}$. The improvement in the remaining lifetime at GD k by the bth transition ($b \geq 1$) is equal to $\frac{E_k(\boldsymbol{x},\boldsymbol{y},\boldsymbol{t})}{\rho_k} - T_{work}$, where $(\boldsymbol{x},\boldsymbol{y},\boldsymbol{t})$ is the UAV loop trajectory during this transition.

However, this improvement can only be achieved if the battery limits at GD k are not exceeded. If GD k already has a high lifetime before the transition, the lifetime improvement will be smaller. In the worst case, where GD k is saturated at the start of the charging process and harvests little energy while the UAV is away, the remaining lifetime of GD k becomes $\frac{\alpha_k}{\rho_k} - T_{work}$ after the transition. For RNL, we can obtain

$$
\begin{aligned}
&\min_{k\in\mathcal{K}}\left\{\frac{\alpha_{k,b}}{\rho_k}\right\} \geq \min_{k\in\mathcal{K}}\left\{\min\left\{\frac{\alpha_{k,b-1}}{\rho_k} + \frac{E_k(\boldsymbol{x},\boldsymbol{y},\boldsymbol{t})}{\rho_k} - T_{work}, \frac{\chi_k}{\rho_k} - T_{work}\right\}\right\} \\
&\geq \min\left\{\min_{k\in\mathcal{K}}\left\{\frac{\alpha_{k,b-1}}{\rho_k}\right\} + \min_{k\in\mathcal{K}}\left\{\frac{E_k(\boldsymbol{x},\boldsymbol{y},\boldsymbol{t})}{\rho_k} - T_{work}\right\}, \min_{k\in\mathcal{K}}\left\{\frac{\chi_k}{\rho_k}\right\} - T_{work}\right\} \\
&\geq \kappa_{b-1} + \min\{T_\Delta, \hat{T}_\Delta\},
\end{aligned} \tag{3.23}
$$

where both T_Δ and $\hat{T}_\Delta$ are functions of T_{work},

$$
T_\Delta = \min_{k\in\mathcal{K}}\left\{\frac{E_k(\boldsymbol{x},\boldsymbol{y},\boldsymbol{t})}{\rho_k}\right\} - T_{work}, \tag{3.24}
$$

$$
\hat{T}_\Delta = \min_{k\in\mathcal{K}}\left\{\frac{\chi_k}{\rho_k}\right\} - T_{work} - \kappa_{b-1} = \kappa_{\max} - T_{work} - \kappa_{b-1}. \tag{3.25}
$$

Clearly, from (3.23), we have a lower bound $\kappa_{b-1} + \min\{T_\Delta, \hat{T}_\Delta\}$ for RNL after transition b. Thus, for the loop trajectory with a working phase of T_{work}, if both T_Δ and $\hat{T}_\Delta$ are positive, the lower bound of RNL is definitely improved by $\min\{T_\Delta, \hat{T}_\Delta\}$. To enhance the RNL, we address the following optimization problem (P5) in each transition loop by maximizing the improvement of the lower bound of the RNL.

$$\begin{aligned}
\text{(P5)}: \quad & \max_{\boldsymbol{x},\boldsymbol{y},\boldsymbol{t},T_{work},T_{total}} \quad \min\{T_\Delta, \hat{T}_\Delta\} \\
\text{s.t.} \quad & \text{C1}: T_{work} = T_{total}; \\
& \text{C2}: T_{work} \le \kappa_{b-1}; \\
& \text{(P1)-C3,C4,C5,C6.}
\end{aligned} \tag{3.26}$$

Since the RNL before transition b may be unknown, we use the lower bound κ_{b-1} as the limit for T, as indicated in constraint C2 in (P5), to ensure a positive RNL during the bth transition. Due to the non-concave objective function, we can also solve problem (P5) iteratively by applying convex approximation at each local point, which uses the same approach for problem (P1). All constraints in problem (P5) are convex, so applying the convex approximation at each local point requires tightly approximating $E_k(\boldsymbol{x}, \boldsymbol{y}, \boldsymbol{t})$ by the concave function given in (3.18). Thus, problem (P5) can be efficiently solved using an iterative algorithm.

Once the solution for transition loop b is obtained, we will update problem (P5) for next loop $b+1$. Specifically, we update the lower bound of RNL by $\kappa_{b+1} = \kappa_b + \Delta\kappa$ according to (3.23), where $\Delta\kappa$ is the optimum obtained from problem (P5). The solution in transition loop b serves as the initialized feasible point for iteratively solving problem (P5) in loop $b+1$. If the solution to problem (P5) in each transition loop consistently yields a positive objective, the transition approach will ensure that the lower bound of RNL improves loop by loop. Since RNL is always above its lower bound, the improvement in RNL can be ensured.

3.6.3 Detailed Algorithm

Based on the analysis in the previous subsections, the transition-based sustainable design operates differently in the following three cases, depending on the initial RNL.

(1) **Case** $\kappa_0 \ge T^\star$: the initial RNL is high enough for the UAV to directly switch to the steady state.

(2) **Case** $\kappa_0 < T_{\min}$: the initial RNL is extremely low. Due to the location and flying speed limits, the UAV cannot sustain the network (the bottleneck GDs) during the remaining lifetime. In this case, neither the original sustainable design nor the transition-based design can ensure a sustainable network.

(3) **Case** $T_{\min} \le \kappa_0 < T^\star$: This case represents the most common situation and is the main focus of the proposed design with transitions. When the battery limits of GDs are with practical interests (not extremely low), i.e., $\kappa_{\max} > T^\star + T_{\min}$ the

improved RNL after multiple transition loops definitely will reach the threshold $T^\star$. On the other hand, when the battery limits of GDs are extremely low, i.e., $\kappa_{\max} \leq T^\star + T_{\min}$, the improved RNL after multiple transition loops will at most reach the maximum capacity $\kappa_{\max}$. However, the proposed transition-based sustainable design in both situations will always outperform the original design.

The above discussions confirm again that the transition-based sustainable design has at least an equal and more likely a higher performance than the original one.

3.7 Simulation Results

This chapter presents the simulation results for the proposed two designs focused on sustainability. Note that the original sustainable design is a special case of the transition-based design. In the simulations, we primarily focus on the transition-based design, using the original design as a benchmark. The simulation methodology is as follows: First, we analyze the RNL performance of both designs. Next, we explore the UAV trajectory difference in different states of the transition-based design. Finally, we evaluate the average UAV power consumption of both designs.

3.7.1 Remaining Network Lifetime Over Time of Different Designs

We first start with Fig. 3.4 to evaluate the RNL performance of original design over time. It can be observed that with a sufficient initial RNL, the RNL exhibits periodic fluctuations and consistently remains above zero. With the number of ground devices set to 10, we can observe their lifetimes respectively. It is important to note that the UAV follows a periodic schedule, sequentially charging the devices in the network when in task period. As a result, the lifetimes of devices 1 and 10 increase at different intervals. Throughout this process, the RNL consistently acts as the lower bound for all devices in the network. When the UAV returns and switches to sleep mode, both the individual device lifetimes and the network RNL decrease due to the lack of charging. The results confirm that our original design effectively sustains the IoT network.

Then, to investigate the impact of initial RNL, we compare the corresponding RNLs over time in the two proposed sustainable designs in Fig. 3.5. The figure shows that with a lower initial RNL, the original design causes the UAV to continuously repeat the same task period, alternating between charging and sleeping phases. This results in a relatively higher average power consumption by the UAV, measured at 39.8132 W. In contrast, the transition-based design gradually improves the RNL over successive loops. After 10 rounds of enhancements, when the RNL becomes sufficiently large, the UAV adopts a more efficient working cycle with extended

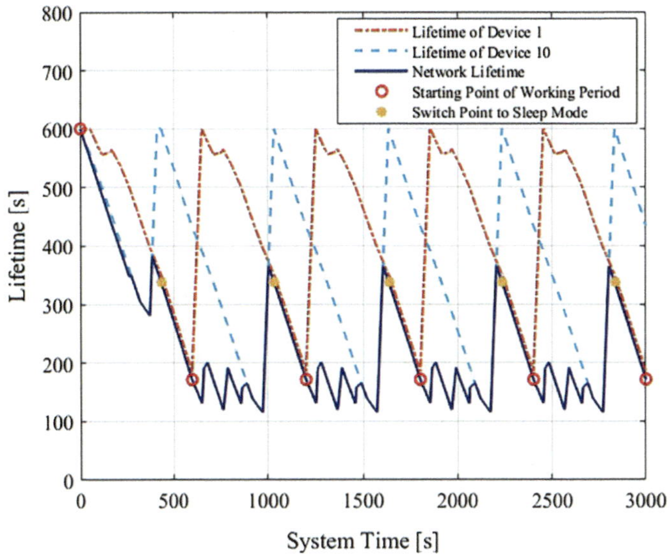

Fig. 3.4 RNL of original design over time

charging and sleeping phases, resulting in a lower average power consumption of 36.3132 W. In the figure, we also mark the threshold T^* for switching to steady state for transition-based design. The results can clearly prove the sustainability and efficiency improvements of the transition-based design, which mitigates the limit of the initial RNL compared to the original design.

Furthermore, we show the RNL behaviors under the transition-based design with varying UAV transmit power in Fig. 3.6. As the transmit power increases, the threshold for achieving steady state decreases. This indicates that the network reaches its steady state configuration more rapidly with higher transmit power. Moreover, fewer transition loops are required because increased transmit power results in higher received RF power at the ground devices (GD). While this higher transmit power reduces RF-DC conversion efficiency, it simultaneously enhances the overall efficiency of RNL improvement. Essentially, the system becomes more effective in optimizing the RNL with fewer iterations due to the higher power levels. The findings can highlight the flexibility and robustness of our analytical model in the transition-based design.

3.7.2 UAV Trajectories of Transition Based Design

We now turn our attention to the UAV trajectory behavior during the transition approach, as illustrated in Fig. 3.7. The figure presents the UAV trajectories in two different transition loops and the steady state. These three trajectories are illustrated

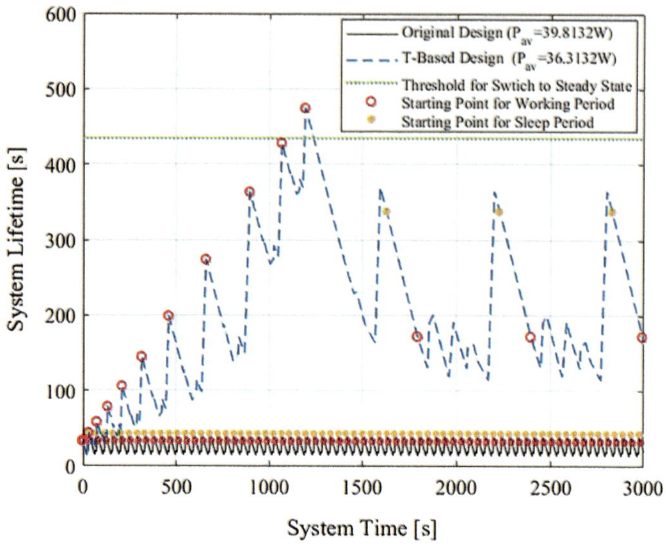

Fig. 3.5 RNL in the proposed two designs over time

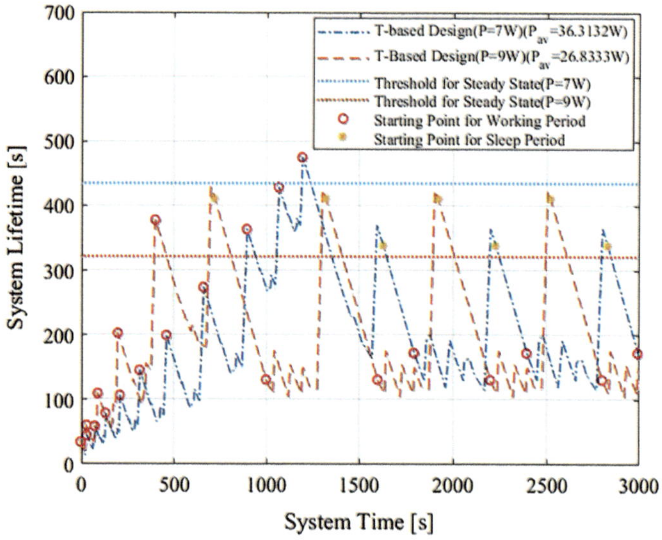

Fig. 3.6 RNL of transition-based design over time

in the first subfigure for representing the differences, at the same time the trajectories are respectively displayed in other subfigures. Clearly, the loop path and the hovering duration in three trajectories are different. This is due to the fact that the individual trajectories are designed based on different RNLs. The UAV performs continuous

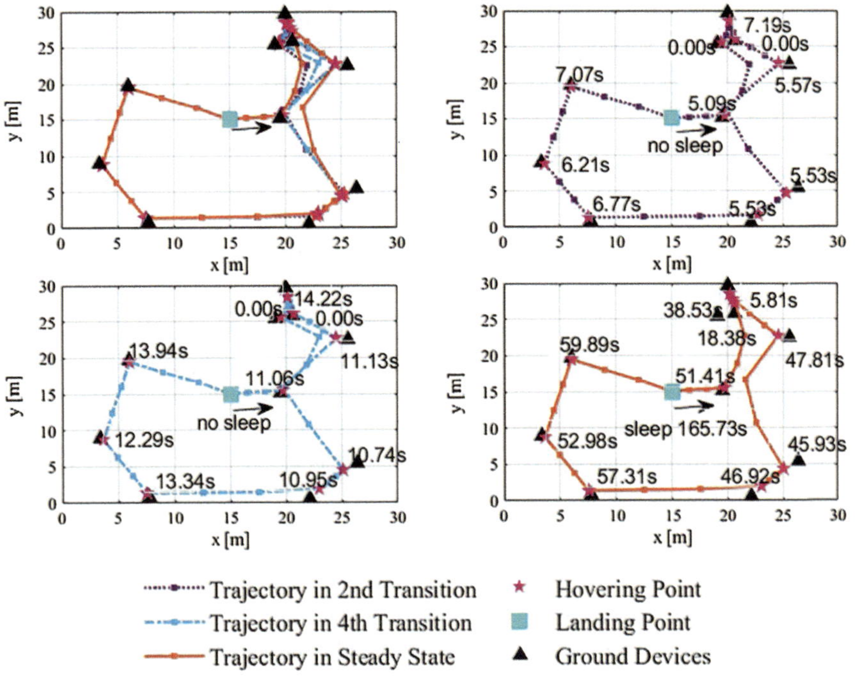

Fig. 3.7 Designed trajectories in transition-based design

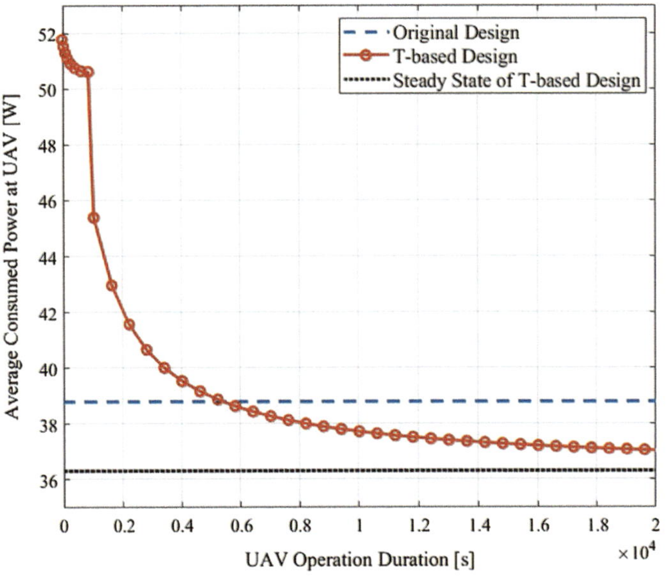

Fig. 3.8 Average UAV consumed power of different designs

charging without entering sleep mode during the transition loops, while it only adopts a sleep mode once the system reaches steady state, allowing for energy conservation. This adaptive behavior, tailored to the RNL fluctuations, ensures efficient energy usage and optimized performance throughout both the transition and steady-state phases.

3.7.3 Average UAV Consumed Power of Different Designs

Finally, to better simulate real-life scenarios, in Fig. 3.8, we observe the average UAV consumed power with operation durations across 20 different distributions of GDs. In a relatively long duration, the average UAV consumed power of transition-based design significantly outperforms the original design. However, in a relatively short duration, the original sustainable design has an advantage in sustaining the network compared to transition-based design. This is due to the fact that the transition-based design includes two states, one is the transition state for improving RNL, and the other is the steady state for saving energy. The core of transition-based design is to sacrificing certain performance to improve RNL during the transition period to achieve long-term energy savings in steady state. The results means that the original sustainable design has an advantage in a relatively short-term sustainable design, while the transition-based design is more advantageous for the long-term sustainable design.

3.8 Conclusion

This chapter considers a network of GDs where the GDs' battery limits and the recent RNL are known prior to design. By adopting the nonlinear WPT model, we utilize a UAV to charge the IoT network for sustainability, aiming to minimize the UAV average power consumption by jointly optimizing its trajectory and mode-switching decisions. Firstly, we formulate a sustainable design problem that considers the initial RNL and battery limits. Then the corresponding algorithm is proposed to successfully address the design in an iterative manner, allowing the UAV to follow a periodic working pattern through a single optimization approach. Since the initial RNL significantly affects performance, we introduce a transition-based sustainable design to address this. In this approach, to fully enhance the RNL, the UAV executes different trajectories for charging in transition loops without sleep. By solving a problem that maximizes the RNL lower bound, the RNL increases after each transition. Once the RNL reaches a predefined threshold, the UAV switches to steady-state operation, repeating the same trajectory with low energy consumption. Simulations confirm that the transition-based design performs as well as or better than the original in the long term. We also observed the steady-state and transition behaviors, demonstrating how the RNL improves through transitions. Finally, the idea of periodic design can be extended to more comprehensive long-term strategies. The modeling approach and transition method can also be adapted to more practical cases, e.g., the multi-UAV collaborative charging to support the entire network.

References

1. Li, M., Li, H., Ma, P., & Wang, H. (2023). Energy maximization for ground nodes in UAV-enabled wireless power transfer systems. *IEEE Internet of Things Journal, 10*, 17096–17109.
2. Vashi, S., Ram, J., Modi, J., Verma, S., & Prakash, C. (2017). Internet of Things (IoT): a vision, architectural elements, and security issues. In: *2017 International Conference on I-SMAC (IoT in Social, Mobile, Analytics and Cloud) (I-SMAC)* (pp. 492–496).
3. Jia, X., Feng, Q., Fan, T., & Lei, Q. (2012) RFID technology and its applications in Internet of Things (IoT). In *2012 2nd International Conference on Consumer Electronics, Communications and Networks (CECNet)* (pp. 1282–1285).
4. Ngu, A. H., Gutierrez, M., Metsis, V., Nepal, S., & Sheng, Q. Z. (2017). IoT middleware: A survey on issues and enabling technologies. *IEEE Internet of Things Journal, 4*, 1–20.
5. Mishra, D., Alexandropoulos, G. C., & De, S. (2018). Energy sustainable IoT with individual QoS constraints through MISO SWIPT multicasting. *IEEE Internet of Things Journal, 5*, 2856–2867.
6. Nguyen, N. T., & Matsuhashi, R. (2019). An optimal design on sustainable energy systems for shrimp farms. *IEEE Access, 7*, 165543–165558.
7. Boukerche, A., Wu, Q., & Sun, P. (2020). Efficient green protocols for sustainable wireless sensor networks. *IEEE Transactions on Sustainable Computing, 5*, 61–80.
8. Hossein Motlagh, N., Taleb, T., & Arouk, O. (2016). Low-altitude unmanned aerial vehicles-based Internet of Things services: Comprehensive survey and future perspectives. *IEEE Internet of Things Journal, 3*, 899–922.
9. Bol, D., et al. (2013). Green SoCs for a sustainable Internet-of-Things. In *2013 IEEE Faible Tension Faible Consommation* (pp. 1–4).
10. Qin, Z., Li, F. Y., Li, G. Y., McCann, J. A., & Ni, Q. (2019). Low-power wide-area networks for sustainable IoT. *IEEE Wireless Communications, 26*, 140–145.
11. Hu, Y., Zhu, Y., Gursoy, M. C., & Schmeink, A. (2019). SWIPT-enabled relaying in IoT networks operating with finite blocklength codes. *IEEE Journal on Selected Areas in Communications, 37*, 74–88.
12. Ku, M.-L., Li, W., Chen, Y., & Ray Liu, K. J. (2015). On energy harvesting gain and diversity analysis in cooperative communications. *IEEE Journal on Selected Areas in Communications, 33*, 2641–2657.
13. Bi, S., Ho, C. K., & Zhang, R. (2015). Wireless powered communication: opportunities and challenges. *IEEE Communications Magazine, 53*, 117–125.
14. Lu, X., Wang, P., Niyato, D., Kim, D. I., & Han, Z. (2015). Wireless networks With RF energy harvesting: A contemporary survey. *IEEE Communications Surveys & Tutorials, 17*, 757–789.
15. Boshkovska, E., Ng, D. W. K., Zlatanov, N., Koelpin, A., & Schober, R. (2017). Robust resource allocation for MIMO wireless powered communication networks based on a non-linear EH model. *IEEE Transactions on Communications, 65*, 1984–1999.
16. Clerckx, B., & Bayguzina, E. (2016). Waveform design for wireless power transfer. *IEEE Transactions on Signal Processing, 64*, 6313–6328.
17. Liu, L., Zhang, R., & Chua, K.-C. (2014). Secrecy wireless information and power transfer with MISO beamforming. *IEEE Transactions on Signal Processing, 62*, 1850–1863.
18. Al-Hourani, A., & Gomez, K. (2018). Modeling Cellular-to-UAV path-loss for suburban environments. *IEEE Wireless Communications Letters, 7*, 82–85.
19. Matolak, D. W., & Sun, R. (2015). Unmanned aircraft systems: Air-ground channel characterization for future applications. *IEEE Vehicular Technology Magazine, 10*, 79–85.

20. Xie, L., Xu, J., & Zhang, R. (2019). Throughput maximization for UAV-enabled wireless powered communication networks. *IEEE Internet of Things Journal, 6*, 1690–1703.
21. Xu, J., Zeng, Y., & Zhang, R. (2018). UAV-enabled wireless power transfer: Trajectory design and energy optimization. *IEEE Transactions on Wireless Communications, 17*, 5092–5106.
22. Mo, X., Huang, Y., & Xu, J. (2020) Radio-map-based robust positioning optimization for UAV-enabled wireless power transfer. *IEEE Wireless Communications Letters, 9*, 179–1830.
23. Hu, Y., Yuan, X., Xu, J., & Schmeink, A. (2019). Optimal 1D trajectory design for UAV-enabled multiuser wireless power transfer. *IEEE Transactions on Communications, 67*, 5674–5688.
24. Yang, Z., Xu, W., & Shikh-Bahaei, M. (2020). Energy efficient UAV communication with energy harvesting. *IEEE Transactions on Vehicular Technology, 69*, 1913–1927.
25. Yuan, X., et al. (2022) UAV trajectory design on completion time minimization of WPT task in UAV-enabled multi-user network. In *2022 IEEE International Conference on Communications Workshops (ICC Workshops)* (pp. 1047–1052).
26. He, Z., et al. (2023). Energy minimization for UAV-enabled wireless power transfer and relay networks. *IEEE Internet of Things Journal, 10*, 19141–19152.
27. Xie, L., Cao, X., Xu, J., & Zhang, R. (2021). UAV-enabled wireless power transfer: A tutorial overview. *IEEE Transactions on Green Communications and Networking, 5*, 2042–2064.
28. Yuan, X., Yang, T., Hu, Y., Xu, J., & Schmeink, A. (2021). Trajectory design for UAV-enabled multiuser wireless power transfer with nonlinear energy harvesting. *IEEE Transactions on Wireless Communications, 20*, 1105–1121.
29. Clerckx, B., et al. (2019). Fundamentals of wireless information and power transfer: From RF energy harvester models to signal and system designs. *IEEE Journal on Selected Areas in Communications, 37*, 4–33.
30. Hu, Y., Yuan, X., Yang, T., Clerckx, B., & Schmeink, A. (2020). On the convex properties of wireless power transfer with nonlinear energy harvesting. *IEEE Transactions on Vehicular Technology, 69*, 5672–5676.

Chapter 4
Joint User Scheduling and UAV Trajectory Design for UAV-Aided Data Collection

Yunfei Gao, Xiaopeng Yuan, and Yulin Hu

Abstract Utilizing the high mobility of UAVs to collaborate with wireless communication networks for data collection is considered an effective technology. This chapter investigates a multi-UAV-assisted data collection system with no-fly zones (NFZs), where UAVs take off from an initial location to collect data from mobile ground users (GUs) while avoiding entering NFZs and preventing collisions. The objective of this work is to minimize the whole data collection time via optimizing the 3D trajectories of UAVs and the scheduling of mobile GUs. On the one hand, the presence of NFZs makes the formulated problem non-convex. In addition, the mobility of users results in a dynamically changing system environment, making it difficult to solve using traditional methods. To tackle the above challenges, we begin by converting the problem into a Markov Decision Process (MDP) and introduce a multi-agent federated reinforcement learning (MAFRL) approach to optimize the long-term goal. Specifically, we implement a multi-step propagation technique alongside a dueling network architecture to enhance the neural network used for agent training, aiming to speed up convergence and increase overall stability. Finally, simulations validate the effectiveness of the method in practical scenarios.

Keywords Date collection · UAV trajectory design · No-fly zones · User scheduling

Y. Gao · Y. Hu (✉)
School of Electronic Information, Wuhan University, Wuhan, China
e-mail: yulin.hu@whu.edu.cn

Y. Gao
e-mail: gaoyunfei@whu.edu.cn

X. Yuan
Chair of Information Theory and Data Analytics, RWTH Aachen University, Aachen, Germany
e-mail: yuan@inda.rwth-aachen.de

Y. Hu et al. (eds.), *Reliable and Efficient Trajectory Design in UAV-Aided Wireless Networks*, https://doi.org/10.1007/978-981-96-5988-3_4

Abbreviations

BCD	Block Coordinate Descent
BS	Base Stations
CSI	Channel State Information
DDQN	Double Deep Q-Network
DNNs	Deep Neural Networks
DRL	Deep Reinforcement Learning
FL	Federated Learning
GDs	Ground Devices
GUs	Ground Users
IRS	Intelligent Reflecting Surfaces
LoS	Line-of-Sight
MAFRL	Multi-agent Federated Reinforcement Learning
MDP	Markov Decision Process
ML	Machine Learning
NFZ	No-fly Zones
NLoS	Non-Line-of-Sight
Mbps	Mbits per second
MC	Monte Carlo
MHz	Megahertz
RL	Reinforcement Learning
SNR	Signal-to-Noise Ratio
TD	Temporal Difference
TDMA	Time Division Multiple Access
UAV	Unmanned Aerial Vehicle

4.1 Introduction

4.1.1 Background

The UAV-aided communication is considered a key technology for the future, providing widespread connectivity with broader and more extensive coverage [1]. Compared to ground-based data collection systems, UAV-aided communication systems offer many advantages. For example, the flexible mobility of UAVs allows them to quickly move to data transmission points, reducing the overall data transmission time [2]. In addition, the flexible mobility of UAVs enables rapid deployment in complex environments such as earthquakes, tsunamis, and fires, ensuring that data transmission is no longer limited by ground-based communication networks [3]. In particular, the higher flight altitude of UAVs allows for a high-probability line-of-sight (LoS) communication channel between the air and the ground. Research on UAV-assisted data collection is especially crucial in fields driven by technology, including smart cities, agricultural monitoring, environmental protection, and disaster response.

4.1.2 Motivation

UAV-aided data collection is of paramount importance in modern technology and decision-making processes across various sectors. Accurate and timely data acquisition provides critical support for scientific research, business analytics, and policy formulation, facilitating process optimization, enhancing efficiency, and driving innovation. However, several challenges persist in data collection. First, the mobility of ground devices (GDs) introduces dynamic changes to the system, which traditional methods struggle to address effectively. Moreover, as concerns over privacy protection and data security intensify, ensuring data integrity and compliance with legal frameworks during the collection process has become a significant challenge. Furthermore, physical constraints in remote or complex environments, high equipment costs, and the substantial time and resources required for data gathering present additional obstacles. Thus, successful data collection not only demands advanced technical capabilities but also necessitates the development of innovative strategies to address these challenges.

In real-world applications, designing efficient 3D UAV trajectories and scheduling of devices while accounting for NFZs, GD mobility, and realistic channel conditions presents a significant challenge. We first model a scenario focused on minimizing data collection time by optimizing both the 3D UAV trajectories and GD scheduling. However, the nature of NFZs renders the problem non-convex, while GD mobility introduces dynamic system changes, making it hard to address using traditional methods like convex optimization and matching theory. Moreover, compared to 2D trajectories, the expanded exploration space in 3D design slows algorithm convergence. Therefore, it is essential to develop algorithms that accelerate convergence and enhance overall system performance.

4.1.3 Main Contributions

Main contributions to this chapter are as follows:

(1) The mobility of GDs, presence of NFZs, UAV collision avoidance, and practical probability LoS channel are all taken into account. To minimize total operation time in a multi-UAV data collection system, we propose a unified approach that jointly optimizes 3D UAV trajectory design and communication scheduling to solve the formulated problem.

(2) The problem we formulated is challenging because of its non-convex structure and the analytical complexity caused by the random mobility of GDs and the large number of interdependent variables. To tackle these challenges, we apply a MAFRL-based method to jointly optimize the UAVs' 3D trajectory and the communication scheduling of GDs. Specifically, this distributed learning approach requires UAVs to only exchange model parameters, reducing communication overhead, maintaining UAVs' privacy, and shortening training time. The problem is then reformulated as a Markov decision process (MDP), where each UAV is treated as an independent agent. Additionally, we employ a multi-step dueling double deep Q-network (DDQN) algorithm to train each agent. Compared to the conventional DDQN method, our approach enhances system performance through the dueling network and accelerates convergence using multi-step bootstrapping with an appropriate step size.

(3) We validated the effectiveness of the proposed algorithm through simulation experiments. In comparison to centralized MAFRL approaches, our method proves to be more robust in dynamic multi-GD environments with NFZs. Furthermore, the system's performance is on par with that of the centralized MARL method and approaches the ideal scenario without NFZs, while outperforming baseline methods. Additionally, the proposed 3D trajectory design is shown to enhance system performance by utilizing the vertical dimension, offering improvements over 2D trajectory designs.

4.1.4 Structural Organization

The work on UAV-aided data collection and trajectory design is introduced in Sect. 4.2, while the system model and problem formulation are presented in Sect. 4.3. Furthermore, Section 4.4 introduces the proposed federated reinforcement learning-based solution. Section 4.5 provides the simulation results and analysis, and finally, the paper is concluded in Sect. 4.6.

4.2 Related Works

4.2.1 UAV-Aided Trajectory Design

The design of UAV trajectories and GD communication scheduling is crucial for UAV-assisted communication networks, enabling UAVs to efficiently reach selected GDs and establish reliable links with minimal time. Several studies highlight the significant advantages of optimizing UAV trajectories and GD communication scheduling in data transmission systems. For example, some focus on maximizing throughput by designing UAV trajectories within communication frameworks [4]. A block coordinate descent (BCD) algorithm has also been proposed to address the non-convex challenges in trajectory design [5]. In scenarios where UAVs are equipped with intelligent reflecting surfaces (IRS), [6] optimizes communication scheduling, IRS phase shifts, and UAV trajectories to maximize the minimum average transmission rate. Additionally, works [7, 8] propose UAV-assisted schemes aimed at improving the energy efficiency of the system by jointly designing UAV 3D trajectories and communication scheduling.

However, the existing literature fails to address more realistic scenarios. For example, studies [4–6, 8] mainly assume a simplified LoS-dominant channel. Additionally, they overlook the mobility of GDs and the presence of no-fly zones (NFZs), assuming GDs remain stationary. In reality, UAV trajectories are heavily influenced by environmental factors, requiring careful design that accounts for altitude variations, multi-dimensional propagation, and the effects of obstacles on communication channels. Moreover, GDs are typically mobile in practical scenarios, and certain areas, e.g., schools and shopping malls, impose flight restrictions on UAVs.

4.2.2 Reinforcement Learning for UAV Application

The presence of NFZs makes the problem non-convex, and the mobility of GDs causes the system to change dynamically, making it challenging to obtain accurate channel state information (CSI). Furthermore, traditional methods such as graph theory [9] and convex optimization [10] techniques struggle to address trajectory optimization and scheduling issues in this practical scenario [11]. To address these challenges, a set of studies have employed artificial intelligence (AI) methods to design the trajectories of UAVs and manage system resources i.e., reinforcement learning (RL), federated learning (FL), and machine learning (ML). In particular, these methods only require continuous optimization of the UAV exploration strategies to achieve long-term benefits, without needing prior knowledge of the system. [12, 13].

For example, the authors of work [14] addressed the problem of multi-task sequence classification through distributed learning. The authors of work [15] utilized a multi-objective reinforcement learning approach to optimize the trajectories and task offloading in multiple UAVs scenario. Additionally, in works [16, 17], the authors applied deep reinforcement learning (DRL) methods to tackle the trajectory and scheduling problems of UAVs. The authors utilized deep reinforcement learning methods based on radio maps to navigate UAVs [18]. Furthermore, the authors of works [19–21] addressed the problem of multi-UAVs trajectory design in perception and coverage scenarios using deep reinforcement learning methods. However, all of the above works are based on a centralized learning approach, which requires uploading model parameters to a server for processing. This can lead to communication overhead and potential data privacy breaches. Furthermore, centralized training involves high time complexity and demands extended training time.

4.3 System Model and Problem Formulation

4.3.1 System Description

In our work, we consider a UAV-aided data collection scenario with L. UAVs and J GDs, with all GDs randomly moving across the ground, as depicted in Fig. 4.1. Additionally, UAVs act as aerial base stations (BSs) deployed to collection data transmitted by GDs within a 3D scenario that includes no-fly zones (NFZs). All UAVs collaboratively train a global trajectory optimization model based on a distributed arning approach. Furthermore, one UAV serves as a parameter server to manage the parameters received from the other UAVs. For a time period T, the time-varying 3D position of UAV l can be expressed as $q_l(t) = \left[x_l(t), y_l(t), z_l(t)\right]$ for $t \in [0, T]$. For convenience, we discretize the time period T into N small, equal-length time slots of duration κ, i.e., $T = \kappa N$. It is important to note that the slot length κ should be small enough, allowing the channel to be considered approximately constant within each time slot. Consequently, we can assume that the change in position within the current time slot is negligible and that the channel state remains stable. Thus, the location of UAV l at time solt n can be denoted as $\boldsymbol{q}_{l,n} = \left[x_{l,n}, y_{l,n}, z_{l,n}\right]$. The location of the j th GD is represented as $\psi_{j,n} = \left[x_{j,n}, y_{j,n}, 0\right]$.

To ensure flight safety, we assume that the minimum altitude at which the UAVs operate is above the maximum height of any buildings., i.e., $z_{min} \leq z_{l,n} \leq z_{max}$. Additionally, we define $\beta_{l,j,n} = \{0, 1\}$ as the data transmission decision variable from the lth UAV to the jth GD. In particular, the case where $\beta_{l,j,n} = 1$ represents that the jth GD decides to send its data to the UAV l, otherwise, if $\beta_{l,j,n} = 0$, the data is not sent. In this study, the communication mode between UAVs and GDs employs

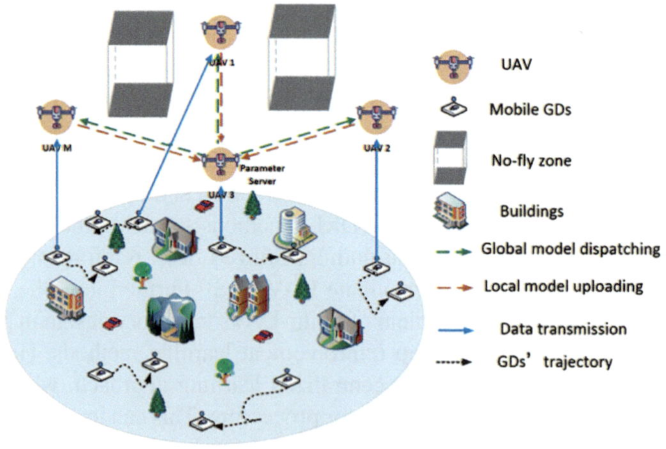

Fig. 4.1 UAVs-aided multi-GDs data collection system

the time division multiple access (TDMA) scheme, meaning that each UAV can only receive signals from one GD in each time slot, i.e., $\sum_{j=1}^{J} \beta_{l,j,n} \leq 1$.

4.3.2 Channel Model

UAV-GD Communication Model

In complex urban environments, the channel model between UAVs and GDs can no longer be represented by a simple LoS model, as electromagnetic waves are affected by scattering, refraction, and other factors. In our work, we consider a probability LoS channel model to represent the channel propagation between UAVs and GDs. In this model, the data transmission rate is primarily determined by both the elevation angle and the distance between the UAV and the GD. In this section, we assume that the probability of a LoS connection between UAV l and GD j is denoted by $a_{l,j,n} = 1$, the probability of LoS is $P^L_{l,j,n}$, and the elevation angle is $\theta_{l,j,n}$. The $P^L_{l,j,n}$ can be expressed as

$$P^L_{l,j,n}\big(a_{l,j,n} = 1\big) = B_3 + \frac{B_4}{1 + e^{-(B_1 + B_2\theta_{l,j,n})}}, \tag{4.1}$$

where $B_1 < 0, B_2 > 0, B_4 > 0$, with B_4 being a constant such that $B_3 + B_4 = 1$. Here, B_1, B_2, B_3, B_4 are constants determined by the surrounding environment. Thus, the corresponding non-line-of-sight (NLoS) probability can be represented as $P^N_{l,j,n}\big(a_{l,j,n} = 0\big) = 1 - P^L_{l,j,n}$. The channel gain can be expressed as

$$h_{l,j,n} = c_{l,j,n}h^L_{l,j,n} + \big(1 - c_{l,j,n}\big)h^N_{l,j,n}, \tag{4.2}$$

where $h^L_{l,j,n}$ and $h^N_{l,j,n}$ represent the channel power gain in the state of LoS and NLoS, respectively. $h^L_{l,j,n}$ and $h^N_{l,j,n}$ are denoted as

$$h^L_{m,k,n} = \alpha_0 d^{-\beta_L}_{m,k,n}, \; h^N_{m,k,n} = \mu\alpha_0 d^{-\beta_N}_{m,k,n}. \tag{4.3}$$

Here, α_0 represents the average channel power gain in the LoS state when the reference distance is unit meter, $\mu < 1$ denotes the additional attenuation factor, and β_L and β_N represent the average path loss exponents for the LoS and NLoS states, respectively, i.e., $2 \leq \beta_L \leq \beta_N \leq 6$. In addition, the Euclidean distance between UAV l and GD j is represented by $d_{l,j,n} = \sqrt{\|\boldsymbol{q}_{l,n} - \boldsymbol{w}_{j,n}\|^2}$. Furthermore, the elevation angle $\theta_{l,j,n}$ is expressed as

$$\theta_{l,j,n} = \frac{180}{\pi} arctan\left(\frac{z_{l,n}}{\|q^{\lambda}_{l,n} - \psi^{\lambda}_{j,n}\|}\right), \tag{4.4}$$

where $q^{\lambda}_{l,n}$ and $\psi^{\lambda}_{j,n}$ denote the horizontal coordinates of UAV l and GD j, respectively, projected onto the ground. It is important to note that the channel power gain $h_{l,j,n}$ is influenced by two sources of randomness: the occurrence of random LoS and NLoS links, and the random small-scale fading. By averaging these two sources of randomness, the expected achievable rate can be expressed as follows:

$$E\left[R_{l,j,n}\right] = P^{L}_{l,j,n}R^{L}_{l,j,n} + P^{N}_{l,j,n}R^{N}_{l,j,n}, \tag{4.5}$$

where $R^{L}_{l,j,n} = B\log_2\left(1 + \frac{\gamma_j}{d^{\alpha_L}_{l,j,n}}\right)$ and $R^{N}_{l,j,n} = Blog_2\left(1 + \frac{\mu\gamma_j}{d^{\alpha_N}_{l,j,n}}\right)$ denote the achievable rates in the LoS and NLoS states. Here, B represents the transmission bandwidth of GD j, and $\gamma_j = \frac{\tau_0 P_j}{\varrho_0 \Re}$, where P_j is the transmit power of the GD j,ϱ_0 represents the receiver noise power, and $\Re > 1$ is the SNR gap between the practical modulation and coding scheme and the ideal Gaussian signaling.

UAV-UAV Communication Model

In our design, since the flying altitude of the UAVs is higher than that of the buildings, we can reasonably assume that the channel between the UAVs and the serving UAV is a LoS channel. Therefore, the data transmission channel for the model parameters exchanged between them can be expressed as:

$$h^{U}_{l,\bar{l},n} = \frac{\Gamma}{\|z_{l,n} - z_{\bar{l},n}\|^2 + \|q^{\gamma}_{l,n} - q^{\gamma}_{\bar{l},n}\|^2}, \tag{4.6}$$

$$h^{D}_{l,\bar{l},n} = \frac{\Gamma}{\|z_{l,n} - z_{\bar{l},n}\|^2 + \|q^{\gamma}_{l,n} - q^{\gamma}_{\bar{l},n}\|^2}, \tag{4.7}$$

where $q^{\gamma}_{l,n}$ and $q^{\gamma}_{\bar{l},n}$ are the horizontal coordinate positions of UAV l and server UAV $\bar{l}$ at time slot n, respectively. Then, the instantaneous transmission rate in the uplink, measured in Mbits per second (Mbps), between UAV l and the server UAV $\bar{l}$ can be expressed as

$$R^{U}_{l,\bar{l},n} = Blog_2\left(1 + \frac{P_l h^{U}_{l,\bar{l},n}}{\phi_0 B}\right) = Blog_2\left(1 + \frac{\chi P_l}{\|z_{l,n} - z_{\bar{l},n}\|^2 + \|q^{\gamma}_{l,n} - q^{\gamma}_{\bar{l},n}\|^2}\right), \tag{4.8}$$

where P_l represents the transmit power of UAV l, ϕ_0 denotes the white Gaussian noise power spectral density in watts/Hz at the server UAV $\bar{l}$, B signifies the total channel bandwidth available for communication in Megahertz (MHz), and the reference signal-to-noise ratio (SNR) is denoted as $\chi = \frac{\Gamma}{\phi_0 B}$ at the reference distance. Let $P_{\bar{l}}$ denote the transmit power of the server UAV $\bar{l}$. Similarly, the transmission rate of model parameters from the server UAV $\bar{l}$ to UAV l can be expressed as

$$R^D_{l,\bar{l},n} = Blog_2\left(1 + \frac{P_{\bar{l}} h^D_{l,\bar{l},n}}{\phi_0 B}\right) = Blog_2\left(1 + \frac{\chi P_{\bar{l}}}{\|z_{l,n} - z_{\bar{l},n}\|^2 + \|q^{\gamma}_{l,n} - q^{\gamma}_{\bar{l},n}\|^2}\right). \tag{4.9}$$

4.3.3 Problem Formulation

As shown in Fig. 4.2, the system operates in a continuous cycle of three stages. First, each local UAV collects data from its selected GDs. Next, the local UAVs upload their model parameters to the server UAV. Finally, the server UAV processes these local parameters and distributes the updated global model to all local UAVs. This process repeats until all GD data is collected. It is assumed that the server UAV is powerful enough to process the local parameters quickly, making computation time negligible. Thus, the total operation time is divided into three phases: data collection, local FL model update, and global FL model broadcast.

- **Data collection time:** Let ω_j represent the total data size in bits that the GD j needs to transmit to the UAV. In particular, the data transmission constraint must be satisfied when UAV l completes its data collection task from GD j. This is expressed as:

$$T^C = N^C \kappa = \sum_{l=1}^{L} N_l^C \kappa, \tag{4.11}$$

  where N^C is the total number of time slots required for all UAVs to complete the data collection task
- **Local FL Model Update Time**: The total data size of the local model parameters that UAV l needs to transmit to the server UAV $\bar{l}$ is denoted by ω_l in bits. In addition, assuming that the UAVs remain stationary during the transfer of model parameters, the time required for UAV l to upload its local model parameters to the server UAV can be expressed as:

$$T^U = N^U \kappa = \upsilon w_l / R^U_{l,\bar{l},n}, \tag{4.12}$$

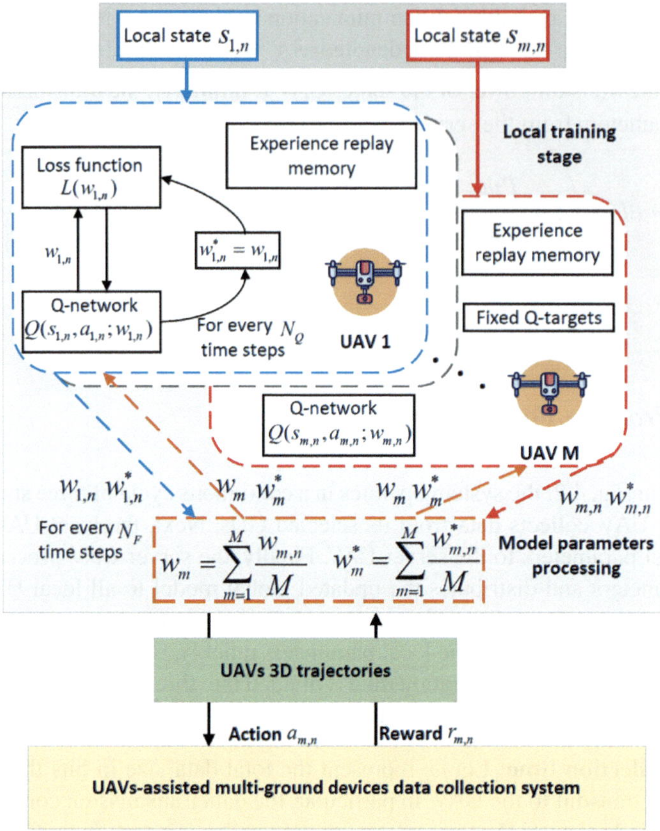

Fig. 4.2 UAVs-assisted multi-GDs data collection system

where υ represents the total number of interactions of model parameters after completing the data collection task, and $R_{l,\bar{l},n}$ denotes the transmission rate of model parameters from the server UAV and the UAV l.

- **Global FL Model Broadcast Time**: Let $w_{\bar{l}}$ denote the total data size of the global model parameters that the server UAV $\bar{l}$ needs to broadcast to other UAVs. The global model parameters broadcast time from the server UAV $\bar{l}$ to the UAV l can be expressed as:

$$T^D = N^D \kappa = \upsilon w_{\bar{l}} / R^D_{l,\bar{l},n} \tag{4.13}$$

where $R^D_{l,\bar{l},n}$ represents the transmission rate of model parameters from the server UAV $\bar{l}$ to the UAV l. Thus, the total operation time can be expressed as $T_{tot} = T^C + T^U + T^D$.

The aim of this design is to minimize the total operation time via jointly optimizing the UAVs' trajectory and GDs' communication scheduling. The formulated problem can be expressed as

$$\begin{aligned}
P1:\quad & \min_{\{\mathbf{q}_{l,n}, v_{l,n}, \alpha_{l,j,n}\}} \quad N_{\text{total}}\,\sigma \\
s.t.\quad & (a)\quad \alpha_{l,j,n} \in \{0,1\}, \\
& (b)\quad \sum_{l=1}^{L} \alpha_{l,j,n} \le 1, \\
& (c)\quad z_{\min} \le z_{l,n} \le z_{\text{maxi}}, \\
& (d)\quad \sum_{t=1}^{L}\sum_{n=1}^{N_l^G} \alpha_{l,j,n} R_{l,j,n} \ge D_j, \\
& (e)\quad \mathbf{q}_{l,n+1} = \mathbf{q}_{l,n} + \Delta_s \vec{v}_{l,n}, \\
& (f)\quad \Delta_s = V\sigma, \\
& (g)\quad \mathbf{q}_{l,n} - \mathbf{q}_{l^*,n} \ge d_{\min}, \forall l \ne l^*, \\
& (h)\quad \mathbf{q}_{l,n} - \mathbf{u}_{m,k} \ge d_{min}.
\end{aligned} \tag{4.14}$$

Constraints (a) and (b) indicate that within a single time slot, a UAV can only communicate with one GD. Constraint (c) represents the UAV's flight altitude limits. Constraint (d) states that the UAV must complete data collection from each GD. Constraints (e) and (f) specify the UAV's movement restrictions, where Δ_s represents the distance the UAV moves within a single time slot, $\vec{v}_{l,n}$ indicates the UAV's flight direction, and V denotes the UAV's flight speed.

4.4 Federated Reinforcement Learning Solution

The problem $\mathcal{P}1$ is difficult to address due to its non-convex feasible region, particularly influenced by the existence of NFZs. Additionally, the mobility of GDs and the constraints imposed by NFZs introduce dynamics and uncertainty into the system. These factors complicate the optimization problem, making it difficult to solve using traditional algorithms [22–24]. To address these challenges, we reformulate the problem as a MDP and subsequently develop a solution method based on Multi-Agent Federated Reinforcement Learning (MAFRL), as illustrated in Fig. 4.2.

4.4.1 Primarily of Reinforcement Learning

RL entails an agent engaging with its environment to improve its policy ι, with the goal of maximizing its long-term reward $\mathbb{E}\left[Y_{l,n}\right]$. The return is defined as the cumulative sum of discounted rewards starting from time step n, where the discount factor $0 < ˥ < 1$. RL is particularly effective for addressing MDP problems, which are characterized by a 4-tuple $\langle \mathcal{S}, \mathcal{A}, \mathcal{P}, r \rangle$. Here, $\mathcal{S}$ denotes the state space, $\mathcal{A}$ represents the action space, $\mathcal{P}$ indicates the state transition probabilities, and r defines the reward function. In particular, at each time slot, the agent chooses an action a_n from the action space $\mathcal{A}$ according to its policy ι, receives a reward r_n, and then transitions to a new state s_{n+1} based on the state transition probabilities ς. DRL is an advancement of Reinforcement Learning (RL) that employs deep neural networks (DNNs) to estimate the Q-value function. This function indicates the expected return of executing an action a_n in a given state s_n. In the DDQN algorithm, the Q-value approximator $Q\left(s_n, a_n | w_{l,n}\right)$, characterized by parameters, $w_{l,n}$, is refined by minimizing the following loss function:

$$(r_{n+1} + \gamma Q(s_{n+1}, \arg\max_{a_n^*} Q(s_{n+1}, a_n^* | w_{l,n}) | w_{l,n}^*) - Q(s_n, a_n | w_{l,n}))^2, \tag{4.15}$$

where $w_{l,n}$ and $w_{l,n}^*$ denote the parameters of the online Q network and the target Q network, respectively. Specifically, the DDQN algorithm utilizes the loss function to update the online Q network, aiming to closely approximate the target Q network, which is updated at regular intervals.

The DDQN network forecasts the state s_{n+1} based solely on the value $Q(s_n, a_n)$ of the current action a_n in the current state s_n. However, if the connection between the action a_n and the state s_n is weak, this prediction may not be reliable. In contrast, the dueling network architecture distinguishes between state and action values, allowing for independent evaluation of each. This approach diminishes the reliance of the state on the action value during the calculation process, resulting in improved stability for the algorithm. Consequently, compared to the traditional DDQN framework, the method utilizing dueling networks enhances both the performance and stability of the system. In particular, we implement multi-step bootstrapping to speed up convergence by taking into account the returns over the next $\wp$ steps. By choosing an appropriate value for $\wp$, the algorithm achieves faster and more effective convergence. The following adjustments have been made to the neural network:

- **Dueling network**. The dueling network architecture introduces two separate streams for estimating the action value. One stream predicts the state value function, which reflects the value of the current state, while the other evaluates the action advantage function, which determines the quality of the actions available in that state. Unlike traditional Q-networks, the dueling network can assess the value of a state without needing to evaluate all possible actions. This structure enhances stability and reduces the likelihood of overestimating the values of specific actions [25]. The Q-function for the dueling network architecture is expressed as follows:

$$Q(s, a|\theta, \phi, \beta) = V(s|\theta, \beta) + A(s, a|\theta, \phi) - \frac{1}{|A|}\sum_{a^{\cdot}} A(s, a^{\cdot}|\theta, \phi), \tag{4.16}$$

Here, $V(s|\theta, \beta)$ represents the state value, which is a measure of how valuable it is to be in state s, $A(s, a|\theta, \phi)$ denotes the action advantage value, which reflects the relative value of taking action a in state s. The parameters β and ϕ correspond to the state value stream and the state-dependent action advantage stream, respectively, while |A| represents the size of the action space. Together, these components form the dueling network architecture, which allows for a more refined evaluation of state and action values by distinguishing between the intrinsic value of being in a particular state and the advantage of selecting specific actions within that state.

- **Multi-step approach**. This approach distinguishes itself from Monte Carlo (MC) and temporal difference (TD) methods by accounting for the return over the future $\wp$ steps. By appropriately choosing the value of $\wp$, the algorithm achieves faster and more efficient convergence compared to traditional methods. The truncated $\wp$-step return is defined as:

$$R_{l,n:n+\wp} = \sum_{i=0}^{\wp-1} \gamma^{i} R_{l,n+i+1}. \tag{4.17}$$

The loss function of the improved method, incorporating the truncated $\wp$-step return and the dueling network architecture, can be expressed as:

$$\left(R_{l,n:n+\wp} + \gamma^{\wp} Q^{*}\left(q_{l,n+\wp}, \vec{v}_{l,n}^{*}; w_{l,n}^{*}\right) - Q^{*}\left(q_{l,n}, \vec{v}_{l,n}; w_{l,n}\right)\right)^{2}, \tag{4.18}$$

where $\vec{v}_{l,n}^{*} = \underset{\vec{v}_{l,n} \in A_{l,n}}{\arg max}\, Q^{*}\left(q_{l,n+\wp}, \vec{v}_{l,n}; w_{l,n}\right)$.

4.4.2 MAFRL Method for Solving the Problem of (P1)

To solve problem of $\mathcal{P}1$, we propose a solution based on MAFRL that simultaneously optimizes the 3D trajectories of UAVs and the scheduling of GDs. In this approach, each UAV operates as an independent agent, learning an optimal policy for its trajectory and communication scheduling decisions. By leveraging federated learning, the agents can collaborate and share model updates while preserving privacy, which is crucial in dynamic environments with mobility constraints and NFZs. The proposed MAFRL framework enables the UAVs to effectively balance exploration and exploitation, resulting in efficient coordination among UAVs to minimize the total operation time for data collection and model transmission tasks.

The MAFRL framework is depicted in Fig. 4.2, and the whole operation is divided into four key phases: local training by each UAV, transmission of local model parameters from individual UAVs to the server UAV, sending of these parameters at the server UAV, and finally, dissemination of the global model parameters to each UAV. The UAVs then engage with the environment and train their neural networks through the following steps:

- **Local UAVs Training**. During the local training phase, each UAV operates independently, interacting with its environment. At the start of each episode, each UAV schedules GDs based on its current position, deciding which GDs will be assigned for that time slot. Afterward, the UAV collects data from its selected GD, identified by ID_j. Once data collection for the current slot is finished, the UAV selects an action from its action space using the ε-greedy strategy, which can be described as:

$$a_{l,n} = \begin{cases} \text{randomly selected from A, with} \varepsilon \text{ probability;} \\ \underset{a_{l,n} \in A_m}{\arg\max}\, Q\left(s_{l,n}, a_{l,n} | w_{l,n}\right), \text{ with } 1 - \varepsilon \text{ probability.} \end{cases} \tag{4.19}$$

  To balance exploration and exploitation, ε gradually decreases from a higher to a lower value throughout the episode, boosting the likelihood of choosing the optimal Q-value action as training progresses. Once the action is taken, the UAV receives a reward and moves to the next state based on the state transition model (equation). The φ-step truncated return is computed, and the tuple $\left(s_{\mathfrak{J}}, a_{\mathfrak{J}}, R_{\mathfrak{J}:min(\mathfrak{J}+\varphi,N)}, s_{min(\mathfrak{J}+\varphi,N)}\right)$ is stored in the replay memory C. A mini-batch transition is then randomly sampled from C, and gradient descent is applied to update the parameter $\mathrm{w}_{\mathrm{l,n}}$ of the online network. The online network's parameter $\mathrm{w}_{\mathrm{l,n}}$ is copied to the target network every N_{Λ} steps.
- **Local Model Parameters Uploaded**. Once n/N_{Λ} becomes an integer, each local UAV transmits its current online model parameter $w_{l,n}$ and target model parameter $w^*_{l,n}$ to the server UAV for aggregation.
- **Federated Aggregation**. When the server UAV has received the model parameters from all local UAVs, it aggregates these parameters to compute the global model parameter, which can be represented as follows:

$$w_l = \sum_{l=1}^{L} \frac{w_{l,n}}{L}, w^*_l = \sum_{l=1}^{L} \frac{w^*_{l,n}}{L}, \tag{4.20}$$

  where $\mathrm{w}_{\mathrm{l,n}}$ and $\mathrm{w}^*_{\mathrm{l,n}}$ denote the model parameters of the on-line Q-network and the target Q-network, respectively. Additionally, w_l and w^*_l represent the global model parameters of the aggregated on-line Q-network and the global model parameters of the target Q-network, respectively.
- **Global Model Parameters Distribution**. After the server UAV processes the model parameters received from the local UAVs, it transmits the global model

parameters back to each local UAV. Subsequently, each local UAV updates its networks by copying the global model parameters every N^{Agg} time slots, such that $w_{l,n+1} = w_l$ and $w^*_{l,n+1} = w^*_l$.

Moreover, the MDP of a local transition tuple (S_l, A_l, π_l, r_l) encompasses the fundamental components for the lth UAV, defined as follows:

- **State Space S_l:** At every time slot n, the state $s_{l,n}$ of each UAV l that characterizes the environment consists of its 3D coordinates $\left[x_{l,n}, y_{l,n}, z_{l,n}\right]$. Specifically, the state space S_l includes all possible locations within the area of interest.
- **Action Space A_l:** In our design, all agents share the same action space A_l. At every time slot n, each agent selects its corresponding action $a_{l,n}$ based on its observed state $s_{l,n}$. The action $a_{l,n}$ represents the flight direction of the UAVs and can be expressed as $[a_1, a_2, a_3, a_4, a_5, a_6]$. Specifically, $a_1 = [0, 1, 0]$ indicates that UAV moves forward, $a_2 = [0, -1, 0]$ signifies the UAV moving backward, $a_3 = [1, 0, 0]$ represents a leftward movement, $a_4 = [-1, 0, 0]$ means the UAV moves to the right, $a_5 = [0, 0, 1]$ indicates the UAV is ascending, and $a_6 = [0, 0, 1]$ denotes the UAV is descending. Furthermore, $a_{l,n}$ is the action taken by the individual agent at time slot n, and all agents act independently.
- **State Transition Function η_l:** Let $\eta_l(s_{l,n+1}|s_{l,n}, a_{l,n})$ represent the transition probability of UAV l. This denotes the probability of transitioning to a new state after executing an action $a_{l,n}$ in the current state $s_{l,n+1}$. The state transition function for UAV l is primarily governed by its motion displacement function, as described in Eq. (4.14e).
- **Reward Function r_l:** To minimize the overall time required for data collection, we define a local reward for each UAV l, denoted as r_l, which is expressed as

$$r_l = \begin{cases} \vartheta, & \text{if outbound/fly into NFZs/collision;} \\ \varpi + \xi R_{l,n}, & \text{else;} \end{cases} \quad (4.21)$$

In this formulation, ϑ is a negative value that serves as a penalty for undesirable behaviors such as flying out of bounds, entering NFZs, or colliding with other UAVs. Meanwhile, ϖ is also a negative value aimed at reducing the time slots required for UAV flights. On the other hand, ξ is a positive value intended to accelerate the data collection process for the UAVs.

4.5 Simulation Results and Analysis

In this section, we assess the overall operational time cost performance of the proposed MAFRL method for the joint design of UAV trajectories in 3D space and the scheduling of GDs. The network comprises two UAVs and six GDs, with each GD requiring the transmission of 40 Mbit of data. The GDs are randomly distributed throughout a 3D urban environment with a side length of 1000 m, and the duration of each discrete time slot, denoted as κ is set to 0.5 s.

Typically, the mobility of the GDs is modeled using a Random Walk approach, where the GDs move randomly within a designated area and frequently change their directions $[0, 2\pi]$, adhering to an angle constraint of and maintaining a speed of 5 m/s. The UAVs, on the other hand, operate at a constant speed of $V = 20$ m/s. This random walk scenario for GDs has numerous practical applications, including situations where GDs roam randomly to sense nearby data. Additionally, the initial positions of both the UAVs and GDs are generated randomly, with the UAVs flying at an altitude ranging from $z_n \in [50, 100]$ meter.

In addition, we present a comparison between the proposed MAFRL algorithm and other widely adopted baseline methods in UAV data collection systems. The comparison strategies are outlined as follows:

No Dueling MAFRL: Proposed MAFRL algorithm without utilizing the dueling architecture.
Without NFZs MAFRL: To validate the applicability of the proposed algorithm in the area, we assume there are no NFZs. In other words, the UAVs are able to perform their tasks anywhere within the designated area. Proposed MAFRL algorithm without utilizing the dueling architecture.
Random Scheduling MAFRL: During the scheduling process, each UAV randomly scheduling GDs for data collection.
Centralized MARL: A centralized DRL approach creates a unified policy for all agents within the system, instead of allowing each agent to make individual decisions at every time slot.

4.5.1 Hyperparameters Selection

In this subsection, we evaluate the effect of different hyperparameters on the performance of the proposed method and identify the optimal values. The simulations assume that the UAVs and GDs start from the same initial positions. The results are presented as the average over 30 episodes, calculated across a total of 1000 episodes.

Figure 4.3 illustrates the relationship between the learning rate (lr) and total operation time. It shows that a learning rate of 0.0005 leads to faster convergence compared to 0.005 and 0.00005, while also offering better stability and less oscillation. This highlights the importance of selecting an appropriate learning rate for effective model training. A learning rate that is too high can cause the model to bypass important local minima, resulting in unstable optimization. Conversely, a learning rate that is too low may cause the model to become stuck in a local minimum, preventing it from reaching the global optimum. As a result, the higher oscillations and worse convergence performance observed with learning rates of 0.005 and 0.00005 make 0.0005 the most suitable choice for this paper.

Figure 4.4 depicts the relationship between total operation time and the size of $\wp$. The plot shows that, compared to $\wp = 64$ and $\wp = 256$, using $\wp = 128$ yields a lower

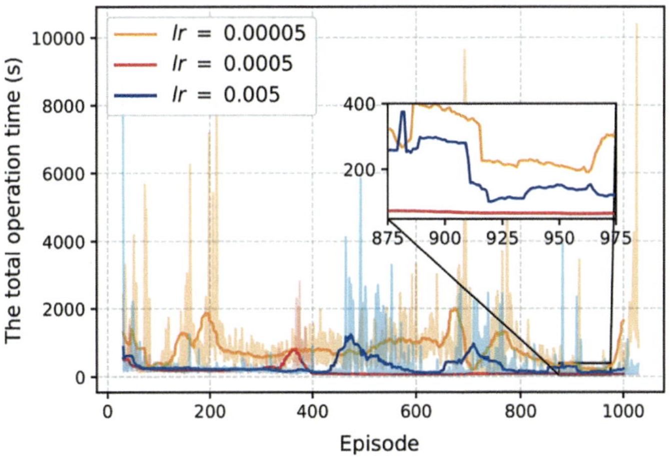

Fig. 4.3 Influence of different learning rate on total completion time

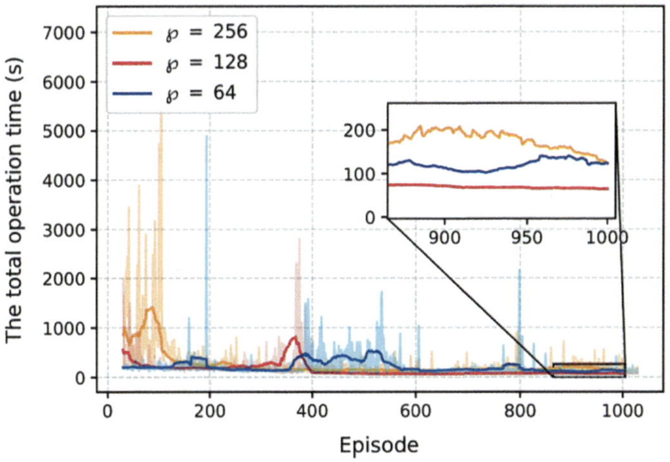

Fig. 4.4 Influence of different ℘ on total completion time

final convergence value, reduced oscillation, and greater stability throughout training. This is because selecting the right value for ℘ is essential for accurately estimating the optimal Q-value during network training. A small ℘ limits the number of states used to update the Q-value, leading to inaccurate predictions. Conversely, a large ℘ slows convergence and may prevent reaching an optimal final value. Therefore, $\wp = 128$ was selected for our simulations to ensure optimal performance.

Figure 4.5 illustrates the relationship between total operation time and mini-batch size. The results show that a mini-batch size of 512 leads to the fastest convergence and shortest operation time. This is because a larger batch size allows the UAVs to

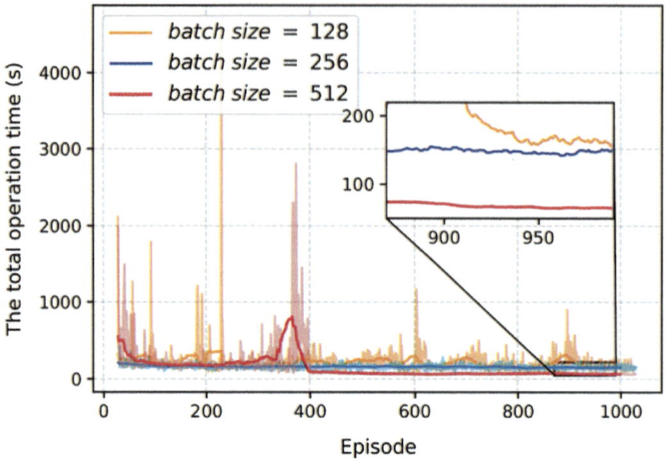

Fig. 4.5 Influence of different *batch size* on total completion time

make better use of the sample data stored in the replay buffer, aiding in the identification of optimal action-selection strategies during environmental interactions. In contrast, smaller batch sizes (such as 128 or 256) result in greater oscillations early in training, as they underutilize stored samples, leading to sub-optimal action selection and fluctuations in overall operation time. Thus, a batch size of 512 was selected to optimize the performance of our method.

4.5.2 Performance Evaluation

Figure 4.6 compares the performance of the proposed method with several baseline approaches. The plot of total operation time shows noticeable fluctuations in the early training stages, which decrease and stabilize as training progresses. This is because UAVs lack experience initially, making it harder to find optimal strategies. As they gain experience, they discover more efficient strategies. Our proposed method performs similarly to the centralized MARL and MAFRL approaches without NFZs, while outperforming the no-dueling MAFRL and random scheduling MAFRL methods. Several factors contribute to this. The centralized MAFRL approach avoids model aggregation, leading to better performance, but it requires UAVs to share data (e.g., position and channel gains) with the server UAV, increasing communication overhead and potentially compromising privacy. The method without NFZs allows UAVs to take shorter paths, improving task completion times. In contrast, the no-dueling MAFRL lacks the stability benefits of the dueling architecture, resulting in poorer performance. Lastly, random scheduling of GDs in the random scheduling MAFRL method causes significant oscillations, negatively affecting its performance.

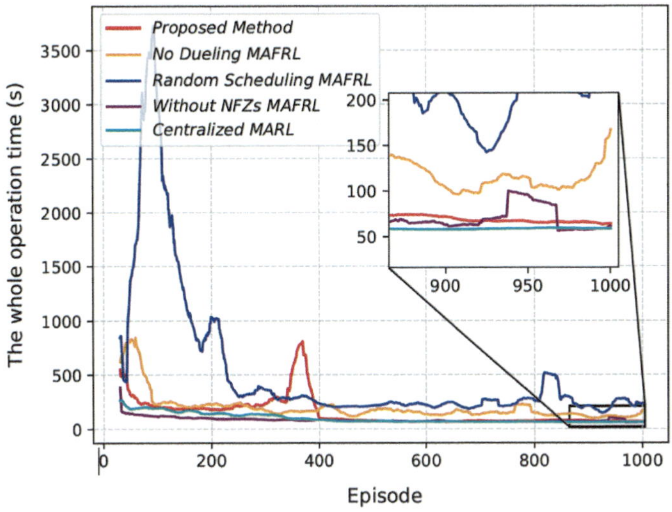

Fig. 4.6 Comparison of total completion time under different schemes

In Fig. 4.7, we compare the 3D trajectories of the UAVs using our proposed method with those of baseline approaches. The trajectories are depicted as 3D curves, with UAV1 and UAV2's starting positions marked by red and blue dots '·', respectively, and the GDs' initial locations marked by 'x' on the ground, showing random movement. From Fig. 4.7a–e, the UAVs exhibit altitude variations while collecting data from the GDs. This behavior reflects a balance between LoS probability and transmission distance. Specifically, when a UAV is at a lower altitude in a NLoS condition, it ascends to achieve a LoS state, and when in LoS, it reduces its Euclidean distance from the GD to improve the transmission rate. Additionally, UAV1 is observed to move closer to GD1, GD3, and GD4 during data collection, while UAV2 approaches GD2, GD5, and GD6. Both UAVs maintain consistent flight paths along their trajectories, which is due to their tendency to approach the assigned GDs to optimize data transmission. By selecting paths that bring them closer to the scheduled GDs, they minimize operational time and maintain consistent flight directions over certain intervals. Moreover, Fig. 4.7d and f show that in the absence of NFZs, UAVs can collect data from previously restricted locations, significantly reducing task completion time compared to scenarios where NFZs are present.

4.5.3 Impact of Different Data Size

Figure 4.8 shows the relationship between total operation time and data size per GD. In Fig. 4.8a and c, it is clear that as the data size per GD increases, both data collection time and overall operation time also rise, as larger data volumes require more time

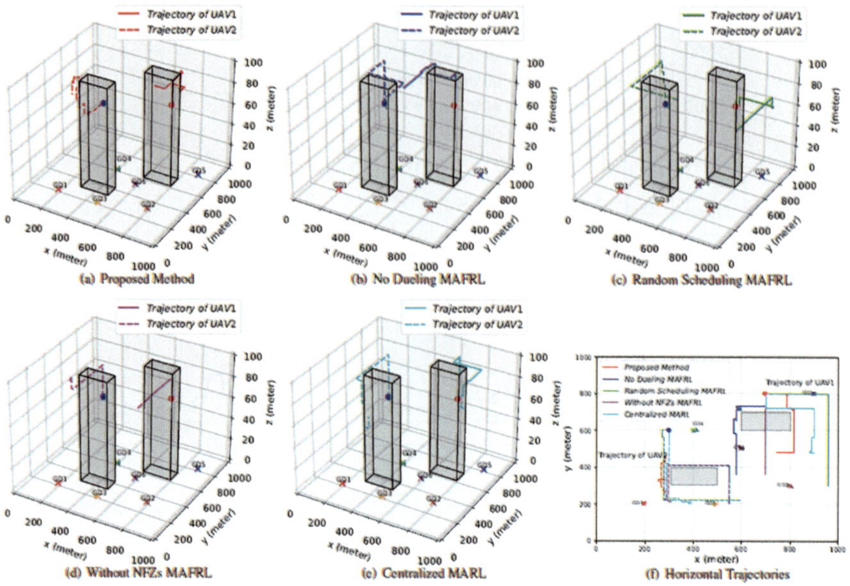

Fig. 4.7 Comparison of UAV trajectories under different schemes

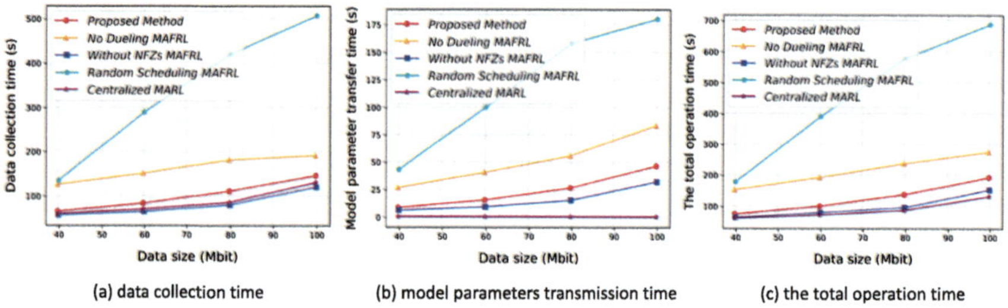

Fig. 4.8 Comparison of time cost under different schemes

for UAVs to complete their tasks. Figure 4.8b indicates that while the centralized MARL method's model parameter transmission time remains constant at zero, other methods show an increase in transfer time as data size grows. This is because the centralized MAFRL method does not require processing model parameters, unlike other approaches that involve more frequent parameter exchanges. However, the centralized MARL method requires UAVs to share sensitive information, such as their positions, raising privacy concerns. The proposed method consistently performs well as data size increases, approaching the performance levels of both the ideal no NFZs MAFRL method and the centralized MAFRL method. Moreover, compared to the random scheduling MAFRL and no-dueling MAFRL methods, our approach shows significantly improved performance, further confirming its effectiveness.

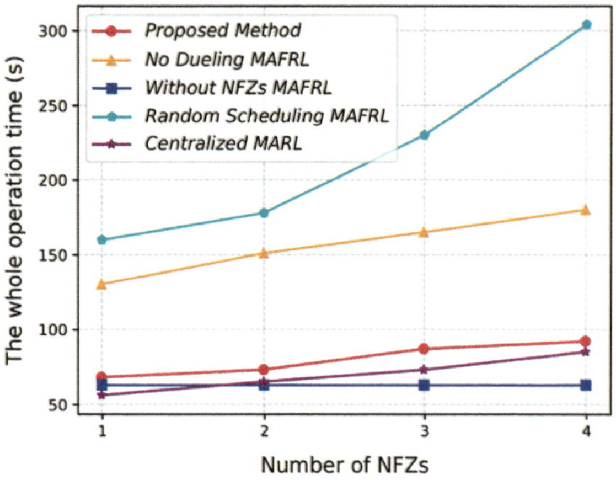

Fig. 4.9 Comparison of performance under different numbers of NFZs

4.5.4 Comparison of Methods with Different Numbers of NFZs

Figure 4.9 illustrates the impact of the number of NFZs on total operation time. The baseline method, which ignores NFZs (MAFRL), shows no change in operation time regardless of the number of NFZs. In contrast, other methods experience an increase in total operation time as the number of NFZs rises. This is due to the added complexity of the environment, forcing UAVs to take longer routes to complete their tasks. Additionally, the figure shows that when the number of NFZs is less than one, the centralized MARL method performs better than the MAFRL method without NFZs in terms of operation time. With fewer NFZs, their impact on system performance is minimal, and operation time mainly depends on the chosen method. As the number of NFZs grows, the advantages of the proposed method become more evident, especially when compared to the benchmark without NFZs and the no-dueling MAFRL approach.

4.6 Conclusion

This paper proposed a learning-based framework for optimizing UAV-assisted data collection in a 3D urban environment with multiple mobile GDs and NFZs. The objective is to minimize total operation time by jointly optimizing the UAVs' 3D trajectories and the communication scheduling for GDs. However, the problem is nonconvex, and the inclusion of GD mobility adds further complexity. Privacy concerns also arise due to risks such as server attacks, fake commands, computational load, and data leakage. To tackle these issues, we modeled the problem as a MDP and proposed a MAFRL method, where each UAV functions as an agent learning optimal policies for trajectory design and scheduling. Simulations were performed to optimize hyperparameters, and the proposed method was compared to traditional approaches. The study further explored the benefits of 3D trajectory design over 2D approaches and examined how varying numbers of NFZs and UAVs influence operation time. Moreover, the proposed method can be extended to more complex systems, including scenarios where GDs move in 3D space or involve multi-task operations. Future research will focus on enhancing the algorithm's effectiveness by designing a continuous reward function that discourages UAVs from entering NFZs or colliding with obstacles.

References

1. Li, B., Fei, Z., & Zhang, Y. (2019). UAV communications for 5G and beyond: Recent advances and future trends. *IEEE Internet of Things Journal, 6*, 2241–2263.
2. Zeng, Y., Zhang, R., & Lim, T. J. (2016). Wireless communications with unmanned aerial vehicles: Opportunities and challenges. *IEEE Communications Magazine, 54*, 36–42.
3. Zeng, Y., Wu, Q., & Zhang, R. (2019). Accessing from the sky: A tutorial on UAV communications for 5G and beyond. *Proceedings of the IEEE, 107*, 2327–2375.
4. Gao, Y., Yang, W., Wu, Y., Cui, Z., & Yu, Y. (2020). Robust trajectory and scheduling design for turning angle-constrained UAV communications. In *2020 WCSP* (pp. 962–967).
5. Hu, J., Cai, X., & Yang, K. (2020). Joint trajectory and scheduling design for UAV aided secure backscatter communications. *IEEE Wireless Communication Letters, 9*, 2168–2172.
6. Song, X., Zhao, Y., Wu, Z., Yang, Z., & Tang, J. (2022). Joint trajectory and communication design for IRS-assisted UAV networks. *IEEE Wirel. Commun. Lett., 11*, 1538–1542.
7. Sun, C., Xiong, X., Ni, W., & Wang, X. (2022). Three-dimensional trajectory design for energy-efficient UAV-assisted data collection. In *ICC 2022—IEEE ICC* (pp. 3580–3585).
8. Xiong, X., Sun, C., Ni, W., & Wang, X. (2023). Three-dimensional trajectory design for unmanned aerial vehicle-based secure and energy-efficient data collection. *IEEE Transactions on Vehicular Technology, 72*, 664–678.
9. Zhong, C., Zheng, G., Zhang, Z., & Karagiannidis, G. K. (2015). Optimum wirelessly powered relaying. *IEEE Signal Processing Letters, 22*, 1728–1732.
10. Li, Q., Zhao, J., Gong, Y., & Zhang, Q. (2019). Energy-efficient computation offloading and resource allocation in fog computing for internet of everything. *China Communications, 16*, 32–41.
11. Nie, Y., Zhao, J., Liu, J., Jiang, J., & Ding, R. (2020). Energy-efficient UAV trajectory design for backscatter communication: A deep reinforcement learning approach. *China Communications, 17*, 129–141.
12. Yang, Z., et al. (2021). AI-driven UAV-NOMA-MEC in next generation wireless networks. *IEEE Wireless Communications, 28*, 66–73
13. Xiao, Z., et al. (2024). Densely knowledge-aware network for multivariate time series classification. In *IEEE Transactions on Systems, Man, and Cybernetics: Systems* (pp. 1–13).
14. Xing, H., Xiao, Z., Qu, R., Zhu, Z., & Zhao, B. (2022). An efficient federated distillation learning system for multitask time series classification. *IEEE Transactions on Instrumentation and Measurement, 71*, 1–12.
15. Song, F., et al. (2023). Evolutionary multi-objective reinforcement learning based trajectory control and task offloading in UAV-assisted mobile edge computing. *IEEE Transactions on Mobile Computing, 22*, 7387–7405.
16. Wang, Y., et al. (2022). Trajectory design for UAV-based internet of things data collection: A deep reinforcement learning approach. *IEEE Internet of Things Journal, 9*, 3899–3912.

17. Li, Y., Aghvami, A. H., & Dong, D. (2022). Path planning for cellular-connected UAV: A DRL solution with quantum-inspired experience replay. *IEEE Wireless Communications, 21*, 7897–7912.
18. Zeng, Y., Xu, X., Jin, S., & Zhang, R. (2021). Simultaneous navigation and radio mapping for cellular-connected UAV with deep reinforcement learning. *IEEE Wireless Communications, 20*, 4205–4220.
19. Wu, F., Zhang, H., Wu, J., & Song, L. (2020). Cellular UAV-to-device communications: Trajectory design and mode selection by multi-agent deep reinforcement learning. *IEEE Transactions on Communications, 68*, 4175–4189.
20. Zhang, W., Wang, Q., Liu, X., Liu, Y., & Chen, Y. (2021). Three-dimension trajectory design for multi-UAV wireless network with deep reinforcement learning. *IEEE Transactions on Vehicular Technology, 70*, 600–612.
21. Liu, X., Liu, Y., & Chen, Y. (2019). Reinforcement learning in multiple-UAV networks: Deployment and movement design. *IEEE Transactions on Vehicular Technology, 68*, 8036–8049.
22. Xu, X., Feng, G., Qin, S., Liu, Y., & Sun, Y. (2023). Joint multi-UAV deployment and resource allocation based on personalized federated deep reinforcement learning. In *ICC 2023* (pp. 5677–5682).
23. Wu, S., Xu, W., Wang, F., Li, G., & Pan, M. (2022). Distributed federated deep reinforcement learning based trajectory optimization for air-ground cooperative emergency networks. *IEEE Transactions on Vehicular Technology, 71*, 9107–9112.
24. Rezwan, S., Chun, C., & Choi, W. (2023). Federated deep reinforcement learning-based multi-UAV navigation for heterogeneous NOMA systems. *IEEE Sensors J., 23*, 29722–29732.
25. Wang, Z., et al. (2016). Dueling network architectures for deep reinforcement learning. In *Proceedings of ICML, 1995–2003.*

Chapter 5
Time Minimization in Delay Constrained UAV Sensing

Kaitao Meng

Abstract The unmanned aerial vehicle (UAV) offers inherent mobility and advanced communication capabilities, making it suitable for agile deployment and adaptable observation compared to traditional fixed sensing systems. In various sensing applications, particularly those requiring low transmission delays, multiple distributed tasks must be efficiently managed. Optimizing the mission completion time in such delay-constrained multi-task UAV-based sensing scenarios necessitates careful design of inter-task trajectories, UAV-base station (BS) associations, and sensing sequences. However, achieving an optimal sensing scheme is challenging due to the interdependencies and non-convex constraints involved in these sub-problems. To address these challenges, this chapter introduces a space pruning-based trajectory search (SPTS) algorithm that leverages geometric properties to expedite the convergence compared to conventional methods like the Ployblock algorithm. Secondly, an optimal UAV-BS association algorithm is developed to efficiently determine the optimal BS for each sensing task. Finally, integrating these algorithms with the Lin-Kernighan-Helsgaun (LKH) algorithm enables computation of a lower bound for mission completion time and derivation of near-optimal solutions. Numerical simulations validate the effectiveness of the proposed approach, demonstrating significant performance enhancements in UAV-based sensing applications.

Keywords UAVs · Trajectory optimization · Transmission delay · UAV-BS association · Sensing sequence

Abbreviations

AoI	Age of Information
BS	Base Station
LKH	Lin-Kernighan-Helsgaun
LoS	Line-of-Sight

K. Meng (✉)
Department of Electronic and Electrical Engineering, University College London, London, UK
e-mail: kaitao.meng@ucl.ac.uk

Y. Hu et al. (eds.), *Reliable and Efficient Trajectory Design in UAV-Aided Wireless Networks*, https://doi.org/10.1007/978-981-96-5988-3_5

MO	Monotonic Optimization
NLoS	Non-Line-of-Sight
SNR	Signal-to-Noise Ratio
SPTS	Space Pruning-based Trajectory Search
TSP	Traveling Salesman Problem
UAV	Unmanned Aerial Vehicle

5.1 Introduction

5.1.1 Background

Unmanned aerial vehicles (UAVs), known for their flexibility in deployment and cost-effectiveness, are poised to transform existing fixed-deployed sensing systems [1, 2]. Specifically, using UAVs connected to cellular networks to conduct sensing and relay sensory data is expected to deliver quicker mission responses and reduce task completion times compared to traditional static sensing systems [3, 4]. However, in many sensing scenarios, such as disaster relief and real-time target monitoring, there are strict requirements for minimizing transmission delays [5–7]. After UAVs complete sensing, the data captured by their onboard sensors must be transmitted to the base station (BS) within a specific time frame to enable timely analysis and prevent data loss due to expiration. As a result, developing effective UAV-based sensing techniques is crucial for minimizing task completion time under transmission delay constraints [8, 9].

Research in UAV-assisted wireless communications has shown that trajectory optimization is a key method for enhancing communication latency [10–12]. For instance, in [13], a data collection strategy for time-sensitive IoT devices was optimized to maximize the number of devices served before a deadline, with UAV trajectories adjusted within a fixed time. In contrast, this study focuses on optimizing task completion time, with transmission delay defined as the period from UAV sensing to successful data upload. This difference introduces new challenges in minimizing task completion time in delay-sensitive sensing applications. A critical trade-off arises for UAVs in such scenarios: they must balance two conflicting objectives. On one side, the UAV should minimize travel time between sensing locations by following direct flight paths. On the other side, it may need to approach the BS to achieve higher data transmission rates and satisfy delay constraints. Although some prior research has tackled this problem, the solutions typically result in sub-optimal UAV trajectories. The problem of optimizing the sensing order in UAV-based sensing missions has also received considerable attention. However, in time-constrained sensing applications, UAVs may need to follow non-linear paths, rendering the Euclidean distance an unreliable measure for trajectory planning.

This chapter formulates the problem of optimizing UAV trajectories to minimize task completion time under transmission delay constraints. The primary challenges lie in the non-convexity of the delay constraints and the absence of a closed-form expression for the total data transmission. To overcome these difficulties, this work shows that the problem can be transformed into a monotonic optimization (MO) problem. However, directly applying the Polyblock algorithm to this MO problem leads to excessive complexity. To address this, we propose a novel Space Pruning-based Trajectory Search (SPTS) algorithm, which leverages the geometric characteristics of the optimal UAV trajectory to bypass unnecessary computations and achieve significantly faster convergence compared to standard Polyblock-based solutions.

5.1.2 Motivation

Optimizing trajectory design has proven to be an effective method for enhancing communication latency in traditional UAV-assisted wireless communication systems. Previous studies have explored approaches like heuristic search algorithms and successive convex optimization techniques to tackle the UAV trajectory optimization problem. However, these works primarily focus on communication performance without considering the sensing process itself, such as the order of sensing tasks or the generation of sensory data. Recent advancements have begun addressing the trajectory optimization problem specifically for UAV-based sensing. Nevertheless, these studies often yield sub-optimal solutions and fail to address the inaccuracies in UAV positioning that may arise from imperfect control mechanisms. In the realm of UAV-assisted wireless sensor networks, it has been demonstrated that optimizing the visiting sequence of sensor nodes can significantly improve data collection performance. While these efforts have focused mainly on data collection in wireless networks, they have not fully addressed transmission delay optimization, which is critical for time-sensitive UAV-based sensing missions. However, these studies typically assume a fixed wireless sensor network, with UAVs collecting data from pre-deployed sensors.

5.1.3 Main Contributions

Main contributions to this chapter are as follows:

(1) Present a delay-constrained multi-task UAV-based sensing problem aimed at minimizing overall mission completion time. It is demonstrated that this can be achieved through optimizing the inter-task UAV trajectory, determining the UAV-BS association for each task, and arranging the sensing order.
(2) Introduce a novel Space Pruning-based Trajectory Search (SPTS) algorithm to reduce flight time between task areas while accounting for control errors. This algorithm improves efficiency by leveraging the geometric properties of the optimal UAV trajectory, thereby avoiding redundant searches.

(3) Develop a new Optimal UAV-BS Association (OUBA) algorithm that eliminates sub-optimal base stations by incorporating auxiliary trajectories and virtual base stations, allowing for a more efficient identification of the optimal UAV-BS pairings.
(4) Propose the OASPTS sensing strategy, which not only calculates a lower bound for the mission completion time but also generates a near-optimal sensing order.

5.1.4 Structural Organization

The remainder of the chapter is organized as follows: Sect. 5.2 provides discussion on related works, and Sect. 5.3 presents the problem formulation and preliminaries. Section 5.4 analyzes the task completion time minimization problem and its conversion into a MO problem. Section 5.5 discusses the UAV-BS association. Section 5.6 describes the sensing task order design. Section 5.7 provides simulation results, and conclusions are drawn in Sect. 5.8.

5.2 Related Works

Trajectory design is a proven method to enhance communication latency in conventional UAV-assisted wireless communications [14, 15]. Various approaches, including heuristic search [16] and successive convex optimization algorithms [17, 18], have been applied to optimize UAV trajectories. However, these methods often overlook key aspects of the sensing process, such as determining sensing order and generating sensory data. Recently, several pioneering studies have focused on optimizing trajectories specifically for UAV-based sensing tasks. For instance, the authors in [19] introduced an online sensing and navigation algorithm that adjusts the UAV's path within a defined flight corridor using the Bessel curve model. Another study [20] proposed an iterative trajectory optimization approach aimed at maximizing the worst-case secrecy rate given uncertain adversary locations. Additionally, Zhanget al. [21] presented an optimized turning line algorithm that iteratively adjusts the sensing positions and trajectory turning points based on a pre-defined sensing order. Deep learning methods have also been explored, as seen in [11, 22], where algorithms use a return function to align with sensing goals, maximizing neural network performance through training. Nonetheless, these methods generally yield only suboptimal inter-task trajectories and do not account for positional inaccuracies in UAVs due to control limitations.

Current research on UAV-assisted wireless sensor networks has shown that data collection can be enhanced by optimizing the visiting order of sensor nodes [23–25]. Most of these studies focus on UAV-aided data collection in wireless networks. However, optimizing transmission delay for time-sensitive UAV-based sensing remains underexplored. Additionally, another line of research addresses the age of information (AoI), defined as the elapsed time since data was generated at source nodes, indicating data freshness, in UAV-aided data collection for pre-deployed wireless sensor networks [26–28]. These AoI-focused studies often rely on a fixed wireless sensor network for data acquisition, whereas in UAV-based sensing, onboard sensors enable more flexible data gathering. Instead of AoI, our work considers the total completion time of multi-task sensing missions to more accurately represent UAV travel times between task zones. Additionally, optimizing the sensing order for UAV-based applications has gained attention. For example, Zhang et al. [29] proposes a method to minimize UAV energy consumption by iteratively adjusting each task's sensing position. In the above work, the task visit order is determined by solving a traveling salesman problem (TSP) based on the Euclidean distances between task areas. However, given transmission delay constraints in these sensing applications, UAVs may need to fly along curved paths between task zones, where Euclidean distance alone may not accurately reflect trajectory length.

5.3 System Model and Problem Formulation

To examine the UAV-based sensing process, we first introduce the system model, followed by the formulation of the mission completion time minimization problem. Specifically, we consider a UAV sensing network connected to a cellular system, as illustrated in Fig. 5.1. The UAV is tasked with performing a set of sensing operations, denoted as $K = \{1, ..., K\}$. Given the real-time nature of these sensing tasks, the sensory data collected from each task k must be transmitted to a specific BS, indexed by $m \in M = \{1, ..., M\}$, within a predefined deadline τ_k.

5.3.1 UAV Flight Model

In this scenario, it is assumed that all targets are located on the ground, with their horizontal positions represented by $\boldsymbol{u}_k$ for $k \in K$. When the UAV is linked to the mth BS and moving towards the next task target j, the total mission completion time $T_{k,j,m}$

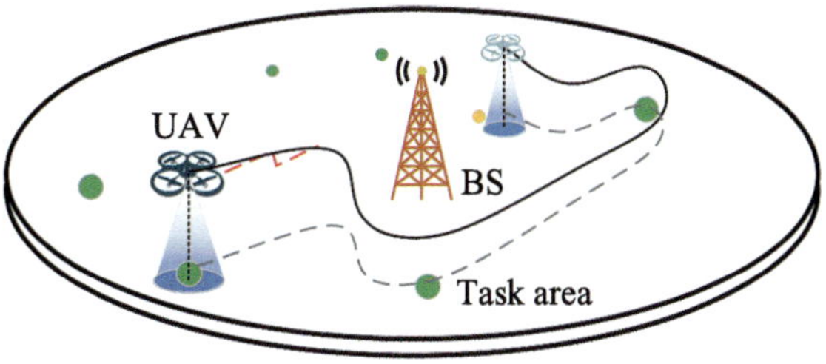

Fig. 5.1 System architecture

for task k is comprised of several components: the sensing time T_k^s, the transmission time $T_{k,j,m}^t$, the non-transmission time $T_{k,j,m}^n$, and the mode adjustment time T^a:

$$T_{k,j,m} = T_k^s + T_{k,j,m}^t + T_{k,j,m}^n + T^a. \tag{5.1}$$

During the sensing phase, the UAV hovers over the target location $\boldsymbol{u}_k$ at a fixed altitude h and collects data. As detailed before, the sensing time is determined by the required data quality. The transmission phase involves the UAV flying towards a designated BS to transmit the collected data. After the data has been transmitted, the non-transmission time refers to the remaining flight time before reaching the next task area. Additionally, the mode adjustment time accounts for the UAV's transition between flight and hovering modes by adjusting its altitude and velocity. Consequently, the total flight time from $\boldsymbol{u}_k$ to $\boldsymbol{u}_j$ can be expressed as:

$$T_{k,j,m}^f = T_{k,j,m}^t + T_{k,j,m}^n. \tag{5.2}$$

For simplicity, the continuous UAV trajectory between two task areas $\boldsymbol{u}_k$ and $\boldsymbol{u}_j$ is approximated by a series of discrete points $\boldsymbol{s}_{k,j} = \{\boldsymbol{s}_{k,j}[1], ..., \boldsymbol{s}_{k,j}[N_{k,j}]\}$, which are sampled along the trajectory at intervals of δ seconds. Each point $\boldsymbol{s}_{k,j}[n] = (x_k[n], y_k[n])$ represents the UAV's position at the nth time frame in the sequence. The length of the discrete trajectory, $N_{k,j}$, depends on the flight time. Given the UAV's maximum speed v_{max}, the positions of the UAV must satisfy:

$$\boldsymbol{s}_{k,j}[n] - \boldsymbol{s}_{k,j}[n-1] \leq v_{max}\delta. \tag{5.3}$$

However, the actual UAV location $\tilde{\boldsymbol{s}}_{k,j}$ may differ from the target location $\boldsymbol{s}_{k,j}$ due to imperfect control, and is described by:

$$\tilde{x}_k[n] = x_k[n] + \Delta x_k[n], \quad \text{and} \quad \tilde{y}_k[n] = y_k[n] + \Delta y_k[n], \tag{5.4}$$

where $(\Delta x_k[n], \Delta y_k[n])$ accounts for the location error between the actual and target positions. Additionally, the UAV's actual altitude is represented by $\tilde{\mathrm{h}} = \mathrm{h} + \Delta \mathrm{h}$. The error in the UAV's position is bounded by:

$$\Delta x_k[n]^2 + \Delta y_k[n]^2 + (\Delta \mathrm{h})^2 \leq \varepsilon_e^2, \tag{5.5}$$

where ε_e^2 represents the error variance. The UAV's starting location (such as a control centre or charging station) is represented by $\boldsymbol{s}_0$, and it is assumed that the UAV must return to this location after completing all tasks. The flight time from the starting location to the first task area $\boldsymbol{u}_k$ is denoted as $T_{0,k,m}^f$, and is given by:

$$T_{0,k,m} = T_{0,k,m}^f = \frac{\boldsymbol{s}_0 - \boldsymbol{u}_k}{v_{max}}. \tag{5.6}$$

5.3.2 Communication Model

The horizontal coordinates of the mth BS are represented by $\boldsymbol{u}_m^{BS}$, assuming that all BSs are positioned at the same altitude, denoted by h^{BS}. When the UAV transitions from the kth task area to the next area $\boldsymbol{u}_j$, the distance from the UAV to the mth BS at a given time frame $1 \leq n \leq N_{k,j}$ can be expressed as:

$$d_{k,m}[n] = \sqrt{s_{k,j}[n] - \boldsymbol{u}_m^{BS2} + \left(\mathrm{h} - \mathrm{h}^{BS}\right)^2} \tag{5.7}$$

Considering uncertainties in the UAV's location, the actual distance between the UAV and the BS is modified as:

$$\tilde{d}_{k,m}[n] = d_{k,m}[n] + \Delta d_{k,m}[n]. \tag{5.8}$$

From the location error model described in (5.5), this uncertainty is bounded by $\Delta d_{k,m}[n] \leq \varepsilon_e$. Assuming that the communication between the UAV and the BS relies heavily on the line-of-sight (LoS) component [30], the channel power gain is modelled following [31] as:

$$\mathrm{h}_{k,m}[n] = \rho_0 \tilde{d}_{k,m}^{-\alpha}[n], \tag{5.9}$$

where ρ_0 refers to the power gain at a reference distance, and $\alpha \geq 2$ is the path-loss exponent. Consequently, the signal-to-noise ratio (SNR) at the mth BS is:

$$\gamma_{k,m}[n] = \frac{P_S \cdot \mathrm{h}_{k,m}[n]}{\sigma^2}, \tag{5.10}$$

where P_S represents the UAV's transmit power, and σ^2 is the noise power. The transmission rate to the mth BS during the nth frame, for data gathered from the kth task area, is then given by:

$$R_{k,m}[n] = B \cdot \log_2\left(1 + \gamma_{k,m}[n]\right), \tag{5.11}$$

where B stands for the bandwidth. To ensure that the data collected from the kth task area is transmitted to the BS within the predefined deadline τ_k, the transmission rate must satisfy the following condition:

$$\sum_{n=1}^{\psi(\tau_k)} R_{k,m}[n] \cdot \delta \geq C_k, \forall k, \tag{5.12}$$

where $\psi(\cdot) = \frac{\cdot}{\delta}$ is the ceiling function, and C_k represents the data size for task k. Furthermore, it is assumed that the transmission deadline τ_k is much shorter compared to the UAV's flight time $T^{f}_{k,j,m}$ between task areas. Hence, the UAV will complete transmitting the sensory data before reaching the next task area.

5.3.3 Sensing Model

The probabilistic sensing model commonly used in the literature [32, 33] is applied in this work. According to this model, the probability of successful sensing by the UAV is described by $p(z) = e^{-\nu \cdot z}$, where ν is a parameter that measures sensing performance [32], and z is the distance between the UAV and the target area, calculated as:

$$z = \sqrt{\Delta x^2 + \Delta y^2 + (\mathrm{h} + \Delta \mathrm{h})^2}, \tag{5.13}$$

based on the uncertain location model. To ensure that the sensing success probability meets or exceeds a predefined threshold Pth, the UAV may need to conduct multiple sensing attempts. The minimum sensing time required is:

$$T^{s}_{k} = t^{s} \cdot \frac{\ln\left(1 - P^{\mathrm{th}}\right)}{\ln(1 - p(\mathrm{h} + \varepsilon_e))}, \tag{5.14}$$

where t^s represents the time interval between consecutive sensing operations.

5.3.4 Problem Formulation

Then, the mission completion time minimization problem can be formulated as:

$$(\mathrm{P1}) \min_{\{s_{k,j}\},\{b_{k,m}\},\pi} T = \sum_{k=0}^{K} \sum_{m=1}^{M} b_{k,m} \cdot T_{\pi(k),\pi(k+1),m},$$

$$s.t \ (5.3) \text{ and } (5.13) \tag{5.15}$$

$$b_{k,m} \cdot \sum_{n=1}^{\psi(\tau_k)} R_{k,m}[n] \cdot \delta \geq b_{k,m} \cdot C_k, \forall m, k, \tag{5.15a}$$

$$\sum_{m=1}^{M} b_{k,m} = 1, \forall k, \tag{5.15b}$$

$$b_{k,m} \in \{0, 1\}, \tag{5.15c}$$

$$(\pi(1), \ldots, \pi(K)) \in \Pi, \tag{5.15d}$$

$$\pi(K+1) = \pi(0) = 0, \tag{5.15e}$$

$$\left\| s_{k,j}[1] - u_k \right\| \leq v_{max} \cdot \delta, \forall k, \tag{5.15f}$$

$$s_{k,0}\left[N_{k,0}\right] = s_0, \forall k, \tag{5.15g}$$

$$s_{k,j}\left[N_{k,k'}\right] = u_{k'}, \forall k, j, j \tag{5.15h}$$

Here, $\underline{R}_{k,m}[n]$ is defined as the worst-case transmission rate, given by:

$$\underline{R}_{k,m}[n] \triangleq \min_{(\Delta x_k[n], \Delta y_k[n], \Delta \mathrm{h})} R_{k,m}[n]. \tag{5.16}$$

In this formulation, the variable $b_{k,m}$ indicates the association between the UAV and the BS for task k: $b_{k,m} = 1$ if the UAV communicates with the mth BS, otherwise $b_{k,m} = 0$. The vector $\boldsymbol{\pi} = (\pi(0), ..., \pi(K+1))$ denotes the sequence in which the UAV performs the sensing tasks. Constraint (P1a) guarantees that the worst-case communication rate satisfies the transmission delay requirement, even considering UAV location deviations. Constraint (P1b) ensures that data from each task is transmitted to only one BS. Constraint (P1g) ensures that the UAV starts from and returns to its predefined location s_0.

5.4 Optimal Inter-Task Trajectory Design

In this section, we delve into the design of inter-task UAV trajectories with a predefined UAV-BS association.

5.4.1 *Optimal UAV Trajectory*

Here, we demonstrate that the optimal inter-task UAV trajectory can be classified into two scenarios. The subproblem is subsequently transformed into a monotonic optimization by introducing auxiliary variables and proving its inherent monotonicity.

Given a UAV-BS association, the trajectory optimization problem between two task areas, denoted as u_k and u_j, is expressed as:

$$\text{(P2)}\ \min_{s_{k,j}} T_{k,j,m} = T_k^s + T_{k,j,m}^f,$$
$$s.t\ \ (5.3),\ \ (5.14), \tag{5.17}$$

$$\sum_{n=1}^{\psi(\tau_k)} \underline{R}_{k,m}[n] \cdot \delta \geq C_k, \tag{5.17a}$$

$$\left\| s_{k,j}[1] - \boldsymbol{u}_k \right\| \leq v_{max} \cdot \delta, \tag{5.17b}$$

$$s_{k,j}\left[N_{k,j}\right] = u_j. \tag{5.17c}$$

Since the sensing time in (1) is independent of the UAV trajectory, (P2) can be simplified by minimizing the flight time. To further reduce complexity, the optimal inter-task trajectory is categorized into two cases based on transmission delay, with the associated optimization transformed into a monotonic one.

For the UAV to transmit more data efficiently, it should travel at maximum speed towards the BS and hover as close as possible to it. The left-hand side of the inequality in Lemma 5.1 quantifies the transmitted data along this trajectory. Location errors are considered via the factor ε_e in (5.14). Thus, if a feasible solution exists for (P2), the maximum data transmission must exceed C_k.

If a feasible UAV trajectory exists, the optimal inter-task UAV trajectory can only take one of two forms:

Case A: The UAV flies directly from $\boldsymbol{u}_k$ to $\boldsymbol{u}_j$ at maximum speed.

Case B: The UAV flies towards the BS along a curved path, then proceeds to the next task area after transmission.

For Case B, it is not difficult to verify that the following condition applies: $\left(T_{k,j,m}^t - \tau_k\right) \cdot T_{k,j,m}^n = 0$. In Case A, all sensory data can be transmitted within the delay constraint τ_k when the UAV flies directly between task areas $\boldsymbol{u}_k$ and $\boldsymbol{u}_j$. However, if this does not hold, the UAV must follow a curved path towards the BS to enhance the data rate and ensure timely transmission, as described in Case B.

Then, we further provide solution to the optimal UAV trajectory in Case B. In this scenario, the UAV completes data transmission before arriving at the next task area, implying $T_{k,j,m}^t = \tau_k$. This indicates that the flight time equals $\tau_k + T_{k,j,m}^n$. To minimize the non-transmission time $T_{k,j,m}^n$, the UAV travels along a direct path to the next task area once data transmission is complete. Then, $T_{k,j,m}^n = s_{k,j}[\psi(\tau_k)] - \boldsymbol{u}_j / v_{max}$, where $s_{k,j}[\psi(\tau_k)]$ represents the UAV's location at the end of transmission. To determine the optimal inter-task UAV trajectory in Case B, the following subproblem must be solved:

$$\text{(P2.1)}\ \min_{s_{k,j}} T_{k,j,m} = T_k^s + T_{kj,m}^f,$$
$$s.t\ \ (5.3) \tag{5.18}$$

$$\sum_{n=1}^{\psi(\tau_k)} \underline{R}_{k,m}[n] \cdot \delta \geq C_k, \tag{5.18a}$$

$$\left\| \boldsymbol{s}_{k,j}[1] - \boldsymbol{u}_k \right\| \leq v_{max} \cdot \delta, \tag{5.18b}$$

$$\boldsymbol{s}_{k,j}\left[N_{k,j}\right] = u_j. \tag{5.18c}$$

The inter-task UAV trajectory optimization problem is transformed into a monotonic optimization using auxiliary variables in the next subsection.

5.4.2 Proposed Accelerated Trajectory Search Algorithm

In this section, we introduce the SPTS algorithm, designed to speed up convergence. The core concept behind SPTS is to narrow the search space for the optimization variables $\boldsymbol{\theta}_{k,m}$, leveraging the geometric properties of the optimal UAV inter-task trajectory.

In the original formulation of the monotonic optimization problem, the range of values for $\theta_{k,m}[n]$ is expressed as:

$$\left[0, \left(\left(h - h^{BS}\right)^2 + \max_{i=k,j} \left\| u_i - u_m^{BS} \right\|^2\right)^{\frac{\alpha}{2}}\right]. \tag{5.19}$$

The UAV's position $s_{k,j}[n]$ in each frame n is limited by its maximum velocity and must lie within the overlap of two spheres $\Theta_1(u_k, n \cdot \delta \cdot v_{max})$ and $\Theta_2\left(u_j, \left(N_{k,j} - n\right) \cdot \delta \cdot v_{max}\right)$, as illustrated in Fig. 5.2. Here, the first argument represents the sphere's centre, while the second denotes its radius. Then, the range of $\theta_{k,m}[n]$ can be constrained based on the distance from the overlapping region $\Theta_1 \cap \Theta_2$ to the mth base station.

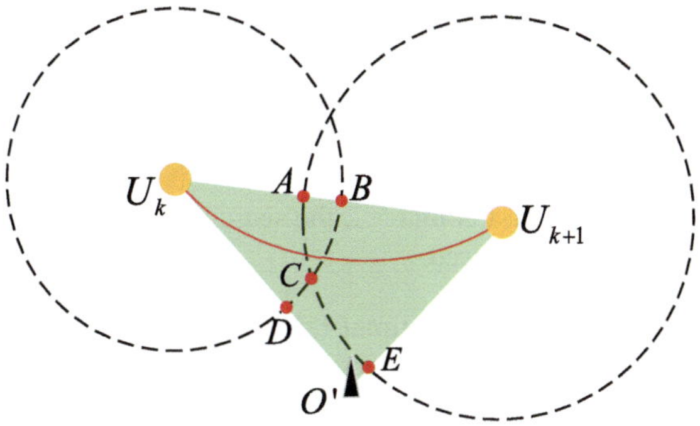

Fig. 5.2 Optimal UAV trajectory confined within a triangular region

Lemma 5.1 The optimal UAV trajectory $\boldsymbol{s}_{k,j}^{*} = \left\{\boldsymbol{s}_{k,j}^{*}[1], \ldots, \boldsymbol{s}_{k,j}^{*}\left[N_{k,j}\right]\right\}$ connecting the task areas $\boldsymbol{u}_k$ and $\boldsymbol{u}_j$ must lie within the triangular region $\Delta_{k,j,m}$ formed by the points $u_k = (\boldsymbol{u}_k, \mathrm{h})$, $u_j = \left(\boldsymbol{u}_j, \mathrm{h}\right)$, and $O' = \left(u_m^{BS}, \mathrm{h}\right)$, as depicted by the blue region in Fig. 5.1. The lower bound $\underline{\theta}_{k,m}[n]$. corresponds to the nearest point to the base station inside the intersection $\Theta_1 \cap \Theta_2 \cap \Delta_{k,j,m}$, reprented by one of the intersection points C, D, or E in Fig. 5.1. The upper bound $\overline{\theta}_{k,m}[n]$ is set by the furthest points, A or B.

Proof Assume that in the optimal UAV trajectory $\boldsymbol{s}_{k,j}^{*} = \left\{\boldsymbol{s}_{k,j}^{*}[1], \ldots, \boldsymbol{s}_{k,j}^{*}\left[N_{k,j}\right]\right\}$ from task area $\boldsymbol{u}_k$ to task area $\boldsymbol{u}_j$, there exists a sub-trajectory $s_{k,j'} = \left\{s_{k,j'}[n_1], \ldots, s_{k,j'}[n_2]\right\}$ outside the triangular area $\Delta_{k,j,m}$. In the following, it will be proved that there exists another trajectory all within the triangular area, whose performance is no worse than $\boldsymbol{s}_{k,j}^{*}$. The path point belonging to $\boldsymbol{s'}_{k,j}$ can be projected vertically onto the nearest boundary of $\Delta_{k,j,m}$. Obviously, the projection point is closer to the BS than the original path point and thus the transmission rate at the location of the projection point is also better than that of the corresponding original path point. Besides, the projection path can be improved further to decrease the task completion time. Hence, the optimal inter-task UAV trajectory must be within the horizontal triangular area $\Delta_{k,j,m}$. Also, limited by the UAV's maximum moving speed, the location $\boldsymbol{s}_{k,j}^{*}[n]$ of the UAV in each frame n must be within the intersection of the two balls $\Theta_1(\boldsymbol{u}_k, n \cdot \delta_t \cdot v_{max})$ and $\Theta_2\left(\boldsymbol{u}_j, \left(N_{k,j} - n\right) \cdot \delta_t \cdot v_{max}\right)$. Then the path point $\boldsymbol{s}_{k,j}[n]$ in the optimal inter-task UAV trajectory must be within the intersection region $\Omega_{k,j} = \Theta_1 \cap \Theta_2 \cap \Delta_{k,j,m}$.

It can be readily verified that the lower bound $\underline{\theta}_{k,m}[n]$ and the upper bound $\overline{\theta}_{k,m}[n]$ of the variable $\theta_{k,m}[n]$ is determined by the feasible distance $d_{k,m}[n]$ from the intersection area $\Omega_{k,j,m}$ to the associated BS. Then, it will be proved that the closest position from $\Omega_{k,j,m}$ to the BS is one of the intersection points C, D, and E, and the farthest position to the BS is one of the points A and B (c.f. the red dot in Fig. 5.2). It can be seen that the closest points from O′ to the two balls Θ_1 and Θ_2 are the points D and E, respectively. If one of the points D and E is within the region $\Omega_{k,j,m}$, the closest point to the BS is the corresponding point within the region. Otherwise, if both points D and E are outside the region,$\Omega_{k,j,m}$ it is not difficult to verify that the closest point is C according to the geometric relationship between O′ and $\Omega_{k,j,m}$. As a result, $\underline{\theta}_{k,m}[n]$ is determined by the closer one of the points C, D, and E inside the region $\Omega_{k,j,m}$. Similarly, since the farthest point in the intersection area must be within the interval between the points A and B, $\overline{\theta}_{k,m}[n]$ is determined by the farther one of the points A and B.

By utilizing Lemma 5.1, the search space for the optimal solution to (P2) can be significantly reduced.

Lemma 5.2 In the optimal UAV trajectory $\boldsymbol{s}_{k,j}^{*}$ between task areas $\boldsymbol{u}_k$ and $\boldsymbol{u}_j$, the variables $\boldsymbol{\theta}_{k,m}$ satisfy a convexity condition:

$$\theta_{k,m}[\beta n_1 + (1-\beta)n_2] \leq \beta\theta_{k,m}[n_1] + (1-\beta)\theta_{k,m}[n_2], \tag{5.20}$$

where n_1, n_2, $\beta n_1 + (1-\beta)n_2 \in \left\{1, ..., \psi\left(T^t_{k,j,m}\right)\right\}$, and $\beta \in [0, 1]$, ensuring that $\beta n_1 + (1-\beta)n_2$ is an integer.

Proof In the optimal UAV path $\boldsymbol{s}^*_{k,j}$ from task location $\boldsymbol{u}_k$ to $\boldsymbol{u}_j$, let points A and B represent any two positions on this path, occurring at times t_1 and t_2 (where $t_1 < t_2$), respectively. Let O denotes the location of the associated BS m, with distances from points A and B to this BS represented as $d_{k,m}(t_1) = |AO|$ and $d_{k,m}(t_2) = |BO|$. These distances, denoted $d_{k,m}(t)$, are interpolated from $d_{k,m}[n]$.

To show that a straight flight from A to B at constant speed satisfies the convexity property relative to the BS, consider any time $t = \beta t_1 + (1-\beta)t_2$, where $\beta \in [0, 1]$. The distance from the trajectory to the BS at time t is $d(t) = |CO|$, which fulfills $d(t) \le \beta|AO| + (1-\beta)|BO|$. Thus, the direct path remains convex.

Next, consider the points A' and B' in Fig. 5.3 where $\left|A'O\right| = \beta|AO|$ and $\left|B'O\right| = (1-\beta)|BO|$. Given that $t = t_2 - \beta(t_2 - t_1)$ and the UAV speed is constant, $|AC| = (1-\beta)|AB|$. Thus, $A'C \| OB$ and $B'C \| OA$, forming a parallelogram $OA'CB'$, which implies $\left|CA'\right| = \left|B'O\right|$. By the triangle inequality, $|CO| \le \beta|AO| + (1-\beta)|BO|$.

Obviously, the optimal trajectory lies within the triangular region in Fig. 5.1, making the BS distance on this trajectory closer than on a direct path from A to B. Thus, in the optimal path, the UAV-BS distance at time $t = \beta t_1 + (1-\beta)t_2$ satisfies

$$d_{k,m}(t) \le \beta d_{k,m}(t_1) + (1-\beta)d_{k,m}(t_2) \tag{5.21}$$

For simplify notation, let $x_1 = d_{k,m}(t_1)$ and $x_2 = d_{k,m}(t_2)$, and we now show $d_{k,m}(t)^\alpha \le \beta x_1^\alpha + (1-\beta)x_2^\alpha$ where $\alpha \ge 2$. Define $g(x_1, x_2) = \beta x_1^\alpha + (1-\beta)x_2^\alpha - d_{k,m}(t)^\alpha$, and substituting (5.21) into $g(x_1, x_2)$, we find

$$g(x_1, x_2) \ge \beta x_1^\alpha + (1-\beta)x_2^\alpha - (\beta x_1 + (1-\beta)x_2)^\alpha = \tilde{g}(x_1, x_2) \tag{5.22}$$

Assuming $x_1 \le x_2$, the partial derivative of $\tilde{g}(x_1, x_2)$ with respect to x_1 yields

$$\begin{aligned}\frac{\partial \tilde{g}}{x_1} &= \alpha\beta x_1^{\alpha-1} - \alpha\beta(\beta x_1 + (1-\beta)x_2)^{\alpha-1} \\ &\le \alpha\beta\left(x_1^{\alpha-1} - (\beta x_1 + (1-\beta)x_1)^{\alpha-1}\right) = 0.\end{aligned} \tag{5.23}$$

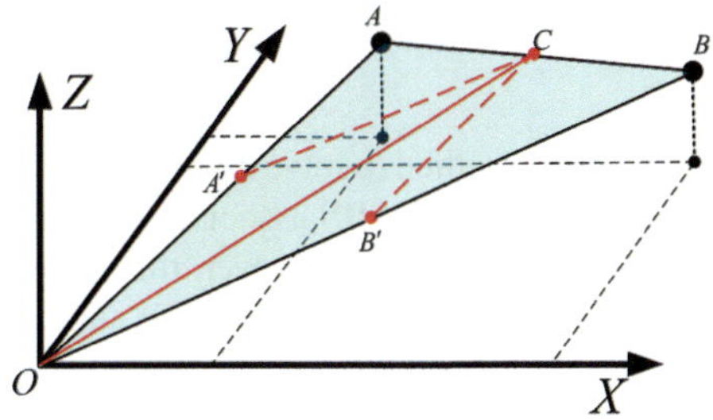

Fig. 5.3 Illustration of proof of Lemma 5.2

Thus, $\frac{\partial \tilde{g}}{\partial x_1} \le 0$, implying $\tilde{g}(x_1, x_2) \ge \tilde{g}(x_2, x_2) = 0$, and therefore $g(x_1, x_2) \ge 0$. Consequently, the optimal UAV trajectory maintains

$$d^{\alpha}_{k,m}(\beta t_1 + (1-\beta)t_2) \le \beta d^{\alpha}_{k,m}(t_1) + (1-\beta) d^{\alpha}_{k,m}(t_2) \tag{5.24}$$

The constraints in (5.18b) hold at the optimal solution, so $d^{\alpha}_{k,m}[n] = \theta_{k,m}[n]$. And $\theta_{k,m}[n]$ satisfies $\theta_{k,m}[\beta n_1 + (1-\beta)n_2] \le \beta\theta_{k,m}[n_1] + (1-\beta)\theta_{k,m}[n_2]$.

Utilizing Lemma 5.2, each new copolyblock $Q^{(r+1)}$ in the $(r+1)$th iteration of the Polyblock algorithm can be narrowed down by removing solutions that violate (5.17).

5.5 UAV-BS Association

In this section, we examine the UAV-BS association and sensing order subproblems of (P1with the objective of minimizing the mission completion time, based on the optimal UAV trajectory determined in the previous section.

The complexity of identifying the optimal UAV-BS association increases with the number of BSs. To reduce this complexity, BSs that are not optimal can be excluded based on a comparison of the UAV's distance to the task area and the maximum achievable data transmission. The UAV-BS association subproblem is then solved by comparing the minimum task completion times for the remaining base stations, referred to as candidate BSs.

As the UAV transitions between task areas $\boldsymbol{u}_k$ and $\boldsymbol{u}_j$, the minimum task completion time for task k is obtained by solving the following subproblem of (P1):

$$\text{(P3)}: \min_{\{s_{k,j}\}, b_{k,m}} \; T_{k,j} = \sum_{\text{m}=1}^{M} \left(b_{k,m} \cdot T_{k,j,m}\right), \tag{5.25}$$

$$s.t \;\; (5.3), \;\; \text{(P2b)}, \;\; \text{(P2c)},$$

$$b_{k,m} \cdot \sum_{n=1}^{\psi(\tau_k)} \underline{R}_{k,m}[n] \cdot \delta \ge b_{k,m} \cdot C_k, \forall m, \tag{5.25a}$$

$$\sum_{m=1}^{M} b_{k,m} = 1. \tag{5.25b}$$

A simple method to solve this problem involves calculating the optimal trajectories for all possible UAV-BS associations and then selecting the one that minimizes the task completion time. However, this approach becomes computationally expensive as the number of BSs increases. To address this, the following lemmas help reduce the candidate BS set $\Upsilon_{k,j}$.

Lemma 5.3 If the distances between the task areas and BSs satisfy the following conditions:

1. $||\boldsymbol{u}_i - \boldsymbol{u}_m^{BS}|| \leq ||\boldsymbol{u}_i - \boldsymbol{u}_{m'}^{BS}||$,
2. $||\boldsymbol{u}_i - \boldsymbol{u}_m^{BS}|| \leq ||\boldsymbol{u}_j - \boldsymbol{u}_{m'}^{BS}||$,

then the m'th BS can be eliminated from the candidate BS set $\Upsilon_{k,j}$.

Proof Condition (1): $||\boldsymbol{u}_i - \boldsymbol{u}_m^{BS}|| \leq ||\boldsymbol{u}_i - \boldsymbol{u}_{m'}^{BS}||$. Assume that the optimal associated BS is the m'th BS, the optimal inter-task UAV trajectory is $\boldsymbol{s}_{k,j}^*$, and the corresponding minimum flight time is $T_{k,j,m'}^{f*}$ In the following, it will be proved that the minimum flight time $T_{k,j,m}^{f*} \leq T_{k,j,m'}^{f*}$, which in turn implies that the m'th BS is non-optimal.

For the ease of analysis, define two regions X_1 and X_2, which are divided by the vertical bisector of the mth BS and the m'th BS (c.f. the green dotted line in Fig. 5.4. According to Lemma 5.1, the optimal inter-task UAV trajectory must be inside the triangle area $\Delta_{k,j,m'}$ when the UAV is associated with the m'th BS. If the optimal inter-task UAV trajectory $\boldsymbol{s}_{k,j}^*$ is inside the region X_2, then the transmission rate under the UAV trajectory $\boldsymbol{s}_{k,j}^*$ to the mth BS is higher than that to the m'th BS. Hence, the UAV trajectory $\boldsymbol{s}_{k,j}^*$ can be further improved to make the flight time shorter when the UAV is associated with the mth BS. Hence $T_{k,j,m}^{f*} < T_{k,j,m'}^{f*}$. If there exists a part of the optimal UAV trajectory $s_{k,j}'$ is inside the region X_1, a new UAV trajectory $\tilde{s}_{k,j}'$ (c.f. the black dotted line in Fig. 5.4) is obtained through symmetrically mapping $s_{k,j}'$ against the green dotted line. Another new trajectory can be obtained by combining the sub-trajectory of $\boldsymbol{s}_{k,j}^*$ in the region X_2, and the trajectory $\tilde{s}_{k,j}'$. When the UAV is associated with the mth BS, the amount of transmitted data under the constructed trajectory is larger than C_k. As a result, the optimal UAV trajectory can be further improved. Hence, the m'th BS can be eliminated from the candidate BS set if condition (1) in Lemma 5.3 holds.

Condition (2): $||\boldsymbol{u}_i - \boldsymbol{u}_m^{BS}|| \leq ||\boldsymbol{u}_j - \boldsymbol{u}_{m'}^{BS}||$. For the ease of discussion, for each BS m', introduce a virtual BS m'', with a location symmetrical to the m'th BS with respect to the vertical bisector of line segment by connecting points k and j. In addition, the optimal inter-task UAV trajectory $\boldsymbol{s}_{k,j}^*$ is symmetrically mapped along the vertical bisector of the line segment to obtain the trajectory $\tilde{\boldsymbol{s}}_{k,j}^*$, whose total flight

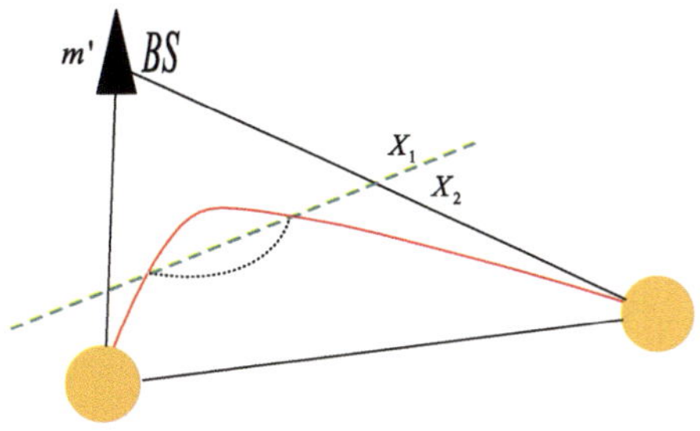

Fig. 5.4 Illustration of proof of Lemma 5.3

time is the same as that of the trajectory $\boldsymbol{s}^*_{k,j}$. Since $T^t_{k,j,m'} = \tau_k$ and $T^n_{k,j,m'} \geq 0$, it is not difficult to find that the amount of transmitted data under the trajectory $\tilde{\boldsymbol{s}}^*_{k,j}$ to the BS m'' is greater than that under the trajectory $\boldsymbol{s}^*_{k,j}$ to the BS m', which in turn implies that the symmetrical trajectory $\tilde{\boldsymbol{s}}^*_{k,j}$ can be further improved to reduce the flight time. In addition, due to $||\boldsymbol{u}_k - \boldsymbol{u}^{BS}_p|| = ||\boldsymbol{u}_j - \boldsymbol{u}^{BS}_q||$ for $p = m''$, $q = m'$ and $p = m'$, $q = m''$. Then $||\boldsymbol{u}_i - \boldsymbol{u}^{BS}_m|| \leq ||\boldsymbol{u}_i - \boldsymbol{u}^{BS}_{m''}||$ for $i = k, j$. Thus, $T^{f*}_{k,j,m} \leq T^{f*}_{k,j,m''}$ by condition (1) in Lemma 5.3. As a result, the m'th BS can be eliminated from the candidate BS set.

Utilizing the results derived above, it is possible to systematically exclude non-optimal base stations from the candidate set $\Upsilon_{k,j}$, leaving only those base stations that contribute to the most efficient UAV trajectory. By eliminating these suboptimal BSs, we reduce the solution space and improve computational efficiency, ensuring that the UAV is primarily connected to base stations that best support its path and operational requirements.

5.6 Sensing Order Optimization

The sensing order subproblem can be reformulated as a Traveling Salesman Problem (TSP) by building a symmetric cost matrix based on the minimum task completion times. With this transformation, the LKH algorithm is applied to obtain a near-optimal solution and an approximate lower bound on the mission completion time.

The sensing order optimization sub-problem of (P1) is expressed as follows:

$$\begin{aligned} \text{(P4)} \quad & \min_{\pi} T = \sum_{k=0}^{K} T_{\pi(k),\pi(k+1)} \\ & s.t. \quad \text{(P1d)}, \quad \text{(P1e)}. \end{aligned} \tag{5.26}$$

As shown, this problem can be restructured as a TSP. Here, the node set $V = \{s_0, \boldsymbol{u}_1, ..., \boldsymbol{u}_K\}$ represents the UAV's starting point and task areas, while the cost matrix $\mathbf{E}$ reflects the corresponding flight times between nodes, expressed by:

$$\boldsymbol{E} = \begin{bmatrix} T_{0,0} & T_{0,1} & \dots & T_{0,K} \\ T_{1,0} & T_{1,1} & \dots & T_{1,K} \\ \vdots & \vdots & \ddots & \vdots \\ T_{K,0} & T_{K,1} & \dots & T_{K,K} \end{bmatrix}, \tag{5.27}$$

where the elements are calculated as:

$$T_{k,j} = \begin{cases} \frac{||s_0 - u_j||}{v_{max}} & k = 0,\ k \neq j \\ T^{f*}_{k,j} + \frac{T^s_k}{2}, & k' = 0,\ k \neq j \\ +\infty, & k = j \\ T^{f*}_{k,j} + \frac{T^s_k}{2} + \frac{T^s_j}{2}, & \text{otherwise} \end{cases} \tag{5.28}$$

Here, $T^{f*}_{k,j}$ represents the minimum flight time between $\boldsymbol{u}_k$ and $\boldsymbol{u}_j$. Thus, the UAV's optimal sensing sequence can be identified by finding the minimum Hamiltonian cycle in graph $G = (V, E)$, where the total mission duration is given by:

$$T = \sum_{k=0}^{K} T^{f*}_{\pi(k),\pi(k+1)} + \frac{T^s_{\pi(k)}}{2} + \frac{T^s_{\pi(k+1)}}{2} = \sum_{k=0}^{K} T_{\pi(k),\pi(k+1)}. \tag{5.29}$$

Due to potential asymmetry in travel times between tasks, where $T^{f*}_{k,j} \neq T^{f*}_{j,k}$, the minimum Hamiltonian cycle is formulated as an asymmetric TSP. To address this, a symmetric TSP is created with node set V' containing $2K + 1$ nodes, where the cost matrix E' is symmetric and sized $2(K + 1) \times 2(K + 1)$. Its elements are defined as:

$$\begin{cases} e'_{K+1+k,j} = e'_{j,K+1+k} = e_{k,j}, & \forall k,\ j,\ k \neq j \\ e'_{K+1+k,k} = e'_{k,K+1+k} = -Q, & \forall k \\ e'_{k,j} = Q, & \text{otherwise} \end{cases}, \tag{5.30}$$

where Q is a sufficiently large constant. According to [35], the optimal solution of the symmetric problem aligns with the asymmetric case.

To solve this symmetric TSP, we employ the LKH algorithm [16], which finds a near-optimal sensing order $\boldsymbol{\pi}'$ efficiently. Specifically, this algorithm iterates on graph $G' = (V', E')$, executing moves to reduce the current trajectory cost $\sum_{k=0}^{K} T_{\pi'(k),\pi'(k+1)}$ until no further improvement can be made. Moreover, the LKH algorithm generates a unique minimum 1-tree. Here, the minimum 1-tree is defined as follows: A 1-tree is a spanning tree on $V \backslash \{1\}$ plus two edges from E incident to node 1. A minimum 1-tree minimizes length among such trees. for graph $G' = (V', E')$ by first obtaining the minimum spanning tree of vertices $V' \backslash \{1\}$ and then adding the two least-cost edges to vertex 1.

Let $\boldsymbol{s}^*_{k,j}$ and $\{b^*_{k,m}\}$ denote the optimal inter-task UAV trajectory and UAV-BS associations derived by our proposed method, respectively, while $\boldsymbol{\pi}'$ represents the sensing order produced by the LKH algorithm.

Proposition 5.1 The total weight $\underline{T}$ of the minimum 1-tree produced from graph G' using the LKH algorithm serves as a lower bound on the optimal objective for (P4). When this minimum 1-tree forms a loop, the sensing strategy $\left\{\left\{\boldsymbol{s}^*_{k,j}\right\}, \left\{b^*_{k,m}\right\}, \pi'\right\}$ yields the optimal solution to (P1).

Proof To address the sensing order sub-problem (P4), it is transformed into a symmetric TSP. After forming the minimum 1-tree of the associated graph $G' =$

(V', E'), the lower bound for the optimal solution to (P1) is computed as $\underline{T}$, representing the aggregate weight of the edges in this 1-tree. The matrix E' is generated through the proposed BS association algorithm and contains the minimum flight times $T_{k,j}$ between task areas, meeting the constraints of sensing quality and transmission delay as specified in (P1). Consequently, finding the minimum weight Hamiltonian cycle in G' is equivalent to minimizing the mission completion time for (P1). Thus, the minimum 1-tree serves as a lower bound to (P1)'s optimal mission completion time, as the symmetric TSP's lower bound reflects this objective. If each node in the minimum 1-tree has precisely two connecting edges, forming a loop, then this 1-tree provides the global optimal solution for (P1).

Therefore, the total mission completion time associated with the sensing strategy $\left\{\left\{s_{k,j}^*\right\}, \left\{b_{k,m}^*\right\}, \boldsymbol{\pi}'\right\}$ approaches the lower bound $\underline{T}$ when the minimum 1-tree from graph G′ is not a loop, as demonstrated in Sect. 5.7. Thus, the sensing strategy $\left\{\left\{s_{k,j}^*\right\}, \left\{b_{k,m}^*\right\}, \boldsymbol{\pi}'\right\}$ is near-optimal.

This subsection provides an analysis of the complexity of the proposed sensing approach. The proposed scheme comprises three components: the trajectory optimization algorithm, the BS association algorithm, and the LKH algorithm. Firstly, determining the complexity of the trajectory optimization algorithm alone remains an open question, as it is based on the Polyblock algorithm [14], whose complexity has not yet been resolved. Therefore, the convergence rate of the SPTS algorithm is numerically compared with that of the Polyblock algorithm. For convenience, the complexity of the proposed trajectory optimization algorithm is denoted by O_{SPTS}. Secondly, the complexity of the BS association algorithm can be verified as $O(M_R \cdot O_{SPTS})$, where M_R is the number of candidate BSs identified by the BS association algorithm. Without using Lemma 5.3, finding the optimal UAV-BS association would have a complexity of $O(M \cdot O_{SPTS})$, with M representing the total BS count. Thus, the complexity is reduced by a factor of M/M_R. Thirdly, generating the $K \times K$ cost matrix has a complexity of $K^2 \cdot O(M_R \cdot O_{SPTS})$, where K is the number of task areas. Additionally, as per [16], the LKH algorithm has a complexity of $O\left(K^2\right)$. Therefore, the overall complexity of the proposed sensing scheme can be expressed as $K^2 \cdot O(M_R \cdot O_{SPTS}) + O\left(K^2\right)$.

5.7 Simulation Results and Analysis

5.7.1 Simulation Setups and Benchmarks

This section presents numerical results to assess the performance of the proposed algorithms. In the simulations, the K task areas are randomly positioned based on a uniform distribution within a 4 km by 4 km^2 region, with BSs distributed uniformly throughout this area. Parameter settings are given as: $P\text{th} = 99\%$, $t^s = 0.01$, $\nu = 0.008$, $B = 10$ MHz, $P_S = 0.1$ W, $v_{\max} = 20$ m/s, $\alpha = 2$, $\rho_0 = -80$ dB,

$h = 50$ m, $h^{BS} = 20$ m. The proposed OASPTS scheme is benchmarked against two baseline approaches: the Euclidean distance-based OTL-LKH (Euclid-OTL-LKH) scheme and the OTL-based LKH (OTL-LKH) scheme. Unlike the proposed method, both baseline schemes assume that for each task area, the UAV associates with the closest BS. Additionally, both schemes utilize the conventional trajectory algorithm to compute an optimized flight path through the task areas, where the UAV first flies toward the BS at maximum speed before moving to the next task area, with the turn point determined by searching along the line between the current task area and the associated BS. Furthermore, the first baseline relies on Euclidean distances, while the second uses OTL distances among task areas to determine the sensing order.

The sensing process of the proposed OASPTS scheme is evaluated against two baseline methods. As shown in Fig. 5.5, there are eight randomly distributed tasks served by four BSs. Red stars represent BSs, blue dots indicate task areas, green lines show UAV-BS associations, and black lines illustrate UAV trajectories. The sensing sequence is marked numerically, with "S" denoting the UAV's starting point. The mission completion times for the Euclid-OTL-LKH and OTL-LKH schemes are 38.2 and 25.1% longer, respectively, compared to the OASPTS scheme. Notably, the UAV trajectories derived from the proposed algorithm (Fig. 5.5c) are smoother and rarely intersect, unlike those of the baselines (Fig. 5.5a and b). This improvement arises from the algorithm's effective use of BS and task area positions to determine an optimized sensing sequence, leading to significantly reduced mission completion times.

5.7.2 Convergence of Inter-Task UAV Trajectory

The convergence rate of the proposed SPTS algorithm is analysed by comparing it with that of the Polyblock algorithm in this subsection. Specifically, Fig. 5.6 displays the convergence of η_k for a given ξ_k, with $\psi(\tau_k) = 5$ and $\varepsilon_\theta = 10^{-3}$. Here, the y-axis represents the transmission ratio η_k for sensory data. As illustrated, the upper bound of η_k progressively diminishes as infeasible solutions are excluded, with the proposed SPTS algorithm reaching 98.7% of this bound after only two iterations. Conversely, the Polyblock algorithm's transmission ratio is still approximately 60% of the upper bound, even after 1000 iterations. This accelerated convergence of η_k significantly enhances the algorithm efficiency, as each binary search step requires fewer iterations.

5.7.3 Number of Candidate BSs

The effectiveness of the OUBA algorithm in reducing suboptimal BSs is verified by examining the sizes of the candidate BS sets obtained using BS selection algorithm. Results for different sizes of initial candidate BS sets are presented in Table 5.1,

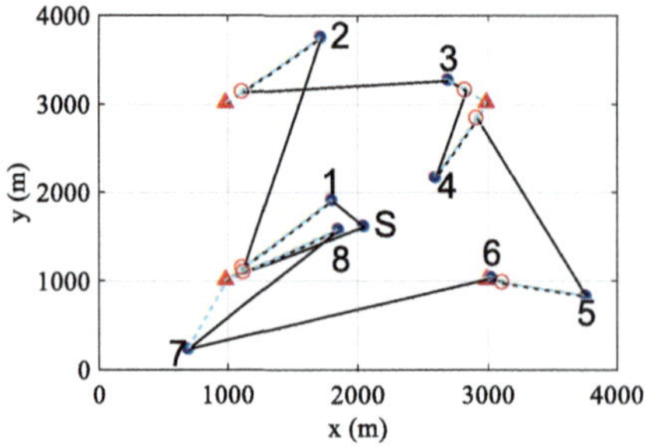

(a) Euclid-OTL-LKH scheme (877.9 s).

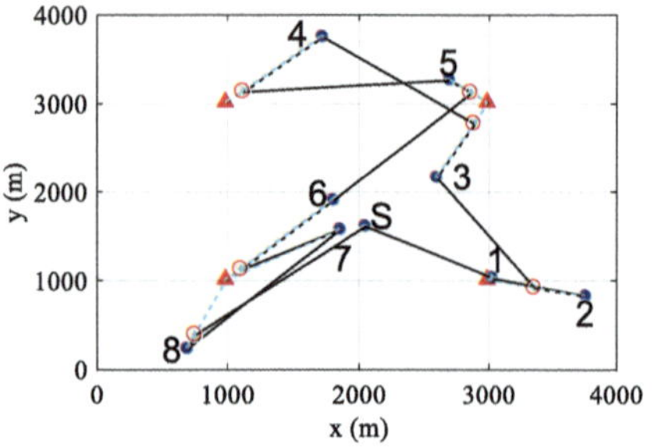

(b) OTL-LKH scheme (794.5 s).

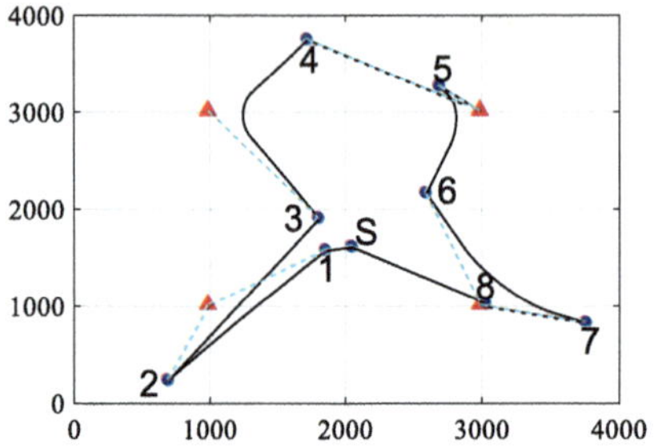

(c) The proposed scheme (635.2 s).

Fig. 5.5 UAV trajectory and task completion time

Fig. 5.6 Convergence comparison

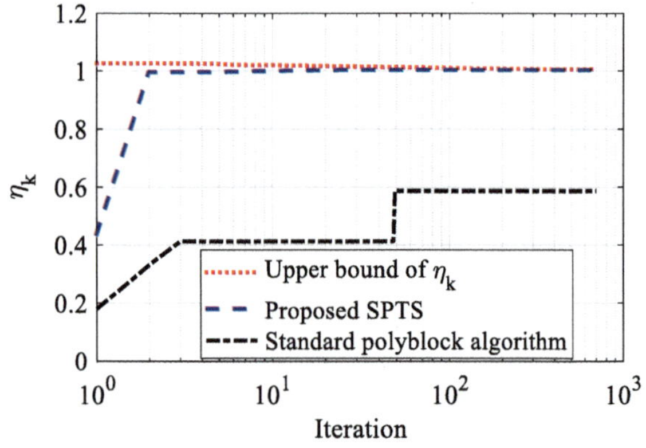

averaged over 100 Monte Carlo trials. The findings show that BS selection algorithm typically narrows the candidate set down to 1–2 BSs, which significantly reduces the computational complexity for determining the optimal UAV-BS association.

Table 5.1 The output BS number of the proposed BS association algorithm

Initial number of BS	Average	Variance
5	1.16	0.32
10	1.42	0.71
15	1.49	1.17
20	1.59	1.31

5.7.4 Mission Completion Time

A comparison between the proposed OASPTS scheme and the two baseline schemes is depicted in Fig. 5.7. The mission completion time achieved by the proposed OASPTS scheme is remarkably close to the lower bound of the global optimal solution, with a difference of <0.2%. For example, with six tasks, the proposed scheme's mission completion time aligns with the global optimum and is 24.5 and 18.5% shorter than that of the Euclid-OTL-LKH and OTL-LKH schemes, respectively. As shown in Fig. 5.7, as the number of tasks increases, so does the total mission time. On average, the OTL-LKH and Euclid-OTL-LKH schemes yield mission times approximately 19 and 31.2% longer than the OASPTS scheme.

Figure 5.8 compares minimum mission completion times across different sensory data volumes, revealing that larger data volumes result in increased mission times. When the sensory data volume is 0 Mbits, the problem simplifies to a TSP, so all three schemes have identical mission completion times. However, as data volume increases, the OASPTS scheme reduces mission time by up to 20 and 43% compared to the OTL-LKH and Euclid-OTL-LKH schemes, respectively.

Mission completion time statistics based on 100 Monte Carlo simulations are shown in Fig. 5.9. Here, the proposed OASPTS scheme's mission times fall between

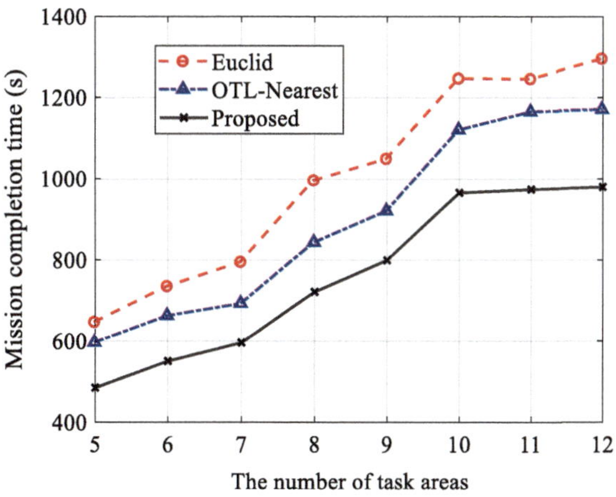

Fig. 5.7 Comparison of mission completion time

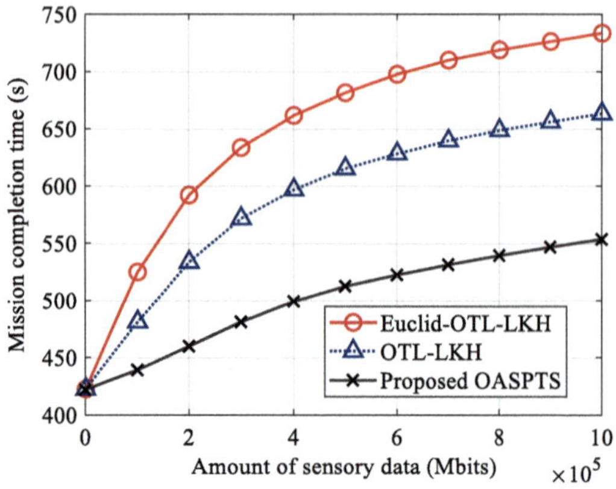

Fig. 5.8 Comparison of mission completion time versus data amount

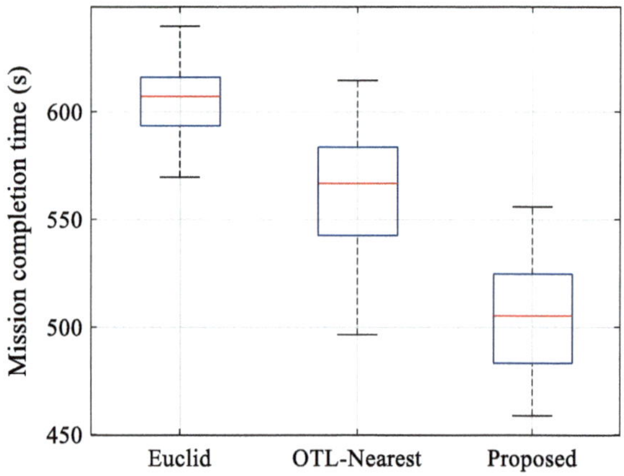

Fig. 5.9 Statistical comparison of mission completion time

454 and 550s, with an average of 500s. In contrast, the Euclid-OTL-LKH and OTL-LKH schemes have average mission times of 607 and 567s, respectively, which are 21.4 and 13.4% longer than the OASPTS scheme.

Figure 5.10 presents a comparison of mission completion times under various transmission delay constraints. It is observed that, as the transmission delay limit increases, the time required to complete the mission gradually decreases. This is primarily because a higher transmission delay limit allows the UAV greater flexibility in selecting a BS and permits a longer transmission duration. When transmission delay limit falls below the lower threshold of 30 s in the given scenario, (P1) lacks a feasible solution (i.e., sensory data cannot be sent to the BS within the specified deadline transmission delay limit), which aligns with the analysis presented in Lemma 5.1. When the transmission delay constraint exceeds 75 s, the mission completion time does not decrease further, as all sensory data is successfully transmitted to the designated BS within a period of 75 s or less.

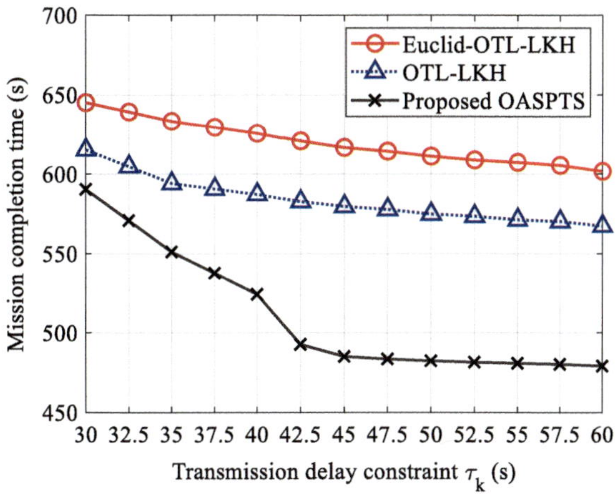

Fig. 5.10 Comparison of mission completion time under different delay constraints

Figure 5.11 shows mission completion times under different BS densities based on 100 Monte Carlo simulations, where BS density refers to the ratio of the number of BSs to the number of tasks. It can be seen that as BS density increases, the average mission completion time decreases. This trend occurs because higher BS density reduces the expected distance between the UAV and its chosen BS. Additionally, Fig. 5.11 demonstrates that, as BS density rises, the proposed OASPTS scheme can reduce mission completion times by up to 16.1 and 28.2% compared to the OTL-LKH and Euclid-OTL-LKH schemes, respectively. However, the improvement is less pronounced at lower BS densities since, for both the proposed method and the baseline schemes, the nearest BS is likely to be optimal when BS density is low.

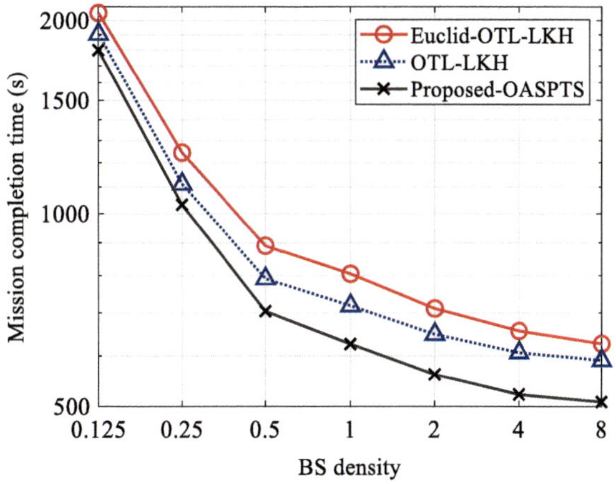

Fig. 5.11 Comparison of mission completion time under different BS densities

5.8 Conclusion

In this chapter, the problem of optimal UAV trajectory design for transmission delay constrained sensing is studied, where the UAV is required to sequentially perform a set of time-critical sensing tasks and transmit the collected sensory data to the BS. By converting this problem into an MO and exploiting the special structure of the formulated problem, a novel SPTS method is proposed. The proposed algorithm can achieve much faster convergence to the optimal UAV trajectory as compared to the standard MO method and thus allow the UAV to complete the sensing tasks with the minimum time.

References

1. Hu, Z., Bai, Z., Yang, Y., Zheng, Z., Bian, K., & Song, L. (2019). UAV aided aerial-ground IoT for air quality sensing in smart city: Architecture, technologies, and implementation. *IEEE Network, 33*(2), 14–22.
2. Meng, K., Wu, Q., Ma, S., Chen, W., & Quek, T. Q. S. (2022). UAV trajectory and beamforming optimization for integrated periodic sensing and communication. *IEEE Wireless Communications Letters, 11*(6), 1211–1215.
3. Zhang, D., Zhao, D., & Ma, H. (2019). On timely sweep coverage with multiple mobile nodes. In *Proceedings of IEEE WCNC*, Marrakech, Morocco.
4. Meng, K., He, X., Wu, Q., & Li, D. (2023). Multi-UAV collaborative sensing and communication: Joint task allocation and power optimization. *IEEE Transactions on Wireless Communications, 22*(6), 4232–4246.
5. Zhang, S., Zhang, H., Di, B., & Song, L. (2018). Joint trajectory and power optimization for UAV sensing over cellular networks. *IEEE Communications Letters, 22*(11), 2382–2385.
6. Mozaffari, M., Saad, W., Bennis, M., Nam, Y., & Debbah, M. (2019). A tutorial on UAVs for wireless networks: Applications, challenges, and open problems. *IEEE Communications Surveys & Tutorials, 21*(3), 2334–2360.
7. Zhan, C., & Zeng, Y. (2019). Completion time minimization for multi-UAV enabled data collection. *IEEE Transactions on Wireless Communications, 18*(10), 4859–4872.
8. Hu, J., Zhang, H., Song, L., Schober, R., & Poor, H. V. (2020). Cooperative internet of UAVs: Distributed trajectory design by multi-agent deep reinforcement learning. *IEEE Transactions on Communications, 68*(11), 6807–6821.
9. Abd-Elmagid, M. A., & Dhillon, H. S. (2019). Average peak age-of-information minimization in UAV-assisted IoT networks. *IEEE Transactions on Vehicular Technology, 68*(2), 2003–2008.
10. Gong, J., Chang, T., Shen, C., & Chen, X. (2018). Flight time minimization of UAV for data collection over wireless sensor networks. *IEEE Journal on Selected Areas in Communications, 36*(9), 1942–1954.
11. Liu, C. H., Chen, Z., & Zhan, Y. (2019). Energy-efficient distributed mobile crowd sensing: A deep learning approach. *IEEE Journal on Selected Areas in Communications, 37*(6), 1262–1276.
12. Zeng, Y., & Zhang, R. (2017). Energy-efficient UAV communication with trajectory optimization. *IEEE Transactions on Wireless Communications, 16*(6), 3747–3760.
13. Samir, M., Sharafeddine, S., Assi, C. M., Nguyen, T. M., & Ghrayeb, A. (2020). UAV trajectory planning for data collection from time-constrained IoT devices. *IEEE Transactions on Wireless Communications, 19*(1), 34–46.

14. Wu, Y., Zhang, B., Yang, S., Yi, X., & Yang, X. (2017). Energy-efficient joint communication-motion planning for relay-assisted wireless robot surveillance. In *Proceedings of IEEE INFOCOM*, Atlanta, GA.
15. Wu, Q., Zeng, Y., & Zhang, R. (2018). Joint trajectory and communication design for multi-UAV enabled wireless networks. *IEEE Transactions on Wireless Communications, 17*(3), 2109–2121.
16. Narayanan, V.K., Miyashita, T., Horikawa, Y., & Hagita, N. (2018). A transient-goal driven communication-aware navigation strategy for large human-populated environments. In *Proceedings of IEEE IROS,* Madrid, Spain
17. Zeng, Y., Zhang, R., & Lim, T. J. (2016). Throughput maximization for UAV-enabled mobile relaying systems. *IEEE Transactions on Communications, 64*(12), 4983–4996.
18. Meng, K., Wu, Q., Ma, S., Chen, W., Wang, K., & Li, J. (2023). Throughput maximization for UAV-enabled integrated periodic sensing and communication. *IEEE Transactions on Wireless Communications, 22*(1), 671–687.
19. Gao, F., Wu, W., Lin, Y., & Shen, S. (2018). Online safe trajectory generation for quadrotors using fast marching method and Bernstein basis polynomial. In *Proceedings of IEEE ICRA*, Brisbane, Australia.
20. Ahmed, S., Chowdhury, M. Z., Sabuj, S. R., & Jang, Y. (2020). Joint worst-case secrecy rate and UAV propulsion energy optimization via robust resource allocation in uncertain adversarial networks. arXiv:2006.15728
21. Zhang, S., Zhang, H., Di, B., & Song, L. (2018). Cooperative sensing and transmission for cellular network-controlled unmanned aerial vehicles. In *Proceedings of IEEE GLOBECOM*, Abu Zaby.
22. Faust, A., Oslund, K., Ramirez, O., Francis, A., Tapia, L., Fiser, M., & Davidson, J. (2018). PRM-RL: Long-range robotic navigation tasks by combining reinforcement learning and sampling-based planning. In M. B. Ghorbel, D. Rodríguez-Duarte, H. Ghazzai, M. J. Hossain, & H. Menouar (Eds.), *Proceedings of IEEE ICRA*, Brisbane, Australia (2018). Joint position and travel path optimization for energy-efficient wireless data gathering using unmanned aerial vehicles. *IEEE Transactions on Vehicular Technology, 68*(3), 2165–2175 (2019).
23. Li, J., Zhao, H., Wang, H., Gu, F., Wei, J., Yin, H., & Ren, B. (2020). Joint optimization on trajectory, altitude, velocity, and link scheduling for minimum mission time in UAV-aided data collection. *IEEE Internet of Things Journal, 7*(2), 1464–1475.
24. Ebrahimi, D., Sharafeddine, S., Ho, P., & Assi, C. (2019). UAV-aided projection-based compressive data gathering in wireless sensor networks. *IEEE Internet of Things Journal, 6*(2), 1893–1905.
25. Liu, J., Wang, X., Bai, B., & Dai, H. (2018). Age-optimal trajectory planning for UAV-assisted data collection. In *Proceedings of IEEE INFOCOM WKSHPS*, Honolulu, Hawaii.
26. Jia, Z., Qin, X., Wang, Z., & Liu, B. (2019). Age-based path planning and data acquisition in UAV-assisted IoT networks. In *Proceedings of IEEE ICC Workshops*, Shanghai, China.
27. Hu, H., Xiong, K., Qu, G., Ni, Q., Fan, P., & Letaief, K. B. (2020). AoI-minimal trajectory planning and data collection in UAV-assisted wireless powered IoT networks. *IEEE Internet of Things Journal*. https://doi.org/10.1109/JIOT.2020.3012835
28. Gao, X., Chen, Z., Pan, J., Wu, F., & Chen, G. (2020). Energy-efficient scheduling algorithms for sweep coverage in mobile sensor networks. *IEEE Transactions on Mobile Computing, 19*(6), 1332–1345.
29. Zhang, S., Zeng, Y., & Zhang, R. (2019). Cellular-enabled UAV communication: A connectivity-constrained trajectory optimization perspective. *IEEE Transactions on Communications, 67*(3), 2580–2604.
30. Ahmed, N., Kanhere, S. S., & Jha, S. (2016). On the importance of link characterization for aerial wireless sensor networks. *IEEE Communications Magazine, 54*(5), 52–57.
31. Shakhov, V. V., & Koo, I. (2017). Experiment design for parameter estimation in probabilistic sensing models. *IEEE Sensors Journal, 17*(24), 8431–8437.
32. Gupta, H. P., Venkatesh, T., Rao, S. V., Dutta, T., & Iyer, R. R. (2017). Analysis of coverage under border effects in three-dimensional mobile sensor networks. *IEEE Transactions on Mobile Computing, 16*(9), 2436–2449.

33. Tuy, H. (2000). Monotonic optimization: Problems and solution approaches. *SIAM Journal on Optimization, 11*(2), 464–494.
34. Zhang, J., Zeng, Y., & Zhang, R. (2018). UAV-enabled radio access network: Multi-mode communication and trajectory design. *IEEE Transactions on Signal Processing, 66*(20), 5269–5284.
35. Jonker, R., & Volgenant, T. (1983). Transforming asymmetric into symmetric traveling salesman problems. *Operations Research Letters, 2*(4), 161–163.

Chapter 6
DRL-Based Joint User Scheduling and UAV Trajectory Design in UAV-Assisted MEC System

Yunfei Gao, Xiaopeng Yuan, Qinwei He, and Yulin Hu

Abstract In this paper, we investigate a DRL-based method for UAV trajectory design and mobile ground user (GU) scheduling in a UAV-assisted mobile edge computing (MEC) system. In this scenario, the UAV first collects data from source GUs, processes it, and then transmits the processed data to the destination GUs. Our objective is to minimize the total operation time, which includes data collection, computation, and decision transmission, by optimizing the UAV's three-dimensional (3D) trajectory and mobile GUs scheduling. In this scenario, considering the mobility of users and obstacle avoidance with 3D buildings, the problem becomes non-convex, and the environment is dynamically changing, making it difficult for traditional methods to address. Therefore, we propose a DRL-based approach for UAV trajectory design and user scheduling. This method only allows the UAV to interact with the environment multiple times, avoiding the challenges of non-convexity. The simulation results demonstrate the superiority of the proposed algorithm. Compared to 2D trajectory design, the 3D trajectory performs better due to the utilization of vertical altitude. Additionally, the simulations confirm the algorithm's robustness across different user mobility patterns.

Keywords 3D trajectory design · User scheduling · MEC · Deep reinforcement learning

Y. Gao · Y. Hu (✉)
School of Electronic Information, Wuhan University, Wuhan, China
e-mail: yulin.hu@whu.edu.cn

Y. Gao
e-mail: gaoyunfei@whu.edu.cn

X. Yuan
Chair of Information Theory and Data Analytics, RWTH Aachen University, Aachen, Germany
e-mail: yuan@inda.rwth-aachen.de

Q. He
Global Energy Interconnection Research Institute Europe GmbH, Berlin, Germany
e-mail: qinwei.he@geiri.eu

Y. Hu et al. (eds.), *Reliable and Efficient Trajectory Design in UAV-Aided Wireless Networks*, https://doi.org/10.1007/978-981-96-5988-3_6

Abbreviations

3D	Three Dimensional
D3QN	Multi-step Dueling DDQN
DRL	Deep Reinforcement Learning
GTs	Ground Terminals
MDP	Markov Decision Process
MEC	Mobile Edge Computing
UAV	Unmanned Aerial Vehicle

6.1 Introduction

6.1.1 Background

Unmanned aerial vehicle (UAV)-assisted communication systems offer several significant advantages. In particular, UAV can be rapidly deployed, making it particularly suitable for post-disaster relief, remote areas, or emergency communication coverage [1, 2]. In such scenarios, the UAV can overcome the limitations of traditional ground-based infrastructure, providing flexible aerial network support. In addition, UAV possess high mobility and flexibility, allowing them to adjust flight altitude and path according to real-time demands, thereby optimizing communication performance [3]. Moreover, by collaborating with ground users or other UAVs, they can extend network coverage, enhance network capacity and performance, and reduce communication latency [4, 5]. This is particularly beneficial in environments with large-scale users or high-density devices. These advantages of the UAV have attracted significant research interest from both academia and industry. For example, the UAV is deployed in tasks such as collecting and transmitting data from ground terminals (GT) [6] or serving as mobile relays [7]. In work [8], the UAV is utilized as flying mobile edge computing (MEC) nodes (with UAVs carrying MEC devices) to assist GTs in completing computational tasks.

6.1.2 Motivation

The UAV-assisted (MEC) system holds significant importance. For instance, it can provide rapidly deployable computing power in environments where traditional computing resources are difficult to access, particularly in scenarios like post-disaster relief and remote areas. In addition, by offloading computational tasks to the UAV, it can effectively reduce the computational load on GTs, decrease latency, and thus enhance the whole computational efficiency and user experience.

However, there are several challenges in UAV-assisted MEC systems that still urgently need to be addressed. First, the increasing number of mobile GTs accessing communication networks in practical scenarios makes the mobility of GTs impossible to ignore. In addition, ignoring the low-altitude flight of the UAV to some extent can result in a loss of system performance as the UAV can not be closer to the GTs in low-altitude areas. Furthermore, the complexity of the physical environment during the UAV missions necessitates obstacle avoidance. Moreover, the mobility and uncertainty of GTs pose significant challenges to dynamic resource allocation and UAV trajectory design, rendering the conventional optimization methods no longer applicable. Finally, accelerating the convergence rate of the algorithm and enhancing the system's performance is essential because 3D trajectory optimization requires simultaneous consideration of the UAV's motion in both horizontal and vertical directions, which increases the complexity of path searching and optimization.

6.1.3 Main Contributions

Main contributions to this chapter are as follows:

(1) Considering the practical aspects of GTs mobility, the need for obstacle avoidance around 3D buildings, the feasible flying zones for UAVs among these structures, and the probabilistic line-of-sight (LoS) channel model (accounting for signal blockage from buildings), we consider a joint design on the 3D UAV trajectory and the GTs' communication scheduling to minimize the whole operation time.
(2) Due to the non-convex nature of the established problem and the dynamic changes in the system caused by GT mobility, the issue becomes challenging to solve. To address these challenges, an effective design method is proposed to optimize the 3D trajectory of the UAV and the communication scheduling of GTs. This includes an improved learning approach for trajectory design and a greedy algorithm for scheduling mobile users. Compared to traditional DDQN algorithms, the proposed multi-step DDQN algorithm accelerates convergence speed and enhances system performance.
(3) We validated the effectiveness of the proposed algorithm through simulation experiments. In comparison to centralized MAFRL approaches, our method proves to be more robust in dynamic multi-GD environments with NFZs. Furthermore, the system's performance is on par with that of the centralized MARL method and approaches the ideal scenario without NFZs, while outperforming baseline methods. Additionally, the proposed 3D trajectory design is shown to enhance system performance by utilizing the vertical dimension, offering improvements over 2D trajectory designs.

6.1.4 Structural Organization

The work on UAV-assisted MEC trajectory design and machine learning is introduced in Sect. 6.2, while the system model and problem formulation are presented in Sect. 6.3. Furthermore, Sect. 6.4 introduces the proposed DRL-based solution. Section 6.5 provides the simulation results and analysis, and finally, the paper is concluded in Sect. 6.6.

6.2 Related Works

6.2.1 UAV-Assisted MEC Trajectory Design

For UAV-assisted MEC systems, previous studies have demonstrated that selecting an optimal deployment location can enhance system performance, such as improving MEC service coverage [9], offloading capacity [10], and energy efficiency [11]. Additionally, optimizing the UAV trajectory can significantly enhance the performance of UAV-assisted MEC systems. On the one hand, designing highly mobile UAV trajectories allows the UAV to move closer to the GTs, reducing transmission distances. On the other hand, an efficient UAV trajectory design can establish a reliable high-probability LoS channel between the UAV and the GTs, facilitating a high-rate offloading channel. Several studies have already explored trajectory design in UAV-assisted MEC systems.

For instance, in work [12], the authors minimized the total energy consumption during the offloading process by designing the UAV's trajectory. Additionally, in work [13], the authors proposed a three-step iterative optimization algorithm to minimize the energy consumption of the MEC system by optimizing the UAV's beamforming vectors, central processing unit (CPU) frequency, and trajectory. In work [14], the authors reduced the total power consumption of the MEC system by optimizing both the UAV's trajectory and resource allocation. Furthermore, the authors in work [17] proposed a UAV-assisted multi-GT framework that maximizes the minimum transmission rate by optimizing the UAV's 3D trajectory. Additionally, the authors presented a 3D trajectory design method to improve energy harvesting and data collection efficiency [18]. In work [19, 20], multiple UAVs are employed to assist multiple GTs, with the UAVs' trajectories designed to enhance system performance. However, the above studies assume that the GTs are stationary, neglecting the mobility of GTs. It is important to note that in practical scenarios, GTs may be mobile, making it essential to account for their mobility.

6.2.2 *Reinforcement Learning for UAV-Assisted MEC Application*

Reinforcement learning (RL) has broad applications in UAV-assisted MEC systems. Through RL algorithms, UAVs can autonomously learn how to optimize flight trajectories, task offloading decisions, and resource allocation to maximize performance in dynamic environments. For instance, deep reinforcement learning (DRL) can help UAVs make real-time decisions under varying channel conditions and GTs mobility patterns, optimizing communication rates, energy consumption, and computational offloading efficiency. Compared to traditional algorithms, RL can adapt to complex environmental changes, gradually improving overall system efficiency and reliability, particularly in multi-UAV and multi-task scenarios.

A set of RL-based trajectory design methods have been proposed for UAV-assisted wireless networks. For example, the authors in [15] present a DRL approach aimed at minimizing UAV mission completion time by jointly optimizing the UAV's trajectory and its association with ground base stations. Meanwhile, a semi-distributed resource management approach is introduced in [16] to reduce the total system power in UAV-assisted MEC systems. Nevertheless, the works of [15, 16] ignores the 3D trajectory design, which will loss the performance of the system. In addition, the mobility of GTs is modeled as random behavior when developing ML-based UAV trajectory designs [21, 22]. In addition, the authors of work [23] propose a RL based multi-UAV MEC framework to optimize quality of service and route planning. In particular, the proposed framework addresses these challenges by modeling user demand and using multi-factor optimization considering user demand, risk, and distance. However, these works overlook obstacle avoidance with respect to 3D buildings and primarily focus on immediate returns.

6.3 System Model and Problem Formulation

6.3.1 *System Description*

In this paper, we study a UAV-assisted MEC system, where the UAV acts as a flying MEC node to assist mobile ground terminals (GTs) in data collection, computation, and decision-making, as shown in Fig. 6.1. Specifically, while performing tasks, the UAV needs to avoid buildings to ensure flight safety. There exists a collection of J source GTs, represented as $\mathcal{J} = \{\mathrm{S}_1, ..., \mathrm{S}_\mathrm{J}\}$, and a collection of K destination GTs, represented as $\mathcal{K} = \{\mathrm{D}_1, ..., \mathrm{D}_\mathrm{K}\}$. The location of source GT and destination GT are represented as $\mathbf{w}_{j,n} = \left[x_{j,n}, y_{j,n}, 0\right], \forall j \in \mathcal{I}, \forall n \in \mathcal{N}$ and $\mathbf{w}_{k,n} = \left[x_{k,n}, y_{k,n}, 0\right], \forall k \in \mathcal{K}, \forall n \in \mathcal{N}$ at time slot n, respectively. The UAV's 3D location is denoted as $\mathbf{q}_n = \left[x_n, y_n, z_n\right], \forall n \in \mathcal{N}$, with an initial location given by $\boldsymbol{q}_0 = \left[x_0, y_0, z_0\right]$. Importantly, the mobility of the GTs influences the performance of UAV trajectory design, as it is assumed that the GTs are in motion while offloading data tasks or receiving

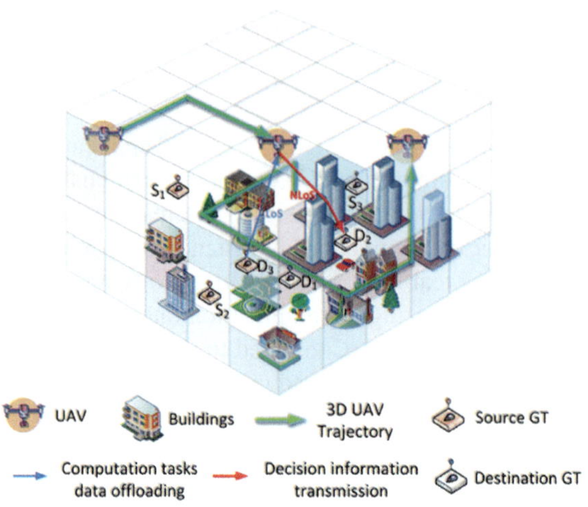

Fig. 6.1 UAVs-assisted multi-GTs data operation system

decision information. This assumption is relevant in practical scenarios such as road inspection, detection, and photography.

The whole operation process is divided into two phases. In the first phase, the UAV collects data from the source GTs, processes it, and makes decisions. In the second phase, the UAV transmits the decisions to the destination GTs. We assume that the data to be collected by source GT j is Ξ_j bits, and the required processing workload is w_j Mcycles. Additionally, the decision information that the UAV needs to transmit to destination GT k is Ξ_k bits. Furthermore, we assume that the total number of time slots for offload data of the GT j is N_j^U, so the total time slots for all source GTs to complete data collection can be represented as $N^U = \sum_{j=1}^{J} N_j^U$. Additionally, we assume that the UAV has sufficient computational capability and can start processing immediately after collecting the data. Given the UAV's computation frequency, the time required to process the data is N^C. Thus, the total number of time slots in the first phase is the sum of the offloading time slots and the time slots for the last task offloading, i.e. $N^U + N^C = \sum_{j=1}^{J} N_j^U + N^C$, as shown in Fig. 6.2. Similarly, we can obtain the total number of time slots for decision transmission as $N^D = \sum_{k=1}^{K} N_k^D$, where N_k^D represents the time slots the UAV takes to transmit the decision to destination GT k. Finally, the total time for the UAV to complete the task is $\left[N^U + N^C + N^D\right]\sigma$, where σ is the duration of a single time slot.

6.3.2 *Channel Model*

In our work, we consider a probability LoS channel model to represent the channel propagation between UAVs and GTs. In this model, the data transmission rate is primarily determined by both the elevation angle and the distance between the UAV and the GT. In this section, we assume that the probability of a LoS connection

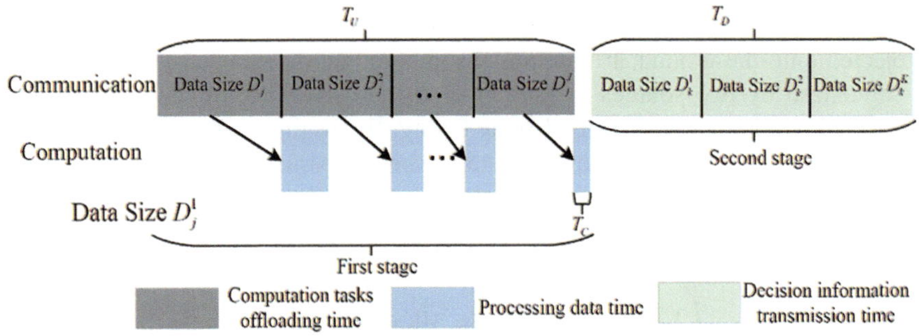

Fig. 6.2 UAVs-assisted multi-GTs data operation process

between the UAV and GD j is denoted by $a_{j,n} = 1$, the probability of LoS is $P_{j,n}^{L}$, and the elevation angle is $\theta_{j,n}$. The $P_{j,n}^{L}$ can be expressed as

$$P_{j,n}^{L}\left(a_{j,n}=1\right)=B_3+\frac{B_4}{1+e^{-\left(B_1+B_2\theta_{j,n}\right)}}, \tag{6.1}$$

where $\mathrm{B}_1 < 0$, $\mathrm{B}_2 > 0$, $\mathrm{B}_4 > 0$, with B_4 being a constant such that $\mathrm{B}_3 + \mathrm{B}_4 = 1$. Here, $\mathrm{B}_1, \mathrm{B}_2, \mathrm{B}_3, \mathrm{B}_4$ are constants determined by the surrounding environment. Thus, the corresponding NLoS probability can be represented as $P_{j,n}^{N}\left(a_{j,n}=0\right)=1-P_{j,n}^{L}$. The channel gain can be expressed as

$$h_{j,n}=c_{j,n}h_{j,n}^{L}+\left(1-c_{j,n}\right)h_{j,n}^{N}, \tag{6.2}$$

where $h_{j,n}^{L}$ and $h_{j,n}^{N}$ represent the channel power gain in the state of LoS and NLoS, respectively. $h_{j,n}^{L}$ and $h_{j,n}^{N}$ are denoted as

$$h_{j,n}^{L}=\alpha_0 d_{j,n}^{-\beta_L},\ h_{j,n}^{N}=\mu\alpha_0 d_{j,n}^{-\beta_N}. \tag{6.3}$$

Here, α_0 represents the average channel power gain in the LoS state when the reference distance is 1 m, $\mu < 1$ denotes the additional attenuation factor, and β_{L} and β_{N} represent the average path loss exponents for the LoS and NLoS states, respectively, i.e., $2 \le \beta_L \le \beta_N \le 6$. In addition, the Euclidean distance between the UAV and GT j is represented by $d_{j,n}=\sqrt{\left\|\boldsymbol{q}_n-\boldsymbol{w}_{j,n}\right\|^2}$. Furthermore, the elevation angle $\theta_{j,n}$ is expressed as

$$\theta_{j,n}=\frac{180}{\pi}arctan\left(\frac{z_{l,n}}{\left\|q_n^{\lambda}-\psi_{j,n}^{\lambda}\right\|}\right), \tag{6.4}$$

where q_n^{λ} and $\psi_{j,n}^{\lambda}$ denote the horizontal coordinates of the UAV and GT j, respectively, projected onto the ground. It is important to note that the channel power gain $h_{j,n}$ is influenced by two sources of randomness: the occurrence of random LoS and NLoS links, and the random small-scale fading. By averaging these two sources of randomness, the expected achievable rate can be expressed as follows:

$$\mathrm{E}\big[\mathrm{R}_{(j,n)}\big] = \mathrm{P}_{(j,n)}^{L}\mathrm{R}_{(l,j,n)}^{L} + \mathrm{P}_{(j,n)}^{N}\mathrm{R}_{(j,n)}^{N} \tag{6.5}$$

where $\mathrm{R}_{j,n}^{L} = \mathrm{B}\log_2\left(1 + \frac{\gamma_j}{d_{j,n}^{\alpha_L}}\right)$ and $\mathrm{R}_{j,n}^{N} = \mathrm{B}\log_2\left(1 + \frac{\mu\gamma_j}{d_{j,n}^{\alpha_N}}\right)$ denote the achievable rates in the LoS and NLoS states. Here, B represents the transmission bandwidth of GD j, and $\gamma_j = \frac{\tau_0 P_j}{\varrho_0 \Re}$, where P_j is the transmit power of the GD j, ϱ_0 represents the receiver noise power, and $\Re > 1$ is the SNR gap between the practical modulation and coding scheme and the ideal Gaussian signaling. Similarly, the instantaneous achievable rate from the UAV to destination GT k is given by,

$$\mathrm{R}_{k,n} = \mathrm{B}\log 2\left(1 + \frac{h_{k,n}P}{\sigma^2\Gamma}\right) = c_{k,n}\mathrm{R}_{k,n}^{L} + \left(1 - c_{k,n}\right)\mathrm{R}_{k,n}^{N}, \tag{6.6}$$

where B represents the UAV's transmission bandwidth, and P denotes the UAV's transmission power.

6.3.3 Computing Model and Obstacle Avoidance Model

Once a computation task is offloaded from the source GTs, the UAV can begin processing it immediately. In particular, once the UAV receives data from the (j + 1) th source, it immediately processes the data from the j th source. We assume that the workloads of the tasks are very small, significantly shorter than the offloading time. As shown in Fig. 6.2 the total data processing time depends only on the size of the last source GT's data. Assuming the workload of the last source GT is w_j, the total processing time slots can be expressed as

$$N^{C} = \left\lceil \frac{C_{\beth} W_i^{*}}{f_{\beth}\sigma} \right\rceil, \tag{6.7}$$

where $\lceil\cdot\rceil$ rounds up to the nearest integer, $C_{\beth}$ represents the number of CPU cycles the UAV requires to process one sample, and $f_{\beth}$ denotes the UAV's data processing capability, primarily determined by its CPU's rotation speed in rounds per second.

In this work, we define the total 3D spatial area as Γ, and the 3D space occupied by buildings as Υ. Therefore, the feasible region for the UAV can be expressed as

$$\Psi \triangleq C_{\Gamma}\Upsilon \tag{6.8}$$

which represents the complement of Γ in the total 3D spatial geographic area. Recall that the GTs move within a 2D spatial area $\mathcal{D}$, which represents the projection of the 3D space Ψ onto the ground.

6.4 Problem Formulation

The aim of this paper is to minimize the total operation time via jointly optimizing the UAV's trajectory and the communication scheduling of offloading transmissions from sources GTs and destination GTs. In particular, we assume that the UAV can obtain the historical location information of the GTs through feedback from the GTs, but the current and future positions remain unknown. The time minimization problem can be formulated as follows:

$$\begin{aligned}
\mathcal{P}1: \quad & \min_{\{\mathrm{q_n}, \vec{\mathrm{v}}_\mathrm{n}, \alpha_{\mathrm{i,n}}, \alpha_{\mathrm{j,n}}\}} \left[N^U + N^C + N^D\right]\sigma \\
\text{s.t.} \quad & (a) \quad \alpha_{\mathrm{i,n}} \in \{0, 1\} \\
& (b) \quad \alpha_{\mathrm{j,n}} \in \{0, 1\}, \\
& (c) \quad \sum_{\mathrm{i=1}}^{\mathrm{I}} \alpha_{\mathrm{i,n}} + \sum_{\mathrm{j=1}}^{\mathrm{J}} \alpha_{\mathrm{j,n}} \leq 1, \\
& (d) \quad 0 \leq \mathrm{z_n} \leq \mathrm{H_{max}}, \\
& (e) \quad \mathrm{q_{n+1}} = \mathrm{q_n} + \Delta_\mathrm{s} \vec{\mathrm{v}}_\mathrm{n}, \\
& (f) \quad \Delta_\mathrm{s} = \mathrm{V}\sigma, \\
& (g) \quad \mathrm{q_n} \in \Psi, \\
& (h) \quad \mathrm{w_{i,n}} \in \mathcal{D}, \ \mathrm{w_{j,n}} \in \mathcal{D}, \\
& (i) \quad \sum_{\mathrm{n=1}}^{\mathrm{N}} \alpha_{\mathrm{i,n}} \mathrm{R_{i,n}} \geq \Xi_\mathrm{i}, \\
& (j) \quad \sum_{\mathrm{n=1}}^{\mathrm{N}} \alpha_{\mathrm{j,n}} \mathrm{R_{j,n}} \geq \Xi_\mathrm{j},
\end{aligned} \tag{6.8}$$

where (6.8a–c) represents the scheduling constraint, and it is important to note that $\left[N^U + N^C + N^D\right]$ denotes the total number of time slots in which UAV completes the task. Constraint (6.8d) defines the range of altitudes for UAV flights. Constraints (6.8e and f) outline the displacement limits for the UAV. Moreover, constraint (6.8) ensures that the UAV avoids obstacles to maintain safe flight, while (6.8h) indicates the valid operational area for GTs on the ground. Additionally, constraints (6.8i and j) guarantee that each source GT's task is completely offloaded and that decision information is transmitted successfully to the corresponding destination GTs.

6.5 Proposed DRL-Based UAV Trajectory Design Scheme

The problem $\mathcal{P}1$ is difficult to address due to it is a mixed-shaping non-convex problem. Additionally, due to the randomness of small-scale channel effects, problem $\mathcal{P}1$ is difficult to solve analytically. Furthermore, the random mobility of GTs causes dynamic changes in the system, and traditional methods require real-time location information, making the problem harder to solve. To address problem $\mathcal{P}1$, we propose a DRL-based approach that eliminates the need to handle non-convexity and enables the UAV to learn an optimized trajectory by continuously interacting with the environment. To improve system performance and accelerate algorithm convergence, we enhanced the DDQN framework by incorporating multi-step propagation techniques and a dueling network architecture.

6.5.1 Modified DDQN Network Framework

Reinforcement Learning (RL) involves an agent interacting with its environment to improve its policy ι, aiming to maximize the long-term reward, $\mathbb{E}\left[Y_{l,n}\right]$. The return is defined as the cumulative sum of discounted rewards from time step n n, with a discount factor $0 < \urcorner < 1$. RL is particularly well-suited for solving Markov Decision Process (MDP) problems, represented by a 4-tuple $\langle \mathcal{S}, \mathcal{A}, \mathcal{P}, \mathrm{r} \rangle$. In this context, $\mathcal{S}$ is the state space, $\mathcal{A}$ is the action space, $\mathcal{P}$ represents the state transition probabilities, and r defines the reward function. At each time step, the agent selects an action a_{n} from the action space $\mathcal{A}$ based on its policy ι, receives a reward r_n, and transitions to a new state s_{n+1} s_{n+1}. Deep Reinforcement Learning (DRL) builds on RL by using deep neural networks (DNNs) to estimate the Q-value function, which represents the expected return of taking action a_{n} a_{n} in state s_n. In the Double Deep Q-Network (DDQN) algorithm, the Q-value approximator $Q\left(s_n, a_n|w_{l,n}\right)$, with parameters $w_{l,n}$, is refined by minimizing the following loss function:

$$\left(r_{n+1} + \gamma Q\left(s_{n+1}, \arg \underset{a_n^*}{max}\, Q\left(s_{n+1}, a_n^*|w_{l,n}\right)|w_{l,n}^*\right) - Q\left(s_n, a_n|w_{l,n}\right)\right)^2, \quad (6.9)$$

where $w_{l,n}$ and $w_{l,n}^*$ represent the parameters of the online Q-network and the target Q-network, respectively. In particular, the DDQN algorithm updates the online Q-network by minimizing the loss function, ensuring it closely approximates the target Q-network, which is updated periodically.

The DDQN network predicts the next state s_{n+1} based solely on the value $\mathrm{Q}(s_{\mathrm{n}}, a_{\mathrm{n}})$ of the current action a_n in state s_n. However, if the connection between the action a_{n} and the state s_{n} is weak, this prediction may not be reliable. In contrast, the dueling network architecture separates the evaluation of state and action values, enabling independent assessments of each. This reduces the dependency of the state on the action value during calculations, leading to improved algorithm stability. As a result, the dueling network-based approach enhances both the performance and stability of the system compared to the traditional DDQN framework. Additionally, we apply multi-step bootstrapping to accelerate convergence by considering the returns over the next $\wp$ steps. By selecting an appropriate value for $\wp$, the algorithm converges more quickly and effectively. On the one hand, the dueling network architecture employs two distinct streams to estimate action values. One stream predicts the state value function, representing the overall value of the current state, while the other estimates the action advantage function, which quantifies the relative quality of available actions in that state. Unlike conventional Q-networks, the dueling network can evaluate the value of a state without examining each possible action. This design improves stability and mitigates the risk of overestimating specific action values. The Q-function in the dueling network is formulated as follows:

$$Q(s, a|\theta, \phi, \beta) = V(s|\theta, \beta) + A(s, a|\theta, \phi) - \frac{1}{|A|}\sum_{a} A(s, a^{'}|\theta, \phi), \tag{6.10}$$

where $V(s|\theta, \beta)$ represents the state value, indicating how valuable it is to be in state s, while $A(s, a|\theta, \phi)$ denotes the action advantage value, which measures the relative benefit of taking action a in state s. The parameters β and ϕ correspond to the state value stream and the state-dependent action advantage stream, respectively. |A| represents the size of the action space.

On the other hand, the approach differentiates itself from Monte Carlo (MC) and temporal difference (TD) methods by incorporating the return over the next $\wp$ steps. By selecting an appropriate value for $\wp$, the algorithm converges more rapidly and efficiently than traditional approaches. The truncated $\wp$-step return is defined as:

$$R_{n:n+\wp} = \sum_{i=0}^{\wp-1} \gamma^i R_{n+i+1}. \tag{6.11}$$

The loss function of the improved method, incorporating the truncated $\wp$-step return and the dueling network architecture, can be expressed as:

$$\left(R_{n:n+\wp} + \gamma^{\wp} Q^*\left(q_{n+\wp}, \vec{v}_n^*; w_n^*\right) - Q^*(q_n, \vec{v}_n; w_n)\right)^2, \tag{6.12}$$

where $\vec{v}_n^* = \arg\max_{\vec{v}_n \in A_n} Q^*\left(q_{n+\wp}, \vec{v}_n^{'}; w_n\right)$.

6.5.2 $\wp_1$-step D3QN-Based Solution for Solving (P1)

In the DRL model, the agent selects an action based on the current state to interact with the environment, receives a reward, and finally moves to the next position. In our design, the UAV acts as the agent and employs the ε-greedy strategy to balance exploration and exploitation. The ε-greedy strategy is expressed as

$$a = \begin{cases} \underset{a}{\text{argmaxQ}}, \vartheta > \kappa, \\ \text{randomly choose an action}, \ \vartheta \le \kappa, \end{cases} \tag{6.13}$$

where $\vartheta \in [0, 1]$ is a randomly generated number, and $\kappa \in [0, 1]$ represents the exploration probability, which primarily balances exploration and exploitation. In particular, the state, action, state transition function, and reward function are defined as follows:

- **State** s_n: We define the state s_n as the three-dimensional coordinates $[x_n, y_n, z_n]$ of the UAV during its flight.
- **Action** a_n: In our design, the action a_n of the UAV refers to its flight direction,

which includes six possible directions $[a_1, a_2, a_3, a_4, a_5, a_6]$. Specifically, $a_1 = [0, 1, 0]$ indicates that the UAV moves forward, $a_2 = [0, -1, 0]$ signifies that the UAV moves backward, $a_3 = [1, 0, 0]$ denotes a leftward movement, $a_4 = [-1, 0, 0]$ represents a rightward movement, $a_5 = [0, 0, 1]$ indicates that the UAV ascends, and $a_6 = [0, 0, -1]$ signifies that the UAV descends.

- **State Transition Function** χ: Let $\chi(s_{n+1}|s_n, a_n)$ represent the transition probability of the UAV. The state transition model illustrates the probability that when the UAV is in state s_n and selects action a_n based on its policy, it transitions to the subsequent state s_{n+1}. This transition is determined in a deterministic manner as governed by (6.8e).
- **Reward Function** r_l : To ensure that the UAV can fly safely (avoiding 3D buildings) and complete its tasks quickly, we have designed the reward function as follows:

$$r_n = \begin{cases} \Lambda, & \text{if the UAV out of bounds or hit buildings;} \\ \phi * \mu * N^C * \sigma + \lambda * R_{k,n}, k \in J \cup K, & \text{else,} \end{cases} \tag{6.14}$$

Here, ϕ is a binary variable used to determine whether the first stage has concluded, with $\phi = 1$ indicating that the first stage is complete, and $\phi = 0$ otherwise. Additionally, μ is a negative value intended to decrease data processing time, Λ is a negative value representing a penalty for the UAV going out of bounds or colliding with buildings, and λ is the throughput coefficient.

The specific process for UAV trajectory design and communication scheduling is as follows. In the initial phase, various parameters are initialized, such as the replay buffer and the model parameters for both the main and target networks. In particular, the initial positions of the UAV and GTs are generated randomly. Following this, the UAV interacts with the environment. At the beginning of training, the UAV executes a greedy algorithm, selecting the source GT to schedule based on the data transmission rate. The source GT then offloads its data to the UAV, which processes it. After the source or destination GT offloads data or transmits a decision, the UAV selects an action from the action space based on the ϵ-greedy strategy. The UAV then executes the selected action and receives a reward before moving to the next location. A small batch of data is randomly selected from the replay memory, and the network

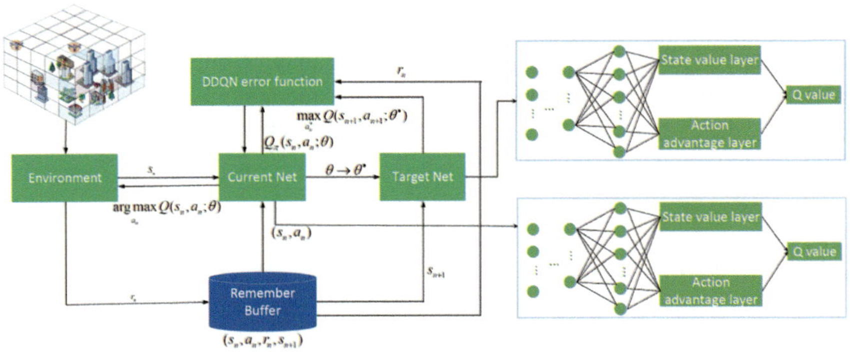

Fig. 6.3 Proposed framework DRL-based UAV trajectory design

parameters are updated using the gradient descent method. The evaluation network's parameters are copied to the target network every M steps. Finally, once the UAV completes its task, the next round of training begins. Figure 6.3 shows the detail framework of the proposed method.

6.5.3 Convergence and Complexity Discussions

To demonstrate the convergence performance of the proposed method, we first establish the convergence of the double Q-learning algorithm and then assess the impact of multi-step bootstrapping and dueling networks on the overall convergence. As shown in Theorem 1 of [24], the double Q-learning algorithm is proven to converge to the optimal value under certain conditions, such as $0 \leq a_n \leq 1$, $\sum_n a_n = \infty$, $\sum_n a_n^2 < \infty$. Additionally, the convergence of the algorithm depends on the proper tuning of hyperparameters. In our proposed method, the UAV's exploration rate decreases with each iteration, and gradient descent is used to update the model parameters θ. After a set number of iterations, the UAV stops exploring the environment, ensuring the stability of the model parameters and guaranteeing convergence. Moreover, improvements in the network architecture, such as the use of multi-step bootstrapping and dueling networks, affect the stability and speed of convergence but do not impact the overall convergence performance, as noted in [25]. Therefore, the convergence of the D3QN method is assured. In addition to theoretical guarantees, we also validate the method's convergence and robustness through simulation results presented in this paper.

The computational complexity of the proposed method primarily arises from the learning phase and the exploration process [21]. The learning process is mainly influenced by the fully connected layers of the neural network, while the UAV's exploration is determined by the size of its search space and step size [21]. Our neural network consists of five fully connected layers. The complexity of the neural network model can be expressed as $\mathcal{O}\left(\sum_{l=1}^{L} f_l f_{l-1}\right)$, where f_l is the number of neural units of the fully connected layer l and L is the number of fully connected layers. Additionally, the complexity of the UAV during exploration and learning is given by $\mathcal{O}(|S||A|)$, where $|S|$ refers to the total number of states the UAV explores with a step size of Δ_D and $|A|$ represents the size of its action space.

6.6 Simulation Results

6.6.1 *Simulation Setting*

In the simulation, we model a densely populated 3D urban environment with dimensions of 1000×1000 m, where the tallest building reaches 50 m. The number of source GTs is set to 3, and there are also 3 destination GTs. The initial positions of both the UAV and the GTs are randomly generated on the ground. The parameters B_1, B_2, B_3 and B_4 are set to $B_1 = -0.4568$, $B_2 = 0.0470$, $B_3 = -0.63$, and $B_4 = 1.63$, respectively. The GTs move at a speed of 5 m/s, while the UAV's speed is 25 m/s, with a maximum flight altitude of $H_{max} = 100$ m. The data sizes that need to be offloaded from the source GTs are $\Xi_{S_1} = 100$ Mbit, $\Xi_{S_2} = 300$ Mbit, and $\Xi_{S_3} = 200$ Mbit. Additionally, the workloads to be processed are $W_1 = 5$ Mcycles, $W_2 = 20$ Mcycles, and $W_3 = 10$ Mcycles, respectively, with the UAV's CPU frequency set at 4 GHz. The size of the decision data that the UAV needs to transmit to each destination GT is set to $\Xi_{D_1} = \Xi_{D_2} = \Xi_{D_3} = 600$ Mbit.

The parameters for the reward function are set as follows: $\mu = -20$, $r_b = -20$, and the throughput coefficient $\lambda = 10$. Both the evaluation network and the target network share the same architecture, each consisting of five hidden layers. The neuron counts in the first four hidden layers are 512, 256, 128, and 128, respectively. The final hidden layer is a dueling layer, containing 7 neurons, one for the state-value estimation and the remaining six representing the action advantages for the six possible actions. ReLU is used as the activation function in the hidden layers, and the Adam optimizer is employed for training. The simulation runs in Python 3.7 using TensorFlow 2.0.0 and Keras 2.3.1.

6.6.2 *Convergence Validation*

In this section, we first examine the algorithm's convergence performance and then choose the hyperparameters. Additionally, we assess the algorithm's effectiveness in terms of overall operation time (the main objective), 3D trajectory optimization, GT scheduling outcomes, and its robustness. It's worth noting that two scenarios with varying random movement behaviors of GTs are simulated. At any given time slot, the UAV is assumed to have access to historical GT position data through feedback from GTs or ground-based sensors, but the future positions of the GTs remain unknown to the UAV. Importantly, our proposed algorithm is independent of any specific GT movement model. The two models described in this section are used solely to illustrate the algorithm's performance, and they are as follows:

Scenario with GTs randomly moving in a **s**mall **a**rea (SA): GTs navigate randomly within a designated area, often altering their directions and speeds. For example, they might be involved in tasks such as capturing images or locating and identifying targets in a limited space.

Scenario with GTs moving along a relatively **l**ong **p**ath (LP): GTs move along a designated ground path, with infrequent changes in direction and speed. For instance, these GTs could be transporting cargo from one point to another while handling data transmission or reception.

We begin with Fig. 6.4a and b to analyze the convergence performance of the proposed method in the SA scenario, focusing on the effects of the moving average return and overall operation time, respectively. The moving average return is calculated by averaging the reward over every 100 episodes across 10,000 episodes. The curves presented are derived from distinct and independent experiments, where the initial positions of the UAV and GTs are randomly assigned. From Fig. 6.4, it is evident that regardless of the initial locations, both the moving average return and overall operation time converge to similar values, demonstrating the proposed algorithm's convergence performance. Furthermore, as seen in Fig. 6.4b, the whole operation time starts at a higher value and gradually decreases before converging. This trend can be attributed to the UAV's initial lack of experience, which hinders effective learning of efficient action strategies during early interactions with the environment. As the UAV gains experience, it refines its action selection strategy, leading to improved efficiency and a reduction in overall operation time. Notably, the final convergence values for the five curves in Fig. 6.4a and b show slight variations, mainly due to differences in the initial locations of the GTs, network parameters, and the dynamic exploration behaviors exhibited by the training curves.

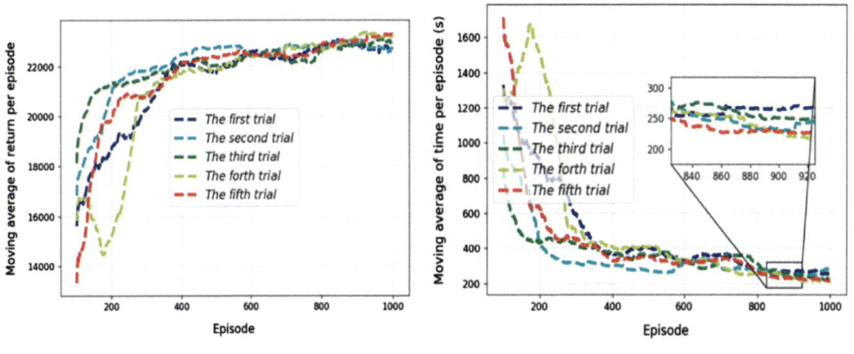

(a) reward of scenario SA (b) whole operation of scenario SA

Fig. 6.4 Convergence investigation

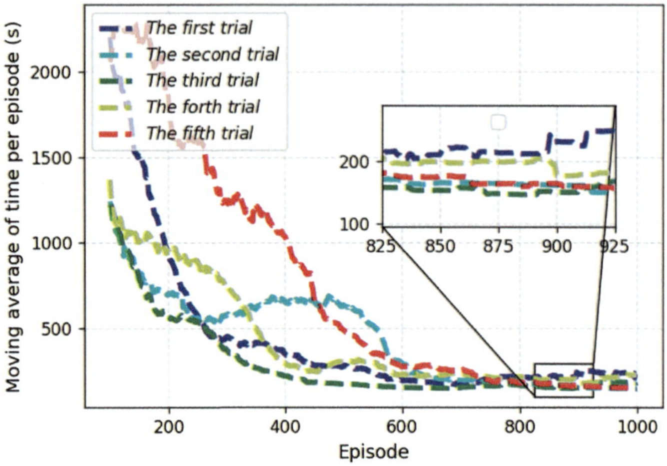

Fig. 6.5 The whole operation time if scenario LP

Next, Fig. 6.5 presents the convergence performance of the overall operation time for UAV 3D trajectory design in the LP scenario. Each curve in the figure corresponds to an independent experiment, with varying initial positions for the UAV and GTs across all trials. In contrast to the SA scenario, the LP scenario demonstrates a quicker convergence of overall operation time. For instance, while the SA scenario reaches convergence at 800 episodes, as indicated in Fig. 6.4b, the LP scenario converges at 700 episodes, as shown in Fig. 6.5. Additionally, the overall operation time in the LP scenario is lower than that in the SA scenario. This difference is primarily attributed to the fact that all GTs move around the central cluster of buildings. Consequently, the UAV only needs to navigate within this area, as depicted in Fig. 6.9. Compared to the SA scenario, the UAV can more swiftly adjust its trajectory in response to the GTs' movements, resulting in a shorter flight distance. This leads to a reduction in the whole operation time and faster convergence.

6.6.3 Performance Evaluation of the Proposed Method

In this section, we assess the performance of the proposed method. Specifically, we compare the 3D trajectory generated by our design with a 2D trajectory that requires the UAV to fly at a constant altitude. This comparative analysis highlights the benefits of allowing the UAV to operate within an unrestricted feasible domain and examines how trajectory behavior varies across two different GT movement scenarios.

In Fig. 6.6, we compare our design featuring a 3D UAV trajectory with the 2D trajectory in scenario SA. The results clearly show that the operational time for our 3D design is shorter than that for the 2D cases, highlighting the efficiency gained from the UAV's adjustable height. This efficiency arises mainly from the enhanced

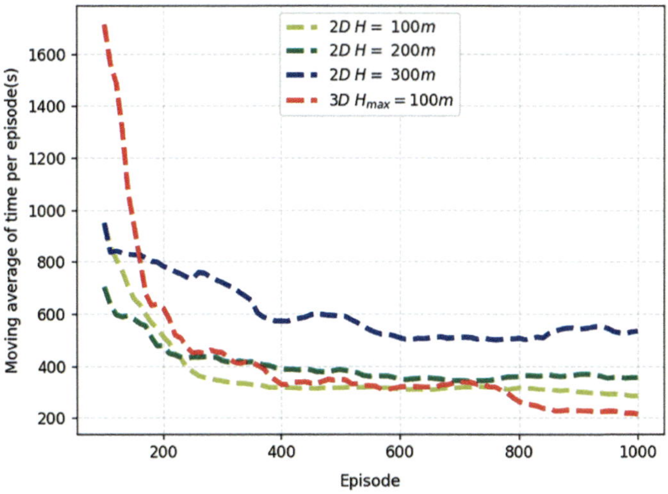

Fig. 6.6 Comparison of 3D trajectory optimization performance with 2D

flexibility of 3D trajectory design, which optimally balances angle and transmission distance. Conversely, our 3D trajectory exploration converges around the 800th episode, which is slower than the 2D design. This slower convergence is attributed to the larger exploration space in the 3D case, necessitating more episodes to find an efficient solution. Additionally, the figure indicates that the overall operation time for the 2D trajectory design tends to increase with a fixed UAV altitude. This occurs because communication connections between the UAV and GTs have a higher LoS) probability at greater altitudes, meaning the transmission rate is primarily affected by the transmission distance.

We examine how the allowed flying altitude ranges of the UAV affect the whole operation time, as shown in Fig. 6.7. Three altitude ranges are analyzed: (a) [0–50] m, where the UAV flies between the ground and the tallest building; (b) [50–100] m, indicating a minimum flying height above building heights; and (c) [0–100] m, allowing flights between and above buildings. Clearly, obstacle avoidance is crucial for the first two cases. The initial positions of the UAV and GTs are consistent across all cases. As seen in Fig. 6.7, the proposed method converges quickly across the three altitude ranges, demonstrating strong adaptability. Notably, under the same initial conditions, case c) achieves the shortest operation time, followed by case (a), while case (b) has the longest. The reasons for this trend are as follows: in case (a), the presence of buildings complicates the UAV's ability to maintain reliable communication with GTs. In case (b), despite higher potential for LoS communication, the longer transmission distances (exceeding the minimum UAV height of 50 m) hinder efficiency. These findings highlight the advantages of a flexible altitude range. Furthermore, case (a) exhibits lower convergence speed and stability compared to cases (b) and (c), with the [0–50] m range showing significant oscillation. This instability stems from the buildings' heights, set at 50 m, forcing the UAV to navigate

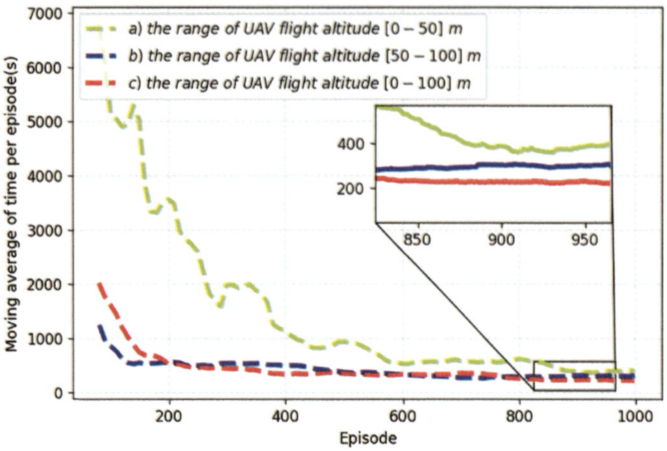

Fig. 6.7 Performance comparison at different flight altitudes

among them and continuously contend with obstacle avoidance constraints. Consequently, the performance of case c) is relatively lower than case b) when episodes are insufficient.

Next, we present the designed UAV 3D trajectories in Figs. 6.8 and 6.9 for the SA and LP scenarios, respectively. The source GTs are labeled as S_1 , S_2, and S_3 , while the destination GTs are represented by D_1, D_2, and D_3. In both figures, the red 3D curves depict the UAV's trajectory, with other curves representing the GTs' paths. The results demonstrate that our method effectively enables the UAV to navigate around 3D buildings during operations with various GT mobility models, highlighting the algorithm's adaptability to complex obstacle avoidance environments. The figures also show that the UAV initially ascends after takeoff before approaching the served GT. For instance, in Fig. 6.8, we observe that the UAV increases its altitude when source GT S_1 offloads tasks, subsequently moving closer to GTs S_3 and S_2 during the first operational stage. This behavior can be explained by the need to balance flight angle and distance from the serving GT to minimize overall operation time. Increasing altitude enhances the LoS probability for servicing GTs, while flying closer to the GT boosts the communication transmission rate.

As depicted in Fig. 6.9, the GTs navigate around the central cluster of buildings, exhibiting a larger movement range compared to scenario SA. Unlike in scenario SA, the UAV primarily flies over the middle cluster of buildings after ascending in the LP scenario. This demonstrates that the UAV adjusts its flight path to follow the moving GTs during the mission. The results from both figures highlight the robustness of our design, showing that our method effectively adapts to varying GT mobility models.

In the system setup illustrated in Fig. 6.10, explore the relationship between the decision regarding the last offloaded task and the UAV's computational efficiency in both the SA and LP scenarios. The figures reveal that as the UAV's CPU frequency increases, the scheduling of the last offloaded task initially changes before stabilizing.

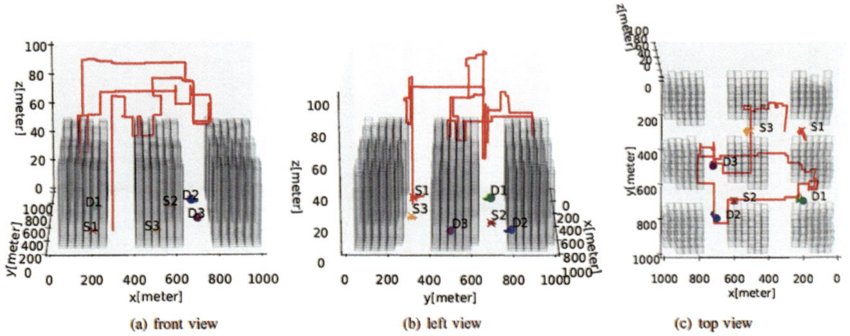

Fig. 6.8 The 3D trajectory of the UAV is displayed under scenario LP

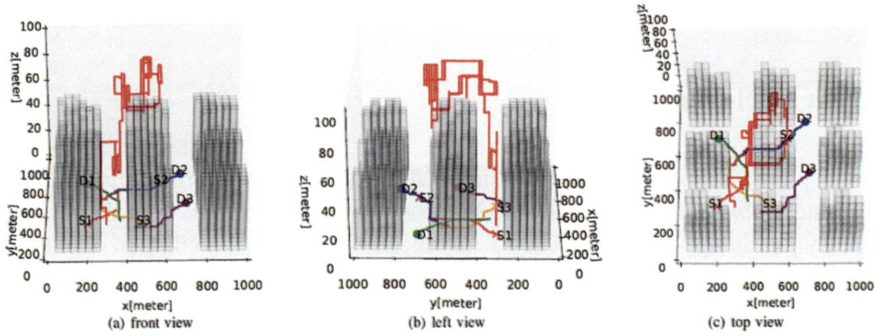

Fig. 6.9 The 3D trajectory of the UAV is displayed under scenario LP

For example, in the SA scenario, when the frequency is below 4 GHz, S_1 is assigned as the last offload task, but this shifts to S_2 when the frequency exceeds 4 GHz. Additionally, we notice a gradual decrease in data processing time for frequencies above 4 GHz across both scenarios. These findings highlight the trade-off among offloading time, processing time of the last offloaded task, and decision transmission time in minimizing overall operation time, which is the sum of these components. Increasing CPU frequency mitigates the impact of the last offloaded task's processing time in this trade-off. At lower frequencies, the processing time is significant, leading the system to favor lighter workload tasks for the last offload. Conversely, if the UAV's CPU is sufficiently powerful, the processing time of the last task becomes less of a bottleneck, and higher CPU frequency no longer affect the scheduling decision.

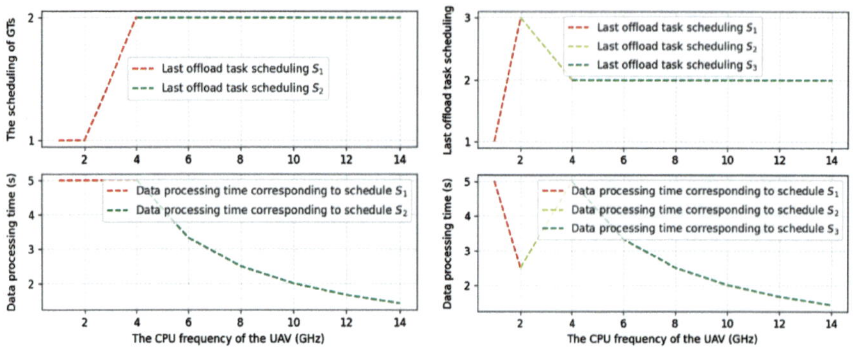

(a) scenario SA **(b) scenario LP**

Fig. 6.10 The scheduling of the last offload task for the GTs and the data processing time throughout the operation under various scenarios

6.7 Conclusion

In this paper, we explored a UAV-assisted MEC network involving multiple mobile GTs and probabilistic LoS channels Our design aims to minimize overall operation time by jointly optimizing the UAV's 3D trajectory and the scheduling of offloading transmissions from source GTs, as well as decision transmissions to destination GTs. Notably, we adopted an unrestricted flight domain for the UAV while ensuring obstacle avoidance around 3D buildings. To tackle this challenge, we transformed the problem into a Markov Decision Process (MDP) and introduced a D3QN-based method, allowing the UAV to function as an agent that learns and explores optimal policies for 3D trajectory and GT scheduling. Through simulations, we confirmed the convergence of our method and examined the effectiveness of various hyper-parameters. Additionally, we assessed the method's performance regarding operation time and robustness, as well as the influence of system parameters, such as GT mobility models and UAV flying altitude, on the designed 3D trajectory and scheduling outcomes.

References

1. Asadpour, M., et al. (2014). Micro aerial vehicle networks: An experimental analysis of challenges and opportunities. *IEEE Communications Letters, 52*, 141–149.
2. Mozaffari, M., Saad, W., Bennis, M., Nam, Y.-H., & Debbah, M. (2019). A tutorial on UAVs for wireless networks: Applications, challenges, and open problems. *IEEE Commun. Surv. Tutor., 21*, 2334–2360.
3. Zeng, Y., Zhang, R., & Lim, T. J. (2016). Wireless communications with unmanned aerial vehicles: Opportunities and challenges. *IEEE Communications Magazine, 54*, 36–42.
4. Ruan, L., et al. (2018). Energy-efficient multi-UAV coverage deployment in UAV networks: A game-theoretic framework. *China Communications, 15*, 194–209.

5. Wang, Z., Duan, L., & Zhang, R. (2019). Adaptive deployment for UAV-aided communication networks. *IEEE Transactions on Wireless Communications, 18*, 4531–4543.
6. Wang, Z., et al. (2022). Robust secure UAV relay-assisted cognitive communications with resource allocation and cooperative jamming. *Journal of the Communications Network, 24*, 139–153.
7. Zeng, Y., Zhang, R., & Lim, T. J. (2016). Throughput maximization for UAV-enabled mobile relaying systems. *IEEE Transactions on Communications, 64*, 4983–4996.
8. Xu, Y., Zhang, T., Yang, D., Xiao, L. (2021). UAV-assisted relaying and MEC networks: Resource allocation and 3D deployment. In *2021 IEEE International Conference on Communications Workshops (ICC Workshops)* (pp. 1–6).
9. Yang, L., Yao, H., Zhang, X., Wang, J., & Liu, Y. (2020). Multi-UAV deployment for MEC enhanced IoT networks. In *2020 IEEE/CIC International Conference on Communications in China (ICCC)* (pp. 436–441).
10. Xu, C., Zhan, C., Liao, J., & Gong, J. (2022). Computation throughput maximization for UAV-enabled MEC with binary computation offloading. In *ICC 2022—IEEE International Conference on Communications* (pp. 4348–4353).
11. Wu, Z., et al. (2022). Joint deployment and trajectory optimization in UAV-assisted vehicular edge computing networks. *Journal of the Communications Network, 24*, 47–58.
12. Xiong, J., Guo, H., & Liu, J. (2019). Task offloading in UAV-aided edge computing: Bit allocation and trajectory optimization. *IEEE Communications Letters, 23*, 538–541.
13. Liu, B., Wan, Y., Zhou, F., Wu, Q., & Hu, R. Q. (2022). Resource allocation and trajectory design for MISO UAV-assisted MEC networks. *IEEE Transactions on Vehicular Technology, 71*, 4933–4948.
14. Xu, D., Sun, Y., Ng, D. W. K., & Schober, R. (2020). Multiuser MISO UAV communications in uncertain environments with no-fly zones: Robust trajectory and resource allocation design. *IEEE Transactions on Communications, 68*, 3153–3172.
15. Chen, Y.-J., & Huang, D.-Y. (2022). Joint trajectory design and BS association for cellular-connected UAV: An imitation-augmented deep reinforcement learning approach. *IEEE Internet of Things Journal, 9*, 2843–2858.
16. Nie, Y., Zhao, J., Gao, F., & Yu, F. R. (2021). Semi-distributed resource management in UAV-aided MEC systems: A multi-agent federated reinforcement learning approach. *IEEE Transactions on Vehicular Technology, 70*, 13162–13173.
17. Duo, B., Wu, Q., Yuan, X., & Zhang, R. (2020). Anti-jamming 3D trajectory design for UAV-enabled wireless sensor networks under probabilistic lOS channel. *IEEE Transactions on Vehicular Technology, 69*, 16288–16293.
18. Sun, C., et al. (2023). Max-min fair 3D trajectory design and transmission scheduling for solar-powered fixed-wing UAV-assisted data collection. *IEEE Transactions on Wireless Communications*, 1–1.
19. Hua, M., Yang, L., Wu, Q., & Swindlehurst, A. L. (2020). 3D UAV trajectory and communication design for simultaneous uplink and downlink transmission. *IEEE Transactions on Communications, 68*, 5908–5923.
20. Zhan, C., et al. (2021). Joint resource allocation and 3D aerial trajectory design for video streaming in UAV communication systems. *IEEE Transactions on Circuits and Systems for Video Technology, 31*, 3227–3241.
21. Zhang, W., Wang, Q., Liu, X., Liu, Y., & Chen, Y. (2021). Three-dimension trajectory design for Multi-UAV wireless network with deep reinforcement learning. *IEEE Transactions on Vehicular Technology, 70*, 600–612.
22. Liu, X., Liu, Y., & Chen, Y. (2019). Reinforcement learning in multiple-UAV networks: Deployment and movement design. *IEEE Transactions on Vehicular Technology, 68*, 8036–8049.

23. Ejaz, M., et al. (2024). Rl-planner: Reinforcement learning-enabled efficient path planning in multi-UAV MEC systems. *IEEE Transactions on Network and Service Management, 21*, 3317–3329.
24. Hasselt, H. V. (2010). *Double Q-learning*. OAI.
25. Xie, H., Yang, D., Xiao, L., & Lyu, J. (2021). Connectivity-aware 3D UAV path design with deep reinforcement learning. *IEEE Transactions on Vehicular Technology, 70*, 13022–13034.

Chapter 7
Joint Trajectories and Resource Allocation for UAV-Enabled Integrated Sensing and Communication

Kaitao Meng

Abstract Integrated sensing and communication (ISAC) technologies have garnered substantial academic and industrial interest. ISAC is recognized for its ability to leverage unified infrastructure and waveforms to simultaneously transmit information and receive echoes, thereby significantly enhancing the efficiency of spectrum, cost, and energy. Unmanned aerial vehicles (UAVs) offer flexible observation and enhanced communication capabilities, poised to revolutionize ISAC systems. Existing ISAC research often overlooks practical asymmetric sensing and communication needs, leading to suboptimal resource utilization. To address this, we propose an integrated periodic sensing and communication mechanism for UAV-enabled ISAC. Our approach optimizes UAV trajectory, user association, sensing selection, and beamforming to maximize system achievable rates while meeting specific sensing and beam pattern requirements. We derive closed-form solutions for beamforming and propose efficient algorithms to solve the non-convex optimization challenges, demonstrating significant improvements over traditional approaches.

Keywords Integrated sensing and communication · UAV · Periodic sensing · User association · Beamforming · Trajectory optimization

Abbreviations

AO	Alternating Optimization
BS	Base Station
FHF	Fly-Hover-Fly
GPS	Global positioning system
ISAC	Integrated sensing and communication
IPSAC	Integrated periodic sensing and communication
LoS	Line-of-sight

K. Meng (✉)
Department of Electronic and Electrical Engineering, University College London, London, UK
e-mail: kaitao.meng@ucl.ac.uk

Y. Hu et al. (eds.), *Reliable and Efficient Trajectory Design in UAV-Aided Wireless Networks*, https://doi.org/10.1007/978-981-96-5988-3_7

NLoS	Non-line-of-sight
QoS	Quality of service
SF	Straight Flight
TDMA	Time division multiple access
UAV	Unmanned aerial vehicles
UPA	Uniform plane array

7.1 Introduction

7.1.1 Background

There has been increasing interest in the investigation of the integrated sensing and communication (ISAC) technologies in both academia and industry due to the potential for spectrum reuse and the urgent demand for advanced sensing capabilities [1, 2]. Unlike traditional systems where radar sensing and communication share the spectrum but use separate infrastructures [3], ISAC systems utilize the same wireless resources for both transmitting data and retrieving information from signal echoes [4]. This approach not only boosts spectrum efficiency and reduces hardware costs but also significantly enhances the balance between sensing and communication performance [5]. With advances in massive MIMO and mmWave/THz technologies, ISAC base stations (BSs) can achieve higher sensing precision, supporting location-based intelligent applications with strict sensing demands [6]. Terms like radar-communication systems have been used to describe similar research topics [7–12]. Industry players such as Huawei regard ISAC as a key technology for six-generation networks [13–16].

Previous studies have shown that co-designing waveform and beamforming for ISAC systems yields mutual benefits for both communication and sensing [17–21]. For example, beamforming models have been proposed to optimize radar transmission while meeting communication requirements [19]. However, one limitation of traditional ISAC systems is their reliance on line-of-sight (LoS) links, which makes sensing performance highly dependent on the presence of direct paths between targets and transceivers. By contrast, non-LoS (NLoS) links are generally seen as undesirable for sensing, as they introduce interference. In scenarios where targets are far from the BS or blocked by obstacles, the effectiveness of sensing can suffer greatly, limiting the operational range of terrestrial ISAC systems.

The unique characteristics of unmanned aerial vehicles (UAVs), such as flexible deployment and strong LoS links, offer a promising solution for enhancing ISAC systems [20, 21]. UAVs can achieve greater communication quality, flexible coverage, and more dynamic observation by leveraging their high mobility [22, 23]. Traditional UAV-enabled networks, however, have focused on either communication or sensing independently, without considering integrated waveform designs [24, 25]. The UAV-enabled ISAC system introduces new challenges, such as balancing beam pattern constraints, resource allocation, and UAV trajectory to maximize achievable rates. Recent research has begun exploring these complexities, with studies optimizing UAV movement and beamforming to improve communication while ensuring adequate sensing [27–29]. Moreover, deploying multiple UAVs for cooperative tasks has shown promise in extending ISAC network coverage, and ISAC-enabled cellular networks could even track and monitor suspicious UAVs to enhance security [30].

7.1.2 Motivation

Despite these advances, most studies assume that sensing and communication must occur simultaneously throughout the system's operation. However, in real-world scenarios, the requirements for sensing and communication are often asymmetric. For instance, low-speed objects may need less frequent sensing, whereas high-speed targets demand higher sensing rates. Current ISAC literature has yet to account for this variation in sensing frequency. Simultaneously forcing both functions can also lead to inefficiencies, such as wasted spectrum and increased interference, while higher energy consumption remains a concern, especially for UAVs with limited power [31]. Addressing these gaps requires studying the impact of sensing frequency on achievable rates [32], particularly for UAV-enabled ISAC systems, which can benefit from the flexibility of autonomous UAV movement.

In this chapter, we propose a UAV-enabled ISAC system where a UAV performs periodic sensing while delivering downlink communication to several users. Our proposed integrated periodic sensing and communication (IPSAC) mechanism allows for flexible balancing between sensing and communication by optimizing transmit beamforming, user association, sensing time, and UAV trajectory, considering practical sensing frequency requirements. Unlike traditional ISAC methods, which force the UAV to perform both functions simultaneously, our approach offers greater flexibility. Both standalone communication and continuous sensing are treated as special cases in our framework.

7.1.3 Main Contributions

Main contributions to this chapter are as follows:

(1) Proposing a UAV-enabled IPSAC mechanism that balances sensing power, frequency, and communication performance across multi-user and multi-target scenarios. We formulate a periodic ISAC problem aimed at maximizing achievable rates while meeting sensing and beam pattern constraints.
(2) Deriving closed-form solutions for beamforming under any given UAV location and for optimal achievable rates when the number of antennas is infinite,

providing insights for algorithm design. A penalty-based algorithm is introduced to jointly optimize beamforming, user association, sensing time, and UAV trajectory.

7.1.4 Structural Organization

Section 7.2 provides discussion on related works. Section 7.3 introduces the system model and problem formulation of the UAV-enabled IPSAC system. In Sect. 7.4, we derive the closed-form optimal beamforming vector and propose a penalty-based algorithm to address the sum achievable rate maximization problem. Section 7.5 provides numerical results to validate the performance of our proposed mechanism. Section 7.6 concludes this chapter.

7.2 Related Works

As discussed in [20, and 21], UAVs can be strategically positioned to enhance both communication and sensing performance, particularly in environments where terrestrial base stations are less effective. However, traditional UAV-enabled networks have primarily focused on either communication or sensing as independent functions, without addressing the challenges of integrated waveform and beamforming designs [24]. The introduction of UAV-enabled ISAC systems presents new complexities, such as balancing beam pattern constraints, resource allocation, and UAV trajectory to maximize achievable rates. More advanced techniques that optimize UAV trajectory and beamforming design, while considering flexible sensing frequencies, are essential. UAV-enabled ISAC systems provide greater flexibility than fixed-deployment counterparts by allowing real-time adjustments to the sensing schedule, beam patterns, and flight paths. This capability leads to more efficient resource allocation and improved overall system performance, as noted in [21].

Recent studies have explored the dynamic interaction between UAV trajectories and ISAC performance. By jointly optimizing UAV flight paths, beamforming strategies, and sensing intervals, researchers aim to develop more sophisticated mechanisms that can balance sensing and communication tasks more effectively. For instance, in [27], the authors proposed an UAV-enabled ISAC system to achieve enhanced synergy between these functions, providing more versatile and energy-efficient solutions for future applications. However, unlike these studies, which primarily focus on simultaneous sensing and communication [33, 34], our work addresses a critical gap by considering the need for adaptable sensing frequencies Fig. 7.1.

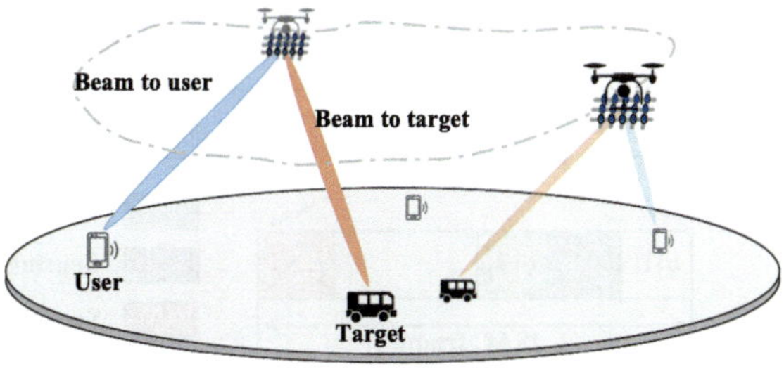

Fig. 7.1 System architecture

7.3 System Model and Problem Formulation

7.3.1 UAV-Enabled ISAC System

We investigate a UAV-based integrated sensing and communication (ISAC) system where the UAV performs sensing of several potential ground targets while also delivering downlink communication to K single-antenna users during a designated flight period of T seconds. The horizontal position of user k is denoted as, which may be determined through either global positioning system (GPS) or uplink signal estimation [35]. The set of the users and that of the prospective targets are denoted by $K = \{1, \cdots, K\}$ and $J = \{1, \ldots, J\}$, respectively. For the targets, the horizontal positions are noted as $\boldsymbol{v}_j = [v_{x,j}, v_{y,j}]^T$, where $j \in J$, and these positions depend on the sensing objectives. For instance, v_j might be set as the estimated position based on prior frames for tracking or uniformly sampled in the region of interest for detection. The entire mission time T is divided into N discrete time slots, each with duration $\delta_t = T/N$, and indexed by $n \in N = \{1, \ldots, N\}$. Each time slot is small enough that the UAV's location can be considered constant to simplify trajectory and beamforming design. The UAV's horizontal coordinates at time slot n are denoted by $\boldsymbol{q}[n] = [q_x[n], q_y[n]]^T$, and it is assumed that the UAV flies at a fixed altitude of H meters under air traffic control constraints [36]. The UAV uses a uniform plane array (UPA) with $M = M_x \times M_y$ antennas, where M_x and M_y are the number of elements along the x-axis and y-axis, respectively. The elements are spaced by $d_x = d_y = \lambda/2$, where λ is the carrier wavelength. The UPA is assumed to be parallel to the ground to simplify technical analysis, as shown in Fig. 5.1.

7.3.2 ISAC Frame Structure

Considering practical timeliness requirements for sensing tasks, we introduce an ISAC mechanism tailored for scenarios involving multiple users and targets, seeking to balance sensing and communication. Each sensing task must be performed at least once in each ISAC frame, as depicted in Fig. 5.2. Let T_L denote the duration of one ISAC frame, with the total number of frames during the flight period being $L = T/T_L$,

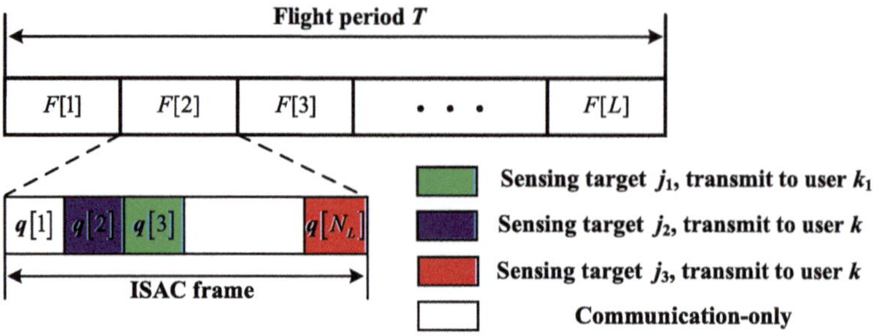

Fig. 7.2 The structure of DNN

and frames are indexed by $l \in L = \{1, \ldots, L\}$.. To avoid interference caused by line-of-sight (LoS) channels, we employ time division multiple access (TDMA), where only one target is sensed in each time slot, minimizing the computational complexity of target estimation. If the UAV is sensing target j during slot n, we set $c_j[n] = 1$; otherwise, $c_j[n] = 0$. Each target is sensed exactly once per frame, as expressed by the following constraints (c.f. Fig. 7.2):

$$\sum_{n=(l-1)N_L+1}^{lN_L} c_j[n] = 1, \forall l \in L, j \in J; \sum_{j=1}^{J} c_j[n] \leq 1, \forall n \in N, \tag{7.1}$$

Additionally, the sensing frequency of each target is given by $1/T_L = 1/(\delta N_L)$.

7.3.3 Communication and Sensing Model

The communication channels between the UAV and users are dominated by a LoS component [37]. The channel gain between the UAV and user k follows the free-space path loss model and is expressed as:

$$\beta_k(\boldsymbol{q}[n], \boldsymbol{u}_k) = \beta_0 d(\boldsymbol{q}[n], \boldsymbol{u}_k)^{-2} = \frac{\beta_0}{H^2 + \|\boldsymbol{q}[n] - \boldsymbol{u}_k\|^2}, \tag{7.2}$$

where β_0 is the channel gain at a reference distance of 1 m. The Doppler effect induced by UAV movement can be fully compensated for at both the communication users and the sensing receiver [38, 39]. The transmit array response vector towards user k is:

$$\boldsymbol{a}^H(\boldsymbol{q}[n], \boldsymbol{u}_k) = \left[1, \ldots, e^{-\frac{j2\pi(M_x-1)d_x\Phi(\boldsymbol{q}[n],\boldsymbol{u}_k)}{\lambda}}\right] \otimes \left[1, \ldots, e^{-\frac{j2\pi(M_y-1)d_y\Omega(\boldsymbol{q}[n],\boldsymbol{u}_k)}{\lambda}}\right], \tag{7.3}$$

where $\Phi(\boldsymbol{q}[n], \boldsymbol{u}_k)$ and $\Omega(\boldsymbol{q}[n], \boldsymbol{u}_k)$ represent the angular positions between the UAV and user k. The baseband equivalent channel from the UAV to user k can be written as:

$$h_k^H(q[n], u_k) = \sqrt{\beta_k(q[n], u_k)}e^{-j\frac{2\pi d(q[n], u_k)}{\lambda}} a^H(q[n], u_k). \tag{7.4}$$

The UAV transmits an information signal s_k to user k at time slot n, where $s_k \sim \mathcal{CN}(0, 1)$ and the signals are uncorrelated. The transmitted signal is:

$$x[n] = w_c[n] \sum_{k=1}^{K} \alpha_k[n] s_k[n], n \in N, \tag{7.5}$$

where $w_c[n]$ is the beamforming vector, and $\alpha_k[n]$ equals 1 if the UAV is serving user k in slot n, otherwise $\alpha_k[n] = 0$ Only one user is served per time slot, as described by the constraint:

$$\sum_{k=1}^{K} \alpha_k[n] \leq 1, \forall, n \in N. \tag{7.6}$$

The received signal at user k is:

$$y_k[n] = h_{c,k}^H(q[n], u_k)\left(w_c[n] \sum_{k=1}^{K} \alpha_k[n] s_k[n] \right) + n_k[n], \forall n \in N, \tag{7.7}$$

where $n_k[n]$ denotes the AWGN noise at user k's receiver. When $\alpha_k[n] = 1$, the corresponding user's SNR is:

$$\gamma_k[n] = \frac{\left| h_{c,k}^H(q[n], u_k) w_c[n] \right|^2}{\sigma_k^2}, \forall n \in N, \tag{7.8}$$

and the corresponding data rate is:

$$R_k[n] = \log_2(1 + \gamma_k[n]). \tag{7.9}$$

The reflected communication signals from targets can also be leveraged for target parameter estimation, with the transmit beam pattern gain toward target j given by:

$$\Gamma\left(q[n], v_j\right) = E\left[\left| a^H\left(q[n], v_j\right)(x[n])\right|^2\right] = a^H\left(q[n], v_j\right)\left(w_r[n] w_r[n]^H\right) a\left(q[n], v_j\right). \tag{7.10}$$

7.3.4 Problem Formulation

This chapter focuses on maximizing the achievable rate by optimizing several factors, including the beamforming vector, user association, sensing time allocation, and UAV trajectory. These optimizations must adhere to constraints related to sensing frequency, sensing power, and quality of service (QoS) standards. Thus, the optimization problem is formulated as follows:

(P1):	$\max_{\boldsymbol{w}_c,\boldsymbol{q}[n],\alpha_k[n],c_j[n]} \frac{1}{N}\sum_{n=1}^{N}\sum_{k=1}^{K}\alpha_k[n]R_k[n],$	(P1)
s.t.	(1), (6)	(P1a)
Sensing constraints:	$c_j[n]\frac{\Gamma(\boldsymbol{q}[n],\boldsymbol{v}_j)}{d(\boldsymbol{q}[n],\boldsymbol{v}_j)^2} \geq c_j[n]\Gamma_j, \forall j \in J, n \in N,$	(P1b)
Allocation variables:	$c_j[n] \in \{0, 1\}, \alpha_k[n] \in \{0, 1\}, \forall j \in J, \forall k \in K, n \in N,$	(P1c)
Communication constraints:	$\frac{1}{N_L}\sum_{n=(l-1)N_L+1}^{lN_L}\alpha_k[n]R_k[n] \geq R_k^{t\hbar}, k \in K, l \in L,$	(P1d)
Transmit power:	Transmit power: $\|\boldsymbol{w}_c[n]\|^2 \leq P_{\max}$	(P1e)
Movement constraints:	$\|\boldsymbol{q}[n] - \boldsymbol{q}[n-1]\| \leq Vt_{\max}$	(P1f)
Start and end point:	$\boldsymbol{q}[1] = \boldsymbol{q}_I, \boldsymbol{q}[N] = \boldsymbol{q}_F.$	(P1g)

The constraint in (P1a) can be extended to scenarios with a path loss exponent of 4, i.e., $c_j[n]\Gamma(\boldsymbol{q}[n], \boldsymbol{v}_j)/d(\boldsymbol{q}[n], \boldsymbol{v}_j)^4 \geq c_j[n]\Gamma_j$, which indicates that the signal-to-noise ratio (SNR) from the reflected target signals must exceed a specific threshold. The case with a 4-exponent path loss can be treated as a mono-static sensing system, while the 2-exponent path loss in (P1a) reflects that the signal power at the target's location must be above the threshold, representing bi-static sensing, where a separate receiver is used for echo analysis. The constraints in (P1b) ensure that the beamforming gain at the target locations meets the required threshold Γ_j. To maintain the required service quality, the minimum achievable rate is constrained by (P1d). The transmit power and UAV speed limits are imposed by (P1e) and (P1f), respectively, while the starting and final locations are set by (P1g).

In some scenarios, a target might require both communication and sensing services (e.g., using sensing data to enhance communication). In such cases, a separate user at the same location could be introduced to handle this dual service. Given that problem (P1) is non-convex and involves integer variables tightly coupled with UAV trajectory and beamforming design, finding a solution is challenging. We tackle this by first deriving a closed-form expression for the optimal beamforming vector and calculating a tight lower bound on the achievable rate. Based on these results, we propose an efficient penalty-based algorithm with a two-layer structure to solve the optimization problem. Moreover, by relaxing the initial and final location constraints, we demonstrate the symmetric structure of optimal solutions across different ISAC frames.

7.4 Penalty-Based Algorithm

In this section, we begin by deriving the optimal closed-form beamforming vector for the proposed IPSAC mechanism, as outlined in Sect. 7.4.1. Next, Sect. 7.4.2 presents a tight lower bound for the original objective value. Based on this, we introduce a penalty-based approach to simultaneously optimize the UAV trajectory, user assignment, and sensing time allocation.

7.4.1 Optimal Closed-Form Beamforming

If $\sum_{j=1}^{J} c_j[n] = 0$ and $\alpha_k[n] = 1$, the optimal beamforming vector for a given UAV location is $\boldsymbol{w}_c^* = \sqrt{P_{max}}\boldsymbol{h}_{c,k}(\boldsymbol{q}[n], \boldsymbol{u}_k)/|\boldsymbol{h}_{c,k}(\boldsymbol{q}[n], \boldsymbol{u}_k)$. Otherwise, if $\mathrm{c_j[n]} = 1$ and $\alpha_k[n] = 1$, the optimal beamforming is highly interdependent with the UAV's flight path. To simplify notation, we represent $\boldsymbol{h}_{c,k}^H(\boldsymbol{q}[n], \boldsymbol{u}_k)$ *as* $\boldsymbol{h}_{c,k}^H$ and $\boldsymbol{a}^H(\boldsymbol{q}[n], \boldsymbol{v}_j)/d(\boldsymbol{q}[n], \boldsymbol{v}_j)$ *as* $\boldsymbol{h}_{r,j}^H$. Maximizing $R_k[n]$ is essentially equivalent to maximizing the received signal strength $\boldsymbol{w}_c^H\boldsymbol{h}_{c,k}\boldsymbol{h}_{c,k}^H\boldsymbol{w}_c$. As a result, the log function is omitted in the objective for simplicity. This transforms the problem into maximizing the signal strength $max\boldsymbol{w}_c^H\boldsymbol{h}_{c,k}\boldsymbol{h}_{c,k}^H\boldsymbol{w}_c$, under certain conditions: $\boldsymbol{w}_c^H\boldsymbol{h}_{r,j}\boldsymbol{h}_{r,j}^H\boldsymbol{w}_c \geq \Gamma_j$ and $\boldsymbol{w}_c[n]^2 \leq P_{max}$.

Proposition 7.1 When $c_j[n] = 1$ and $\alpha_k[n] = 1$, the optimal beamforming vector for any given UAV location is given as:

$$w_c^* = \frac{1}{\lambda_1}\left(\sqrt{\beta_{c,k}}\boldsymbol{h}_{c,k} + \lambda_2\sqrt{\Gamma}\boldsymbol{h}_{r,j}e^{-i\varphi_{k,j}}\right), \textit{if}\ \cos\varphi_{k,j}^2 MP_{max}/d(\boldsymbol{q}[n], \boldsymbol{v}_j)^2 < \Gamma_j. \quad (7.11)$$

$\beta_{c,k} = \left\|\boldsymbol{h}_{c,k}^H\right\|^2/\left\|\boldsymbol{h}_{r,j}\right\|^2\gamma^2$, $\lambda_1 = \gamma\left\|\boldsymbol{h}_{c,k}^H\right\|^2 \sin\varphi_{k,j}/\sqrt{P_{\max}\left\|\boldsymbol{h}_{r,j}\right\|^2 - \Gamma_j}$, $\lambda_2 = r\left\|\boldsymbol{h}_{c,k}^H\right\|^2$ $\sqrt{\Gamma_j - \gamma^2\left\|\boldsymbol{h}_{c,k}^H\right\|\left\|\boldsymbol{h}_{r,j}\right\|\cos\varphi_{k,j}/\left\|\boldsymbol{h}_{r,j}\right\|^2}\sqrt{P_{\max}\left\|\boldsymbol{h}_{r,j}\right\|^2\Gamma_j - \Gamma_j^2}\sin\varphi_{k,j}$, $\varphi_{k,j} = \arccos\frac{\left|\boldsymbol{h}_{c,k}^H\boldsymbol{h}_{r,j}\right|}{\left\|\boldsymbol{h}_{c,k}^H\right\|\left\|\boldsymbol{h}_{r,j}\right\|}$ and $\gamma = \sqrt{\Gamma_j}\cos\varphi_{k,j} + \sqrt{P_{\max}\left\|\boldsymbol{h}_{r,j}\right\|^2 - \Gamma_j}\sin\varphi_{k,j}$.

Proof Proposition 7.1 can be proved by Karush–Kuhn–Tucker (KKT) conditions, and details are omitted due to limited spacing.

In Proposition 7.1, the optimal beamforming vector could be intuitively viewed as two linearly superimposed beams toward user and target, respectively, which directly shows the influencing factors of the associated user's achievable rate. Also, the closed-form beamforming in (7.11) can also hold for arbitrary user channels. For $\cos\varphi_{k,j}^2 MP_{max}/d(\boldsymbol{q}[n], \boldsymbol{v}_j)^2 < \Gamma_j$, the optimal SNR at user k can be obtained as follows:

$$\gamma_{k,j}^* = \gamma_0 \frac{d(\boldsymbol{q}[n], \boldsymbol{v}_j)^2}{d(\boldsymbol{q}[n], \boldsymbol{u}_k)^2} \left(\sqrt{\Gamma_j} \cos\varphi_{k,j} + \sqrt{\frac{MP_{max}}{d(\boldsymbol{q}[n], \boldsymbol{v}_j)^2} - \Gamma_j} \sin\varphi_{k,j} \right)^2, \tag{7.12}$$

where $\gamma_0 = \beta_0/\sigma^2$. In Eq. (7.12), the optimal SNR for the user is primarily influenced by two components: Γ_j and $\sqrt{MP_{max}/d(\boldsymbol{q}[n], \boldsymbol{v}_j)^2 - \Gamma_j}$, along with the channel correlation factor, specifically $\cos\varphi_{k,j}^2$. When $\cos\varphi_{k,j}^2 = 1$, the communication and target channels become linearly dependent, meaning the channel power gain at user k reaches $\beta_0 MP_{max}/d(\boldsymbol{q}[n], \boldsymbol{v}_j)^2$. This scenario only occurs when the user's and the target's positions are identical. In contrast, when $\cos\varphi_{k,j} = 0$, the communication and target channels are orthogonal, reducing the user's channel power gain to $\gamma_0 \frac{d(\boldsymbol{q}[n], \boldsymbol{v}_j)^2}{d(\boldsymbol{q}[n], \boldsymbol{u}_k)^2} \left(\sqrt{\frac{MP_{max}}{d(\boldsymbol{q}[n], \boldsymbol{v}_j)^2}} - \Gamma_j \sin\varphi_{k,j} \right)^2$.

Lemma 7.1 If both $M_x \to \infty$ and $M_y \to \infty$, for any given UAV position, the optimal SNR for user k when sensing target j can be expressed as $\gamma_{k,j}^* = \gamma_0 (MP_{max} - \Gamma_j d(\boldsymbol{q}[n], \boldsymbol{v}_j)^2)/d(\boldsymbol{q}[n], \boldsymbol{u}_k)^2$.

Proof Let $\Delta\Omega = \Omega(\boldsymbol{q}[n], \boldsymbol{v}_j) - \Omega(\boldsymbol{q}[n], \boldsymbol{u}_k)$ and $\Delta\Phi = \Phi(\boldsymbol{q}[n], \boldsymbol{v}_j) - \Phi(\boldsymbol{q}[n], \boldsymbol{u}_k)$. In cases where $\Delta\Omega = 0$ and $\Delta\Phi = 0$, meaning $u_k = v_j$, we have $\gamma_{k,j}^* = \gamma_0 (MP_{max} - \Gamma_j d(\boldsymbol{q}[n], \boldsymbol{v}_j)^2)/d(\boldsymbol{q}[n], \boldsymbol{u}_k)^2$. On the other hand, when $\Delta\Omega \neq 0$ and $\Delta\Phi \neq 0$, the expression for $\cos\varphi_{k,j}$ can be written as follows:

$$\left|\cos\varphi_{k,j}\right| = \left| \frac{\sin M_x \Delta\Phi\pi/2}{M_x \sin\Delta\Phi\pi/2} \right| \left| \frac{\sin M_y \Delta\Omega\pi/2}{M_y \sin\Delta\Omega\pi/2} \right|. \tag{7.13}$$

When $M_x \to \infty$ and $M_y \to \infty$, the expression (7.13) becomes zero, meaning $\cos\varphi_{k,j} = 0$. As per Lemma 7.1, when the antenna count is large, user k's SNR simplifies to $\gamma_{k,j}^*$ because the user channel and the target channel become nearly uncorrelated. However, solving (P1) remains complex due to the intertwined integer variables and non-convex constraints. In the following subsection, we develop a lower bound on the achievable rate using the optimal beamforming vector from Proposition 7.1 to simplify the solution to problem (P1).

7.4.2 Lower Bound on Achievable Rate

Given any user association, sensing time allocation, and UAV trajectory, the optimal beamforming vector can be derived using Proposition 7.1. The achievable rate for user k at the nth time slot is thus given by

$$R_k[n] = \alpha_k[n]\left(1 - \sum_{j=1}^{J} c_j[n]\right)R_k^C[n] + \alpha_k[n]\sum_{j=1}^{J} c_j[n]R_{k,j}^{ISAC}[n], \tag{7.14}$$

where the achievable rate for user k during communication-only time is $R_k^C[n] = log_2\left(1+\gamma_0\frac{MP_{max}}{d(\boldsymbol{q}[n],\boldsymbol{v}_j)^2}\right)$, and during sensing is $\log_2\left(1+\gamma_{k,j}^*\right)$, for $\cos\varphi_{k,j}^2 MP_{max}/d(\boldsymbol{q}[n],\boldsymbol{v}_j)^2 < \Gamma_j$, where $\gamma_{k,j}^*$ is defined in (7.12). Therefore, the sum achievable rate is maximized by optimizing user association, sensing time allocation, and UAV trajectory jointly. Despite this, solving the problem is still challenging due to the non-concave nature. To address this, we derive a tight lower bound as shown below.

Lemma 7.2 The optimal achievable rate of user k while sensing target j satisfies the following condition:

$$R_k^{ISAC}[n] \geq \log_2\left(1+\gamma_0\frac{MP_{max} - d(\boldsymbol{q}[n],\boldsymbol{v}_j)^2}{d(\boldsymbol{q}[n],\boldsymbol{u}_j)^2}\right) = \underline{R}_{k,j}^{ISAC}. \tag{7.15}$$

To prove (7.15), it suffices to show that $\gamma_{k,j}^* \geq \gamma_0(MP_{max} - \Gamma_j d\left(\boldsymbol{q}[n],\boldsymbol{v}_j)^2\right)/d(\boldsymbol{q}[n],\boldsymbol{u}_k)^2$ holds, given the monotonic nature of the log function. When $\alpha_k[n]=1$ and $\sum_{j=1}^{J} c_j[n]=1$ or $\cos\varphi_{k,j}^2 MP_{max}/d(\boldsymbol{q}[n],\boldsymbol{v}_j)^2 \geq \Gamma_j$, maximum ratio transmission (MRT) is the optimal beamforming vector for problem (P1), hence, the inequality in (7.15) holds directly. For the case $\cos\varphi_{k,j}^2 MP_{max}/d(\boldsymbol{q}[n],\boldsymbol{v}_j)^2 < \Gamma_j$, the inequality $\sqrt{\Gamma_j}\cos\varphi_{k,j} + \sqrt{\frac{MP_{max}}{d(\boldsymbol{q}[n],\boldsymbol{v}_j)^2} - \Gamma_j}\sin\varphi_{k,j} \geq \sqrt{\frac{MP_{max}}{d(\boldsymbol{q}[n],\boldsymbol{v}_j)^2} - \Gamma_j}$ holds.

The lower bound of user k's SNR in (7.15) becomes tight as M approaches infinity, as stated in Lemma 7.1. By examining this lower bound in (7.15), we observe that the value of $\cos\varphi_{k,j}$ remains small since $(\sin M_x\Delta\Phi\pi/2)/(M_x\sin\Delta\Phi\pi/2)$ is relatively minimal for $\Delta \geq 1/M$. In particular, we have $R_k^{ISAC}[n] = \underline{R}_{k,j}^{ISAC}[n]$ when $\Delta = 2i/M$. Based on Lemma 7.2, we can reformulate the lower bound of user k's achievable rate as:

$$\underline{R}_k[n] = \alpha_k[n]R_k^C[n] + \sum_{j=1}^{J}\alpha_k[n]c_j[n]\left(\underline{R}_{k,j}^{ISAC}[n] - R_k^C[n]\right). \tag{7.16}$$

Consequently, we introduce problem (P1.1) as a lower bound for the achievable rate maximization problem by defining $R_k[n]$ as $\underline{R}_k[n]$ in (P1). Solving problem (P1.1) provides a high-quality solution to problem (P1), as elaborated in the following sections.

7.4.3 Penalty-Based Problem Transformation

Although the optimal achievable rate of user k is simplified to its tight lower bound, the integer variables $\alpha_k[n]$ and $c_j[n]$ remain coupled in both the objective function and constraints, which complicates the optimization process. To resolve this, we introduce a new variable $e_{k,j}[n] = \alpha_k[n]c_j[n]$ to decouple the integer variables. Thus, the lower bound $R_k[n]$ can be reformulated as follows:

$$\underline{R}_k[n] = \alpha_k[n]R_k^C[n] + \sum_{j=1}^{J} e_{k,j}[n]\Big(\underline{R}_{k,j}^{ISAC}[n] - R_k^C[n]\Big). \tag{7.17}$$

where $e_{k,j}[n]$ is a binary variable, i.e., $e_{k,j}[n] \in \{0, 1\}$. To maintain consistency in the problem (P1.1), new constraints are introduced to replace those in (1), as follows:

$$\alpha_k[n] \geq e_{k,j}[n], \forall k \in \mathcal{K}, j \in \mathcal{J}, n \in \mathcal{N}, \tag{7.18}$$

$$\sum_{n=(l-1)N_L+1}^{lN_L} \sum_{k=1}^{K} e_{k,j}[n] = 1, \forall l \in \mathscr{L}, j \in \mathcal{J}, \tag{7.19}$$

$$\sum_{k=1}^{K} \sum_{j=1}^{J} e_{k,j}[n] \leq 1, \forall n \in \mathcal{N}. \tag{7.20}$$

Constraint (7.18) ensures that $e_{k,j}[n] = 1$ only when $\alpha_k[n] = 1$. We can readily prove that the newly formulated problem, denoted by (P1.2), with the updated constraints (7.18)–(7.20), is equivalent to the original problem (P1.1). Moreover, the beam pattern gain constraints in (P1b) can be rewritten as:

$$\sum_{k=1}^{K} e_{k,j}[n](MP_{max} - d\big(\boldsymbol{q}[n], \boldsymbol{v}_j)^2\Gamma_j\big) \geq 0. \tag{7.21}$$

While converting $\alpha_k[n]$ and $e_{k,j}[n]$ into continuous variables and using a rounding function to derive binary solutions, there is a risk of not satisfying the QoS constraints in (P1d) and the beam pattern gain constraints in (P1b). To address this, we introduce slack variables $\overline{\alpha}_k[n]$ and $\overline{e}_{k,j}[n]$ to transform binary constraints into equivalent equality constraints. Specifically, (P1c) can be expressed as:

$$\alpha_k[n](1-\overline{\alpha}_k[n])=0,\alpha_k[n]=\overline{\alpha}_k[n],\forall k\in\mathcal{K},n\in\mathcal{N},\tag{7.22}$$

$$e_{k,j}[n]\big(1-\overline{e}_{k,j}[n]\big)=0,e_{k,j}[n]=\overline{e}_{k,j}[n],\forall k\in\mathcal{K},j\in\mathcal{J},n\in\mathcal{N}.\tag{7.23}$$

From these constraints, it is evident that $\alpha_k[n]$ and $e_{k,j}[n]$ must be either 1 or 0, ensuring that (P1c) is satisfied. Penalty terms for constraints (7.22) and (7.23) are then added to the objective function of (P1.2), yielding the following minimization objectives:

$$\underline{R}=\frac{1}{N}\sum_{n=1}^{N}\sum_{k=1}^{K}\underline{R}_k[n]-\frac{1}{2\eta}\sum_{n=1}^{N}\sum_{k=1}^{K}\big(|\alpha_k[n](1-\overline{\alpha}_k[n])|^2+|\alpha_k[n]-\overline{\alpha}_k[n]|^2\big),\tag{7.24}$$

where $\eta>0$ is the penalty factor, which penalizes violations of the equality constraints in (7.22) and (7.23). This new problem with the objective function (7.24) is denoted by (P2). Despite relaxing these constraints, it can be shown that the solution will always satisfy the binary conditions of $\overline{\alpha}_k[n]$ and $\overline{e}_{k,j}[n]$ as $1/\eta\rightarrow\infty$. For efficiency, η is initialized with a large value and gradually reduced until a sufficiently small value is reached, ensuring a feasible binary solution is found. To solve the optimization problem efficiently, we apply the alternating optimization (AO) method, which iteratively optimizes the variables in different blocks.

7.4.4 Inner and Outer Layer Iteration

In this subsection, we propose a two-layer penalty-based algorithm. The inner layer divides (P2) into three sub-problems, where association and UAV trajectory are iteratively optimized. In the outer layer, the penalty coefficient is updated to ensure that the constraints in (7.22) and (7.23) are eventually satisfied.

Given the values of $\alpha_k[n]$ and $e_{k,j}[n]$ and $\boldsymbol{q}[n]$, since the slack variables $\overline{\alpha}_k[n]$ and $\overline{e}_{k,j}[n]$ only appear in the objective function, the optimal values for these variables can be derived by setting the derivative of the objective function w.r.t. $\overline{\alpha}_k[n]$ and $\overline{e}_{k,j}[n]$ to zero:

$$\overline{\alpha}_k^{\text{opt}}[n]=\frac{\alpha_k[n]+\alpha_k^2[n]}{1+\alpha_k^2[n]},\forall k,n,\overline{e}_j^{\text{opt}}[n]=\frac{e_{k,j}[n]+e_{k,j}^2[n]}{1+e_{k,j}^2[n]},\forall j,k,n.\tag{7.25}$$

For fixed values of $\overline{\alpha}_k[n]$ and $\overline{e}_{k,j}[n]$ and $\boldsymbol{q}[n]$, (P2) is convex, with a quadratic objective function and linear constraints, it can be efficiently solved using convex optimization tools like CVX.

Given the values of $\alpha_k[n]$ and $e_{k,j}[n]$ and $\overline{\alpha}_k[n]$ and $\overline{e}_{k,j}[n]$, the UAV trajectory optimization sub-problem is neither concave nor quasi-concave due to the non-convex nature of the objective function and constraints. To address this, we use successive convex optimization, introducing additional slack variables $\{z_{c,k}[n]\}$ and $\{z_{r,j}[n]\}$. The rate expressions $R_k^C[n]$ and $\underline{R}_{k,j}^{ISAC}[n]$ are modified as:

$$\tilde{R}_k^C[n] = log_2\big(1 + \beta_0 P_{max} M / z_{c,k}[n]\big), \tag{7.26}$$

$$\tilde{R}_{k,j}^{ISAC}[n] = log_2\left(1 + \gamma_0 \frac{P_{max}M \ - z_{r,j}[n]\Gamma_j}{z_{c,k}[n]}\right), \tag{7.27}$$

together with:

$$z_{c,k}[n] \geq \|\boldsymbol{q}[n] - \boldsymbol{u}_k\|^2 + H^2, \tag{7.28}$$

$$z_{r,j}[n] \geq \left\|\boldsymbol{q}[n] - \boldsymbol{v}_j\right\|^2 + H^2, \forall k, j, n. \tag{7.29}$$

This reformulation results in problem (P2.4), which is equivalent to (P2.3) because at the optimal solution, the constraints involving $z_{c,k}[n]$ and $z_{r,j}[n]$ are active. During the rth iteration, we approximate $\tilde{R}_{k,j}^{ISAC}[n]$ as:

$$\tilde{R}_k^C[n] \geq log_2\left(1 + \frac{A_k}{z_{c,k}^{(r)}[n]}\right) - \frac{A_k\left(z_{c,k}[n] - z_{c,k}^{(r)}[n]\right)}{\left(z_{c,k}^{(r)}[n]^2 + A_k z_{c,k}^{(r)}[n]\right)\ln 2} = \hat{R}_k^C[n], \tag{7.30}$$

where $A_k = PO_{k\,max}^2$. Furthermore, by introducing a new variable $u_{c,k}[n]$ stratifying $u_{c,k}[n] \leq z_{c,k}[n] + \gamma_0 M P_{\max} - z_{r,j}[n]\Gamma^{th}$, making $\tilde{R}_{k,j}^{ISAC}[n] \geq \log_2\big(u_{c,k}[n]\big) - \log_2\big(z_{c,k}[n]\big)$, and then $\log_2\big(z_{c,k}[n]\big)$ can be approximated as a linear function of *variable* $z_{c,k}[n]$ using Taylor expansion, similar to the approach in (7.30), then $\tilde{R}_{k,j}^{ISAC}[n]$ can be transformed into a concave function w.r.t. $u_{c,k}[n]$ and $z_{r,j}[n]$. Thus, it is a convex problem that can be efficiently solved using convex optimization solvers like CVX.

In the outer layer, the penalty coefficient η is progressively reduced by updating $\eta = z\eta$, where z is a scaling factor such that $0 < z < 1$. A larger value of z can lead to better performance but may require more iterations in the outer layer.

7.4.5 Convergence Analysis and Computational Complexity

To ensure the convergence of the proposed penalty-based algorithm, the termination criterion for the outer layer is defined as follows: $\max(\alpha_k[n](1 - \overline{\alpha}_k[n])|, |\alpha_k[n] - \overline{\alpha}_k[n]|, |e_{k,j}[n] \ |(1 - \overline{e}_{k,j}[n])|, |e_{k,j}[n] - \overline{e}_{k,j}[n]|) \leq \xi$, where ξ is a predefined threshold for accuracy. Within the inner layer, the objective function of (P2) is non-increasing over iterations with the AO method and is bounded above due to the limited flying time and maximum transmit power. Therefore, a stationary point is achieved in the inner layer. As the outer layer iterates and the penalty coefficient decreases, the equality constraints (7.23) and (7.24) are ultimately satisfied.

The computational complexity of the proposed algorithm can be evaluated as follows. In the inner layer, the primary computational cost arises from steps 5 and 6. In step 5, calculating $\alpha_k[n]$ and $e_{k,j}[n]$ has a complexity of $O\big((KN + JKN)^{3.5}\big)$, where $KN + JN$ represents the number of variables. Similarly, in step 6, optimizing the UAV trajectory incurs a complexity of $O\big((2N + KN + JN)^{3.5}\big)$, where $2N + KN + JN$ represents the total number of variables. Hence, the total complexity of the proposed algorithm is $O\big(L_{outer}L_{inner}\big((KN + JKN)^{3.5} + (2N + KN + JN)^{3.5}\big)\big)$, where L_{inner} and L_{outer} represent the number of iterations required for convergence in the inner and outer layers, respectively.

7.5 Simulation Results and Analysis

7.5.1 Simulation Setups

In this section, we present numerical results to evaluate the performance of the proposed periodic sensing and communication design for UAV-based ISAC systems. The simulations are conducted in a square area of 1 km $\times$ 1 km, focusing on $K = 4$ users and $J = 4$ targets within the sensing area of interest. Unless specified otherwise, the system parameters are set as follows: the UAV is equipped with $M = 16$ antennas ($M_x = M_y = 4$), and the beam pattern gain threshold is $\Gamma = 6e^{-5}$. The UAV's maximum horizontal flight speed is $V_{max} = 30$ m/s, with a flight altitude of $H = 40$ m. The channel power gain at the reference distance is $d_0 = 1$ meter, and the noise power at each user is $\beta_0 = -30$ dB and $\sigma^2 = -100$ dB, respectively. The maximum transmit power is limited to $P_{max} = 0.1$ W. The flight period, ISAC frame length, and time slot duration are denoted as $T = 80$ s, $T_L = 20$ s, and $\delta_t = 0.25$ s, respectively. The minimum achievable rate requirement per user is set to 0.25 bps/Hz.

We compare our proposed mechanism against two benchmarks:

- **Straight Flight (SF)**: The UAV flies in a straight line from the initial location to the final location at a constant speed.
- **Fly-Hover-Fly (FHF)**: The UAV first flies at maximum speed in a straight line from the initial location to an optimized location. After hovering at this optimized location, the UAV then flies at maximum speed towards the final destination.

Apart from the UAV trajectory, the corresponding beamforming, user association, and sensing time slots during the flight period for these benchmarks are determined using the proposed algorithm, without updating the UAV trajectory.

7.5.2 Comparison Versus Sensing Power Requirement

In Figs. 7.3 and 7.4, we present the UAV trajectories and the corresponding maximum achievable rates for different beam pattern gain thresholds Γ, comparing our proposed penalty-based algorithm (for solving problem (P1)) with benchmark schemes. From Fig. 7.3, it is evident that as the sensing beam pattern power threshold Γ increases, the UAV's trajectory contracts, transitioning from larger arcs toward the users to smaller arcs between the targets and users. Additionally, the closest distance between the UAV and the users increases, as the UAV is required to perform sensing tasks closer to the targets. Specifically, when $\Gamma = 0$, i.e., the UAV moves sequentially above each user, utilizing its mobility. However, when $\Gamma = 12 \times 10^{-5}$, the UAV remains within a smaller area near the targets due to higher sensing power demands. In this scenario, as the UAV moves closer to the targets, its distance from the communication users inevitably increases. This trade-off results in higher power consumption to satisfy the beam pattern gain requirement, limiting the system's maximum achievable rate. This challenge becomes more pronounced as both the beam pattern gain and the distance between users and targets increase.

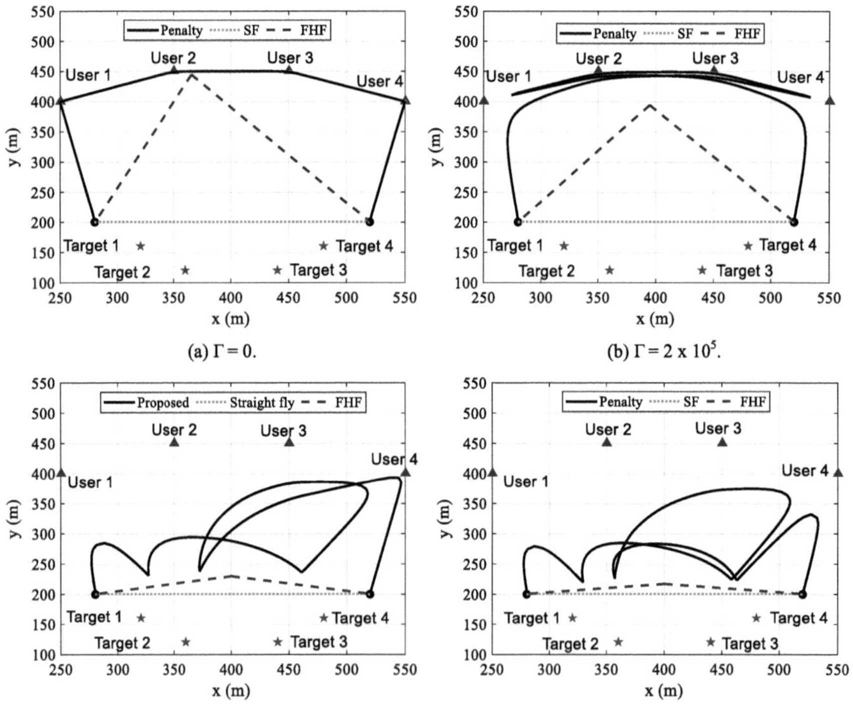

(a) $\Gamma = 0$.
(b) $\Gamma = 2 \times 10^5$.
(c) $\Gamma = 6 \times 10^5$.
(d) $\Gamma = 12 \times 10^5$.

Fig. 7.3 UAV trajectories comparisons among the proposed penalty-based algorithm and benchmarks under different Γ (T = 40 s and TL = 20 s)

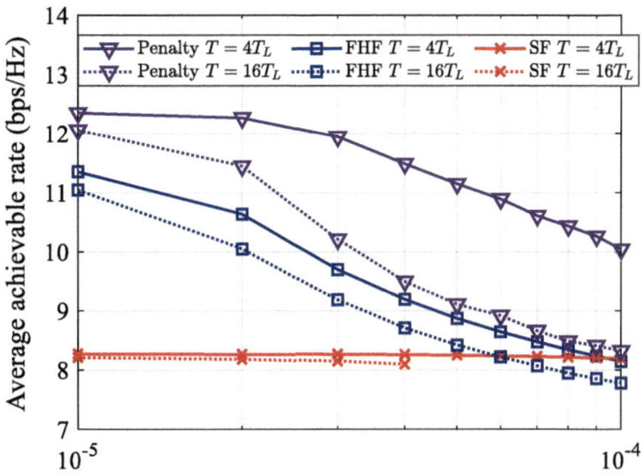

Fig. 7.4 Achievable rate versus beam pattern gain threshold

Figure 7.4 illustrates the impact of the beam pattern gain constraints on the maximum achievable rate. As shown, the rate decreases progressively as Γ increases. Moreover, the performance gain of our proposed scheme over the "SF" benchmark increases as the sensing power requirement decreases, allowing the UAV's trajectory to be optimized within a larger feasible region, thereby improving communication performance. When Γ exceeds 4×10^{-5}, the "SF" scheme becomes infeasible under high sensing power requirements, as it fails to meet both the users' quality-of-service (QoS) constraints and the beam pattern gain requirements for targets without trajectory optimization. Furthermore, under lower sensing frequency conditions, the proposed scheme significantly outperforms the "FHF" benchmark, as more time slots within each ISAC frame can be allocated for communication, improving overall performance.

Figure 7.5 compares the relationships between sensing power requirements, sensing frequency, and achievable rates. As the sensing frequency increases, the achievable rate decreases for all schemes. Additionally, under higher Γ, the achievable rate for the proposed algorithm degrades more rapidly compared to lower thresholds. This is because a higher beam pattern gain threshold forces the UAV to perform sensing tasks nearer the targets, increasing the path loss during the communication-only periods. Finally, Fig. 7.5 demonstrates that as the sensing frequency decreases, the performance advantage of our proposed scheme over the "FHF" and "SF" benchmarks grows, as the UAV has more flexibility to adjust its trajectory to enhance communication performance.

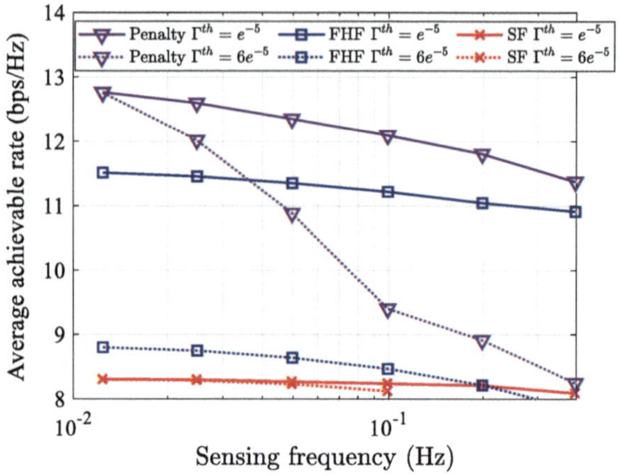

Fig. 7.5 Achievable rate versus sensing frequency requirement

7.5.3 *User-Target Association and Beam Pattern*

The association between users and targets during sensing time slots is presented in Fig. 7.6, with parameters set to $T = T_L = 40$ seconds and $\Gamma = 10^{-3}$. The UAV's flight speed is illustrated in Fig. 7.6, where blue dashed lines indicate user association and target selection. As observed in Fig. 7.6, the UAV tends to serve the user closer to the associated target.

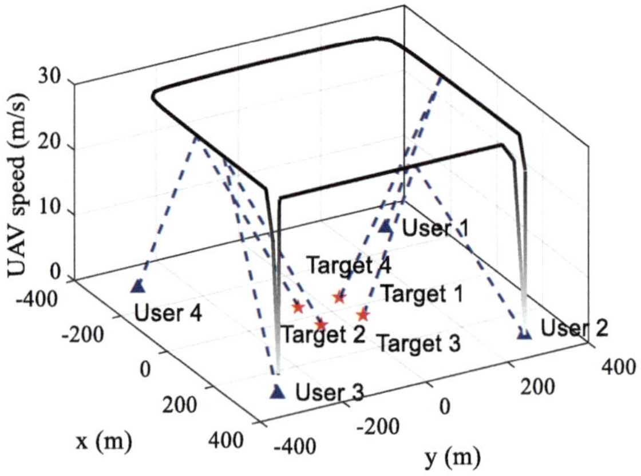

Fig. 7.6 UAV trajectory and its corresponding speed

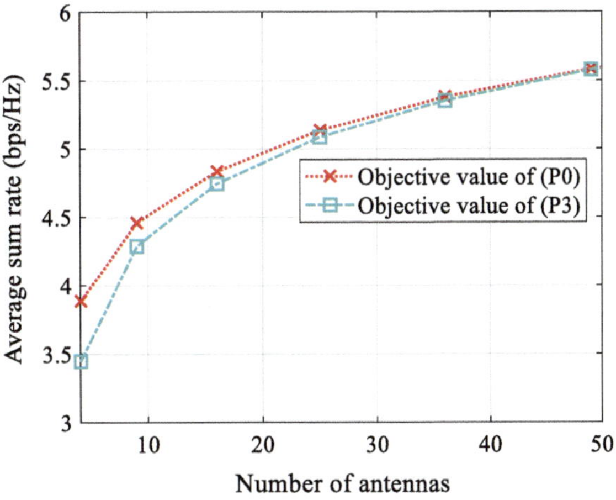

Fig. 7.7 Comparisons under different number of antennas

7.5.4 Gap Between Lower Bound and Actual Rate, and Low-Complexity Method

Since problem (P2) is an approximation of problem (P1), we substitute the optimized solution from the proposed penalty-based algorithm back into the objective function of problem (P1) to determine the actual achievable user rate. As shown in Fig. 7.7, the gap between the average achievable rate during sensing, and the approximate rate decreases as the number of antennas increases. Notably, when the number of antennas M exceeds 16, the original objective value differs from the approximation by $<1\%$, confirming the accuracy of the lower bound derived in Lemma 7.1.

Furthermore, the communication performance of the proposed penalty-based algorithm and the low-complexity method is compared under different flight periods T in Fig. 7.8 The low-complexity algorithm demonstrates a higher performance gain over the two benchmarks as the flight period increases. However, as the flight period lengthens, the performance advantage of the penalty-based algorithm over the low-complexity method decreases. Specifically, the low-complexity algorithm incurs no more than 5% performance loss compared to the penalty-based algorithm when the flight period exceeds 200 s. This is due to the structural characteristics of the optimal solutions across different ISAC frames. For longer flight periods, the time spent traveling from the initial location or toward the final destination constitutes a smaller portion of the overall flight time, allowing the communication rate to approach that achieved without location constraints.

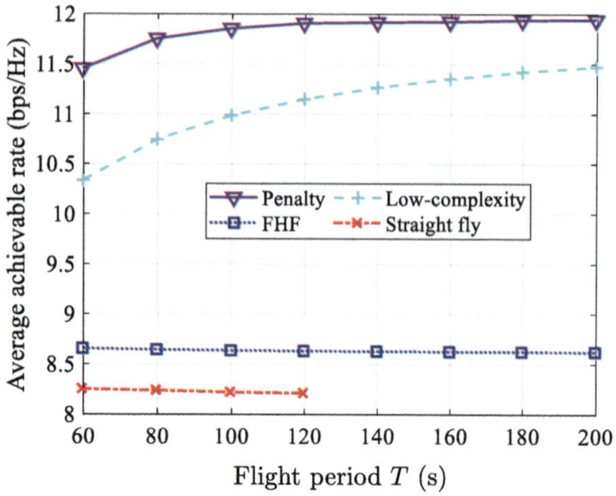

Fig. 7.8 Comparisons versus different flight periods

7.6 Conclusion

This chapter explored a novel UAV-enabled periodic ISAC system, focusing on the joint optimization of beamforming, user association, sensing time allocation, and UAV trajectory to maximize the total achievable rate. A closed-form solution for the optimal beamforming vector was derived, significantly simplifying the beamforming process. Additionally, we provided a lower bound on the achievable rate to aid in the UAV trajectory design. Our numerical results demonstrated the superiority of the proposed approach compared to benchmark schemes and validated the advantages of the periodic ISAC framework.

References

1. Hassanien, A., Amin, M. G., Aboutanios, E., & Himed, B. (2019). Dual-function radar communication systems: A solution to the spectrum congestion problem. *IEEE Signal Processing Magazine, 36*(5), 115–126.
2. Ericsson. (2020). A research outlook towards 6G. Ericsson, Stockholm, Sweden.
3. Li, B., Petropulu, A. P., & Trappe, W. (2016). Optimum co-design for spectrum sharing between matrix completion-based MIMO radars and a MIMO communication system. *IEEE Transactions on Signal Processing, 64*(17), 4562–4575.
4. Yuan, W., Wei, Z., Li, S., Yuan, J., & Ng, D. W. K. (2021). Integrated sensing and communication-assisted orthogonal time frequency space transmission for vehicular networks. *IEEE Journal of Selected Topics in Signal Processing, 15*(6), 1515–1528.
5. Zhang, J. A., et al. (2021). An overview of signal processing techniques for joint communication and radar sensing. *IEEE Journal of Selected Topics in Signal Processing, 15*(6), 1295–1315.
6. Godrich, H., Haimovich, A. M., & Blum, R. S. (2010). Target localization accuracy gain in MIMO radar-based systems. *IEEE Transactions on Information Theory, 56*(6), 2783–2803.
7. Sturm, C., & Wiesbeck, W. (2011). Waveform design and signal processing aspects for fusion of wireless communications and radar sensing. *Proceedings of the IEEE, 99*(7), 1236–1259.

8. Oliveira, L. G., Nuss, B., Alabd, M. B., Diewald, A., Pauli, M., & Zwick, T. (2022). Joint radar-communication systems: Modulation schemes and system design. *IEEE Transactions on Microwave Theory and Techniques, 70*(3), 1521–1551.
9. Liu, F., Zhou, L., Masouros, C., Li, A., Luo, W., & Petropulu, A. (2018). Toward dual-functional radar-communication systems: Optimal waveform design. *IEEE Transactions on Signal Processing, 66*(16), 4264–4279.
10. Liu, R., Li, M., Liu, Q., & Swindlehurst, A. L. (2021). Dual-functional radar-communication waveform design: A symbol-level precoding approach. *IEEE Journal of Selected Topics in Signal Processing, 15*(6), 1316–1331.
11. Mishra, K. V., Shankar, M. R. B., Koivunen, V., Ottersten, B., & Vorobyov, S. A. (2019). Toward millimeter-wave joint radar communications: A signal processing perspective. *IEEE Signal Processing Magazine, 36*(5), 100–114.
12. Zhang, J. A., Huang, X., Guo, Y. J., Yuan, J., & Heath, R. W. (2019). Multibeam for joint communication and radar sensing using steerable analog antenna arrays. *IEEE Transactions on Vehicular Technology, 68*(1), 671–685.
13. Tan, D. K. P., et al. (2021). Integrated sensing and communication in 6G: Motivations, use cases, requirements, challenges and future directions. In *Proceedings of IEEE JC&S*.
14. Wild, T., Braun, V., & Viswanathan, H. (2021). Joint design of communication and sensing for beyond 5G and 6G systems. *IEEE Access, 9*, 30845–30857.
15. Wymeersch, H., et al. (2021). Integration of communication and sensing in 6G: A joint industrial and academic perspective. In *Proceedings of IEEE PIMRC*.
16. IEEE. (2021). *Project IEEE 802.11bf WLAN sensing*. Retrieved March 19, 2021, from https://www.ieee802.org/11/Reports/tgbfupdate.htm
17. Zhang, Q., Wang, X., Li, Z., & Wei, Z. (2021). Design and performance evaluation of joint sensing and communication integrated system for 5G mmWave-enabled CAVs. *IEEE Journal of Selected Topics in Signal Processing, 15*(6), 1500–1514.
18. Cui, Y., Liu, F., Jing, X., & Mu, J. (2021). Integrating sensing and communications for ubiquitous IoT: Applications, trends, and challenges. *IEEE Network, 35*(5), 158–167.
19. Liu, F., Masouros, C., Petropulu, A. P., Griffiths, H., & Hanzo, L. (2020). Joint radar and communication design: Applications, state-of-the-art, and the road ahead. *IEEE Transactions on Communications, 68*(6), 3834–3862.
20. Gupta, L., Jain, R., & Vaszkun, G. (2016). Survey of important issues in UAV communication networks. *IEEE Communications Surveys & Tutorials, 18*(2), 1123–1152.
21. Meng, K., Wu, Q., Ma, S., Chen, W., Wang, K., & Li, J. (2023). Throughput maximization for UAV-enabled integrated periodic sensing and communication. *IEEE Transactions on Wireless Communications, 22*(1), 671–687.
22. Wu, Q., Liu, L., & Zhang, R. (2019). Fundamental trade-offs in communication and trajectory design for UAV-enabled wireless networks. *IEEE Wireless Communications, 26*(1), 36–44.
23. Meng, K., Wu, Q., Ma, S., Chen, W., & Quek, T. Q. (2022). UAV trajectory and beamforming optimization for integrated periodic sensing and communication. *IEEE Wireless Communications Letters, 11*(6), 1211–1215.
24. Meng, K., Li, D., He, X., & Liu, M. (2021). Space pruning-based time minimization in delay-constrained multi-task UAV-based sensing. *IEEE Transactions on Vehicular Technology, 70*(3), 2836–2849.
25. Zhang, S., Zhang, H., Han, Z., Poor, H. V., & Song, L. (2020). Age of information in a cellular internet of UAVs: Sensing and communication trade-off design. *IEEE Transactions on Wireless Communications, 19*(10), 6578–6592.
26. Meng, K., He, X., Li, D., Liu, M., & Xu, C. (2021). Sensing quality-constrained packet rate optimization via multi-UAV collaborative compression and relay. In *Proceedings of IEEE INFOCOM Workshop* (pp. 1–6).
27. Lyu, Z., Zhu, G., & Xu, J. (2021). *Joint maneuver and beamforming design for UAV-enabled integrated sensing and communication*. arXiv preprint arXiv:2110.02857
28. Chen, X., Feng, Z., Wei, Z., Gao, F., & Yuan, X. (2020). Performance of joint sensing-communication cooperative sensing UAV network. *IEEE Transactions on Vehicular Technology, 69*(12), 15545–15556.

29. Wang, Z., Liu, R., Liu, Q., & Han, L. (2021). *QoS-oriented sensing communication-control co-design for UAV-enabled positioning*. arXiv preprint arXiv:2108.09725
30. Wei, Z., Liu, F., Ng, D.W.K., & Schober, R. (2022). Safeguarding UAV networks through integrated sensing, jamming, and communications. In *Proceedings of IEEE ICASSP* (pp. 8737–8741).
31. Wu, Q., et al. (2021). A comprehensive overview on 5G-and-beyond networks with UAVs: From communications to sensing and intelligence. *IEEE Journal on Selected Areas in Communications, 39*(10), 2912–2945.
32. Wu, Q., & Zhang, R. (2018). Common throughput maximization in UAV-enabled OFDMA systems with delay consideration. *IEEE Transactions on Wireless Communications, 66*(12), 6614–6627.
33. Chen, L., Liu, F., Wang, W., & Masouros, C. (2021). Joint radar-communication transmission: A generalized Pareto optimization framework. *IEEE Transactions on Signal Processing, 69*, 2752–2765.
34. Liu, F., Masouros, C., Li, A., Sun, H., & Hanzo, L. (2018). MU-MIMO communications with MIMO radar: From co-existence to joint transmission. *IEEE Transactions on Wireless Communications, 17*(4), 2755–2770.
35. Garcia, N., Wymeersch, H., Larsson, E. G., Haimovich, A. M., & Coulon, M. (2017). Direct localization for massive MIMO. *IEEE Transactions on Signal Processing, 65*(10), 2475–2487.
36. Wu, Q., Zeng, Y., & Zhang, R. (2018). Joint trajectory and communication design for multi-UAV enabled wireless networks. *IEEE Transactions on Wireless Communications, 17*(3), 2109–2121.
37. Wu, Q., Xu, J., & Zhang, R. (2018). Capacity characterization of UAV-enabled two-user broadcast channel. *IEEE Journal on Selected Areas in Communications, 36*(9), 1955–1971.
38. Kang, E., Hwang, H., & Han, D. (2010). A fine carrier recovery algorithm robust to Doppler shift for OFDM systems. *IEEE Transactions on Consumer Electronics, 56*(3), 1218–1222.
39. Xing, M., Jiang, X., Wu, R., Zhou, F., & Bao, Z. (2009). Motion compensation for UAV SAR based on raw radar data. *IEEE Transactions on Geoscience and Remote Sensing, 47*(8), 2870–2883.

Chapter 8
Joint Analog Beamforming and UAV Trajectory Planning for Nonlinear Wireless Power Transfer

Xiaopeng Yuan, Peng Wu, and Yulin Hu

Abstract This chapter investigates an unmanned aerial vehicle (UAV)-enabled nonlinear wireless power transfer (WPT) network which involve multiple ground sensor nodes (SNs). The UAV performs as an energy source to transfer energy wirelessly to these SNs. For efficiently energy harvesting (EH), the UAV is equipped with a three dimensional (3D) uniform linear array (ULA) antenna, facilitating analog beamforming for focused power transfer. Given the UAV energy consumption and adopting a practical nonlinear EH model, we delve into the UAV energy efficiency maximization for WPT operation which leads to the formulation of an efficiency maximization problem through jointly designing analog beamforming, UAV trajectory planning, and transmit power allocation. To tackle the intricate non-convex optimization challenge, we introduce a cosine-based approximation for the complicated 3D ULA antenna pattern which holds certain convexity. Synergizing the convexity of the modified antenna pattern with that of the nonlinear EH model, we construct a convex subproblem centered on given local point where solving this subproblem iteratively ensures a gradual enhancement in the objective. Simulation results confirm the algorithm convergence and performance benefits.

Keywords Unmanned aerial vehicle · Analog beamforming · Trajectory planning · Wireless power transfer (WPT) · Energy efficient design

---

X. Yuan
Chair of Information Theory and Data Analytics, RWTH Aachen University, Aachen, Germany
e-mail: yuan@inda.rwth-aachen.de

P. Wu · Y. Hu (✉)
School of Electronic Information, Wuhan University, Wuhan, China
e-mail: yulin.hu@whu.edu.cn

P. Wu
e-mail: peng.wu@whu.edu.cn

Y. Hu et al. (eds.), *Reliable and Efficient Trajectory Design in UAV-Aided Wireless Networks*, https://doi.org/10.1007/978-981-96-5988-3_8

Abbreviations

3D	Three Dimensional
A2G	Aerial to Ground
EH	Energy Harvesting
IoT	Internet of Things
RF	Radio Frequency
SN	Sensor Node
UAV	Unmanned Aerial Vehicle
ULA	Uniform Linear Array
WPT	Wireless Power Transfer

8.1 Introduction

8.1.1 Background

Radio frequency (RF)-enabled wireless power transfer (WPT), which is a promising solution to provide permanent and cost-effective energy supply for low-power electronic devices, is expected to have rich applications in future wireless networks of the Internet of Things (IoT) [1]. In conventional WPT systems, RF signals are mainly sent to distributed energy receivers such as low-power sensor nodes (SNs) through fixed transmitters [2]. However, due to severe wireless signal pathloss, it is necessary to deploy transmitters in an ultra dense manner to compensate for the low power transmission efficiency which will greatly increase costs of network construction and maintenance [3]. Owing to the excellent performance in mobility and high quality aerial-to-ground (A2G) channel, unmanned aerial vehicle (UAV) can be employed as mobile WPT source to provide wireless energy for ground devices, where the UAV mobility is utilized to improve the WPT efficiency through UAV trajectory design (also called path planning) and transmit power control [4]. In the meanwhile, the combination of UAV trajectory design with transmit power allocation can further deliver RF energy to devices with better efficiency [5, 6]. However, almost all works on UAV-enabled WPT networks use omni-directional antennas for energy transmit in the form of broadcasting, where the energy harvesting (EH) experiences significant attenuation over large transmission distances. Therefore, it is necessary to study channel enhancement technologies especially the antenna beamforming in WPT systems.

8.1.2 Motivation

In UAV-enabled WPT networks, WPT operation efficiency can be improved through power control and trajectory design techniques. However, due to the significant pathloss of RF signals in the A2G channels, the conventional WPT method with omni-directional antenna broadcasting incurs significant transmission losses, making it extremely difficult to apply in practical applications. With the increasing emphasis on energy efficiency in WPT networks, researching channel enhancement schemes to reduce power loss has important practical significance. Among all the channel enhancement technologies, directional antennas beamforming can focus RF power on the orientated devices, reducing power waste compared to broadcast transmission. Therefore, researching effective beamforming algorithms is of great significance for the practical application of UAV-enabled WPT networks.

8.1.3 Main Contributions

Main contributions to this chapter are as follows:

(1) Comparing with conventional WPT work with omni-directional antenna and 1D directional antenna, 3D directional antenna offers unprecedented orientation flexibility and superior power focusing capabilities for wireless power transfer (WPT) applications. This work for the first time investigates 3D ULA-based analog beamforming in UAV WPT networks. To address the intricacies of the 3D antenna gain model, we introduce a novel cosine-based approximation for the antenna pattern, which significantly simplifies the analysis of 3D ULA analog beamforming and enables the development of an efficient iterative solution framework;
(2) To tackle to complicated formulated joint design problem, we mathematically utilize the convexity of nonlinear EH model and modified antenna model to construct a convex subproblem, based on which an iterative algorithm is proposed for converging to a high-quality solution;
(3) Through simulation analyses, we evaluate the proposed WPT energy efficiency design with 3D directional antenna and observe the effect of UAV trajectory and antenna pattern for the efficiency maximization.

8.1.4 Structural Organization

Relevant works on UAV WPT and antenna beamforming are introduced in Sect. 8.2. Section 8.3 shows the system model. Section 8.4 introduces the modified antenna pattern and formulates the considered energy efficiency improvement design problem where an iterative algorithm based on the convex approximation is provided, which is evaluated through simulations in Sect. 8.5. Section 8.6 is the conclusion.

8.2 Related Works

8.2.1 UAV-Enabled WPT

The inclusion of UAV in WPT networks has attracted lots of attention, where by integrating WPT technology on UAVs and deploying them as mobile wireless power sources, the performance of WPT can be significantly enhanced in the WPT network [7], and there have been some existing works on UAV-enabled WPT networks [8–16]. For further improving the WPT performance, authors in [8–10] optimized the UAV trajectory based on the ideal linear EH model. Especially, a low complexity trajectory design scheme for 1D scenario was proposed in [10] after mathematically analyzing the optimal UAV trajectory. In general, the conversion efficiency in the EH process varies significantly with the value of received signal power, especially when the received power increases, the conversion efficiency will first increase and then decrease [11, 12]. The ideal linear EH model cannot accurately characterize this trend, and it is necessary to consider the nonlinear EH model in UAV WPT networks. To characterize the conversion efficiency, two nonlinear EH model have been proposed which are according to the curve fitting on statistic data [13] and analytical model from practical circuit [14]. Utilizing the practical nonlinear EH model, the above works with linear EH model are further extended where the UAV trajectory and transmit power are jointly designed to improve the system performance [15–17].

The above studies have to some extent improved the operation efficiency of UAV from the perspective of jointly design UAV trajectory and transmit power. research and application. However, confirmed by [18], in practical EH scenarios, the significant pathloss of RF signal has limited the harvested power to a large extent, especially when using the omni-directional antenna to transfer energy through broadcasting, a large amount of power is wasted in free space while only very small portion is harvested by the device.

8.2.2 UAV WPT with Directional Antenna

Beamforming with directional antenna seems to be a promising technology to improve channel quality and WPT efficiency [19]. There are generally two main beamforming strategies, namely digital beamforming and analog beamforming [20]. Digital beamforming is based on specific instantaneous channel state information and can flexibly adjust the phase shift of each antenna element, but it requires relatively complex electronic operations [21]. In contrast, based on uniformly linear antenna (ULA) arrays, simulated beamforming can form directional beams with relatively low cost and operational complexity by modifying the uniform phase shift between antenna elements which can be integrated into UAVs to enhance WPT performance [22, 23]. Authors in [24] studied a UAV-assisted WPT network with analog directional antenna containing two devices where a approximated 3dB flat-top model is adopted for helping system analysis. The antenna model is more suitable coverage analysis [25] while it will result in significant estimation errors in the received power and will generate greater bias in evaluating the harvested power. To evaluate the performance of UAV WPT system with directional antenna more preciously, stochastic geometry method is applied in [26] to derive the energy coverage probabilities, while a more accurate antenna model is investigated in [27] to promote the UAV-enabled WPT.

Although these researches have contributed a lot to UAV WPT networks, almost all work is with UAV carried with 1D ULA for analog beamforming, resulting in an annular projection which still wastes RF energy. To harness the benefits of beamforming in UAV-enabled WPT networks more efficiently, a more flexible antenna array architecture, such as a 3D ULA, should be considered for shaping dedicated beams. Nevertheless, the complicated 3D antenna gain introduces significant challenges in application scenarios, where the system modelling of UAV 3D directional antenna WPT, along with the corresponding joint analog beamforming and UAV trajectory planning, remains absent in the literature. Therefore, in this chapter, we focus on the scenario of UAV-enabled WPT networks with 3D antenna, to investigate energy efficiency enhancement strategy with analog beamforming, trajectory planning and power control algorithm.

8.3 System Model

8.3.1 Considered Scenario

We delve into a UAV-enabled wireless sensor network scenario, where a solitary UAV with 3D ULA antenna maintains at a constant altitude H to serves as a wireless energy source for K sensor nodes (SNs) dispersed across a designated ground area as shown in Fig. 8.1, whose location is denoted as $\boldsymbol{s}_k = \left(s_{x,k}, s_{y,k}\right), k \in \mathcal{K} \triangleq \{1, ..., K\}.$. The UAV facilitates the energy to multiple SNs through radio-frequency (RF) signals. The accumulated energy is then utilized by the SNs for their routine functions like encompassing data acquisition, battery replenishment, and wireless communication. The operational time for the UAV's WPT task is predefined as T and to harness the exceptional mobility of UAV, we assume it maneuvering smoothly along the horizontal plane where its horizontal position at any instant $t \in \mathcal{T} \triangleq [0, T]$ can be denoted by $\boldsymbol{q}(t) = (x(t), y(t))$, which traces a continuous UAV trajectory. Additionally, we impose a maximum velocity limit V on the UAV, restricting its speed to

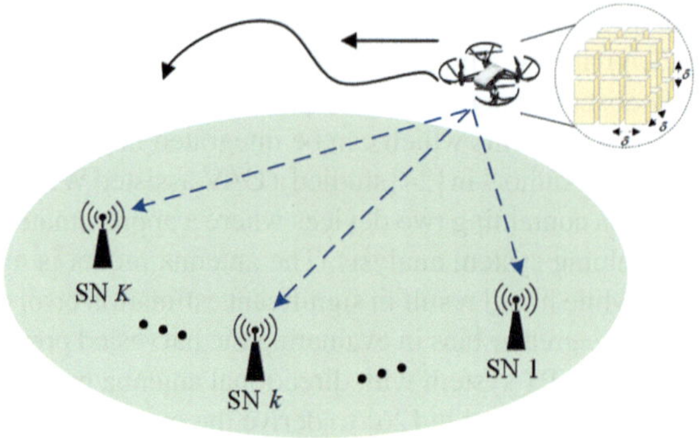

Fig. 8.1 UAV-enabled WPT network

$||(\dot{\boldsymbol{q}}(t)||_2 \leq V, \forall t \in \mathcal{T}$ where $\dot{\boldsymbol{q}}(t)$ denotes the first-order derivative of $\boldsymbol{q}(t)$. In practical scenarios, the UAV often undertakes multiple tasks sequentially. Consequently, besides the designated WPT duration, the starting and ending positions of UAV for the WPT mission might be predetermined by the subsequent and preceding tasks, respectively which are denoted as $\boldsymbol{q}(0) = (x_1, y_1), \boldsymbol{q}(T) = (x_2, y_2)$ respectively.

To tackle the difficulty on infinite trajectory variables $\boldsymbol{q}(t)$ over continuous time, we discretize the trajectory to facilitate the system design. For given discretization slots number N, the slot resolution is $\delta_t = TN^{-1}$. In slot $n, \forall n \in \mathcal{N} \triangleq \{1, ..., N\}$, the UAV position is represented as $\boldsymbol{q}[n] = (x[n], y[n])$ such that $\boldsymbol{q}[1] = (x_1, y_1), \boldsymbol{q}[N+1] = (x_2, y_2)$ and UAV speed limitation is rewritten as $v[n] = ||\boldsymbol{q}[n+1] - \boldsymbol{q}[n]|| \leq V\delta_t$.

8.3.2 3D ULA Analog Beamforming

Furthermore, for enhancing the efficiency of WPT, we postulate that the UAV is equipped with a directional antenna which operates a practical 3D ULA uniform linear array configuration to be optimized for focusing and amplifying transmitted power. Specifically, the 3D ULA comprises numerous isotropic antenna elements that are strategically positioned in a three-dimensional space where the antenna elements number along x, y, z axess is L_x, L_y, L_z and $L = L_xL_yL_z$, each spaced uniformly by a distance δ, as exemplified in Fig. 8.1.

For the specified analog beamforming, it needs a linearly progressive phase from element to element for the 3D ULA where the phase differences between neighbour elements for each axes is $u_x\pi, u_y\pi, u_z\pi$ and $u_x, u_y, u_z \in [-1, 1]$ [28]. With the common antenna spacing where δ is half of wavelength, the antenna gain is:

$$G_A(\theta, \phi) = L\left(\frac{\sin(L_x\varphi_x)}{L_x\sin(\varphi_x)}\right)^2\left(\frac{\sin(L_y\varphi_y)}{L_y\sin(\varphi_y)}\right)^2\left(\frac{\sin(L_z\varphi_z)}{L_z\sin(\varphi_z)}\right)^2 = LG_x(\varphi_x)G_y(\varphi_y)G_z(\varphi_z) \tag{8.1}$$

where θ and ϕ are the elevation and the azimuth angles to denote the certain direction. $\varphi_x = \frac{\pi}{2}(\sin\theta\cos\phi + u_x)$, $\varphi_y = \frac{\pi}{2}(\sin\theta\sin\phi + u_x)$, $\varphi_x = \frac{\pi}{2}(\cos\theta + u_z)$ are the auxiliary variables.

As δ_t is sufficiently small such that the phase differences is invariant in each slot which can be also called the beamforming scheme and denoted by $\mathrm{u}[n] = \left(u_x[n], u_y[n], u_z[n]\right)$. Accordingly, the the auxiliary variables for SN k can be represented by the function of UAV position $(x[n], y[n])$ i.e., $\varphi_{x,k}[n] = \frac{\pi}{2}\left(\frac{s_{x,k}-x[n]}{d_k[n]} + u_x[n]\right)$, $\varphi_{y,k}[n] = \frac{\pi}{2}\left(\frac{s_{y,k}-y[n]}{d_k[n]} + u_y[n]\right)$, $\varphi_{x,k}[n] = \frac{\pi}{2}\left(\frac{H}{d_k[n]} + u_z[n]\right)$, and $d_k[n] = (\boldsymbol{q}[n], H) - (\boldsymbol{s}_k, H)$ is the distance between UAV and SN k. Consequently, the antenna gain (also called the antenna pattern) from UAV to SN k is:

$$G_{A,k}[n] = LG_x\left(\varphi_{x,k}[n]\right)G_y\left(\varphi_{y,k}[n]\right)G_z\left(\varphi_{x,k}[n]\right). \tag{8.2}$$

8.3.3 Channel Model

As the Channel links between UAV and SNs are generally dominated by LoS propagation, we have sufficient motivation to employ the universally accepted free-space path loss model where the channel power gain form UAV to SN k in slot n is $h_k[n] = \beta_0 G_{A,k}[n]d_k[n]^{-2}$ where β_0 is the reference power gain. Denote the UAV transmit power as $P[n]$ which gives recevied RF power at SN k as $\mathrm{Q_k[n]} = \beta_0 P[n]G_{A,k}[n]d_k[n]^{-2}$.

8.3.4 Nonlinear Energy Harvesting

Additionally, pertaining to the EH at the SN k side, it transforms the received RF signal into a direct-current (DC) signal which is an indispensable processed has been confirmed that this conversion process exhibits nonlinearity. Therefore, we adopt a realistic nonlinear EH model, formulated in [11], grounded on a practical diode-facilitated conversion circuitry where the harvested power of SN k can be quantified given a particular received signal power $\mathrm{Q_k[n]}$:

$$P_{EH,k}[n] = \left[\frac{1}{\xi}W_0\left(\xi e^{aI_s}\sum_{j=0}^{\mathrm{n_0}}\kappa_j Q_k[n]^j\right) - I_s\right]^2 R_L \tag{8.3}$$

where $\mathrm{I_s}$ is inverse bias saturation current of diode, $\mathrm{n_0}$ is the truncation order, $\mathrm{W_0}(\cdot)$ is the principal branch of Lambert W function. ξ and κ_j are basic electric components, and $\mathrm{R_L}$ is the load resistance value.

The implicit nature of $W_0(\cdot)$ poses a significant challenge in analyzing the nonlinear EH model. Nevertheless, [11, 14] presents a way forward by leveraging the convexity of $P_{EH,k}[n]$ with respect to reciprocal of received power $Q_k[n]^{-1}$, i.e., $\frac{d_k[n]^2}{\beta_0 P[n] G_{A,k}[n]}$. Above all, the overall harvested energy of SN k is $E_k(\boldsymbol{q}, \boldsymbol{P}, \mathrm{U}) = \sum_{n=1}^{N} P_{EH,k}[n]\delta_t$ where $\boldsymbol{q} = \{\boldsymbol{q}[n]\}$, $\boldsymbol{P} = \{P[n]\}$, $\mathrm{U} = \mathrm{u}[n]$ are UAV trajectory, transmit power and analog beamforming variables.

8.3.5 *UAV Energy Consumption Model*

For a rotary-wing UAV, its overall energy consumption consists of the WPT energy and propulsion energy where the propulsion power consumption can be given by:

$$P_c(v[n]) = \gamma_1 \left(1 + \frac{3v[n]^2}{\mathrm{V}_{tip}^2}\right) + \gamma_2 \left(\sqrt{1 + \frac{v[n]^4}{4v_0^4}} - \frac{v[n]^2}{2v_0^2}\right)^{1/2} + \frac{1}{2} d_0 \rho_0 s_0 A_0 v[n]^3 \tag{8.4}$$

where γ_1 and γ_2 are physical property constants, $\mathrm{V_{tip}}$ is tip speed, $\mathrm{v_0}$ is the mean rotor induced velocity in hovering, d_0, ρ_0, s_0 and A_0 are the fuselage drag ratio, the air density, rotor solidity and the rotor disc area.

Noting that in the analyzed network, in addition to the WPT task, UAV is also tasked with a vital flying objective from the starting point to the ending destination. To prioritize energy efficiency, particularly for the WPT operation, we designate E_0 as the necessary energy expenditure for the flying task, which remains constant and can be derived from work [29], such that the extra energy consumption for WPT operation is $E_{ex}(\boldsymbol{q}, \boldsymbol{P}) = \sum_{n=1}^{N} (P_c(v[n]) + P[n])\delta_t - E_0$.

8.4 Problem Formulation

8.4.1 *Modified Cosine Antenna Pattern*

Before introducing the considered design problem, it is worth noting that the original antenna pattern shown in (8.2) extremely complicated which constrains multiple zero points and counts against subsequent algorithm design. Therefore, in this subsection, we first propose a flexibly modified cosine antenna pattern to facilitate the related design.

As indicated in (8.2), the antenna gain $G_{A,k}[n]$ is the multiplication of independent sinc functions $G_{i,k}(\varphi_i[n])$, $i \in (x, y, z)$ in each axes. According to existing works, we are motivated to approximate the main lobe by the cosine function and the side lobe by a decreasing function, where for each sinc function $G_{i,k}(\varphi_i[n])$, $i \in (x, y, z)$, the proposed modified cosine antenna pattern is

$$\tilde{G}_i(\varphi_{i,k}[n]) = \begin{cases} \cos^2(\frac{L_i}{2}\varphi_{i,k}[n]), & |\varphi_{i,k}[n]| < \frac{\Delta\pi}{L_i} \\ \frac{cos^2(\frac{\Delta\pi}{2})}{\frac{L_i^2}{2\Delta\pi}tan(\frac{\Delta\pi}{2})\left(\varphi_{i,k}[n]^2 - \frac{\Delta^2\pi^2}{L_i^2}\right)+1}, & |\varphi_{i,k}[n]| \geq \frac{\Delta\pi}{L_i} \end{cases} \tag{8.5}$$

where $\Delta \in (0, 1]$ is modification factor and a larger Δ means a better approximation for the main lobe and a smaller value of side lobe which is depicted in Fig. 8.2.

For the modified antenna pattern, we have $\tilde{G}_i(\varphi_{i,k}[n])^{-1}$ is monotonically increasing and convex in $\varphi_{i,k}[n]^2$ and the proof is provided as follows.

When $|\varphi_{i,k}[n]| < \frac{\Delta\pi}{L_i}$, the first order derivative of $\tilde{G}_i(\varphi_{i,k}[n])^{-1}$ as $\frac{\partial\tilde{G}_i(\varphi_{i,k}[n])}{\partial(\varphi_{i,k}[n]^2)} = \frac{1}{\cos^2(\frac{L_i}{2}\varphi_{i,k}[n])} \geq 0$ and the second order derivative can be accordingly given by $\frac{\partial^2\tilde{G}_i(\varphi_{i,k}[n])}{\partial(\varphi_{i,k}[n]^2)^2} = \frac{1}{8\varphi_{i,k}[n]^3\cos^4(\frac{L_i}{2}\varphi_{i,k}[n]^2)}(L_i^2\varphi_{I,k}[n] - L_i\sin(L_i\varphi_{I,k}[n])L_i^2\varphi_{i,k}[n](1 - \cos(L_i\varphi_{i,k}[n]))) \geq 0$ since the two parts $L_i^2\varphi_{i,k}[n] - L_i\sin(L_i\varphi_{i,k}[n]) \geq 0$ and $1 - \cos(L_i\varphi_{i,k}[n])) \geq 0$.

When $\varphi_{i,k}[n]| \geq \frac{\Delta\pi}{L_i}$, $\tilde{G}_i(\varphi_{i,k}[n])^{-1} = \cos^{-2}\left((\frac{\Delta\pi}{2})\frac{L_i^2}{2\Delta\pi}\tan(\frac{\Delta\pi}{2})\left(\varphi_{i,k}[n]^2 - \frac{\Delta^2\pi^2}{L_i^2}\right)+1\right)$, which is an affine in $\varphi_{i,k}[n]^2$. Therefore, $\tilde{G}_i(\varphi_{i,k}[n])^{-1}$ is monotonically increasing and convex in $\varphi_{i,k}[n]^2$.

According to (8.4), the antenna gain under the modified antenna pattern is:

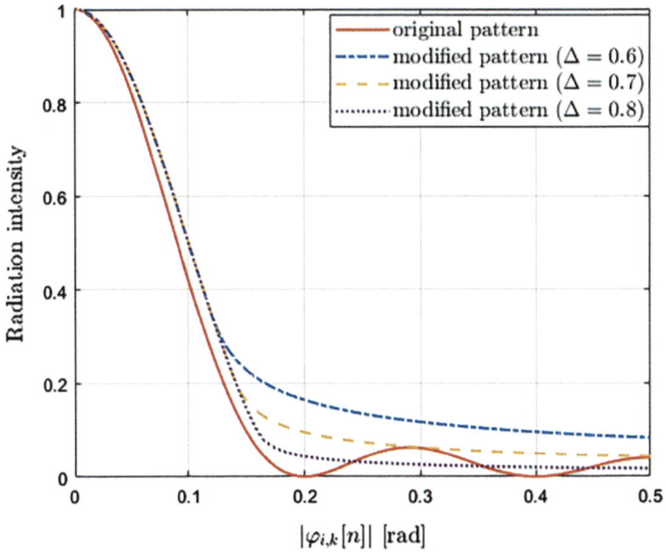

Fig. 8.2 Comparison of original antenna pattern and modified antenna pattern

$$\tilde{G}_{A,k}[n] = L\tilde{G}_x(\varphi_{x,k}[n])\tilde{G}_y(\varphi_{y,k}[n])\tilde{G}_z(\varphi_{x,k}[n]). \tag{8.6}$$

8.4.2 Problem Formulation

To equitably enhance the UAV-enabled EH performance while concurrently minimizing the energy consumption of the UAV, an energy efficient design is studied in this chapter. Particularly, denoting the minimal harvest energy among SNs $E_H(\boldsymbol{q}, \boldsymbol{P}, \mathrm{U}) = \min_k \tilde{E}_{H,k}(\boldsymbol{q}, \boldsymbol{P}, \mathrm{U}) = \min_k \sum_{n=1}^{N} \mathcal{F}\left(\frac{\beta_0 P[n]\tilde{G}_{A,k}[n]}{d_k[n]^2}\right)\delta_t$ as EH performance where $\mathcal{F}(\cdot)$ is the nonlinear implicit conversion function defined in (8.2) and is convex in $\frac{d_k[n]^2}{\beta_0 P[n]\tilde{G}_{A,k}[n]}$ according to [11]. The WPT task energy efficiency is formulated as $\eta(\mathbf{q}, \mathbf{P}, \mathrm{U}) = \frac{E_H(\boldsymbol{q}, \boldsymbol{P}, \mathrm{U})}{E_{ex}(\boldsymbol{q}, \boldsymbol{P})}$ which is to be maximized through jointly design UAV trajectory $\boldsymbol{q}$, transmit power $\boldsymbol{P}$ and analog beamforming $\mathbf{U}$, i.e., the problem is formulated as:

$$(\mathrm{OP}) : \max_{\boldsymbol{q}, \boldsymbol{P}, \mathrm{U}} \eta(\mathbf{q}, \mathbf{P}, \mathrm{U}) \tag{8.7}$$

$$\|\boldsymbol{q}[n+1] - \boldsymbol{q}[n]\| \leq V\delta_t, \forall n \in \mathcal{N} \tag{8.7a}$$

$$\mathbf{q}[1] = (x_1, y_1), \mathbf{q}[N+1] = (x_2, y_2) \tag{8.7b}$$

$$0 \leq P[n] \leq P_{max}, \forall n \in \mathcal{N} \tag{8.7c}$$

$$u_x[n], u_y[n], u_z[n] \in [-1, 1], \forall n \in \mathcal{N} \tag{8.7d}$$

where $P_{\max}$ is the UAV maximum transmit power. Alhouth all the constraints in problem (OP) are convex, the objective function $\eta(\mathbf{q}, \mathbf{P}, \mathrm{U})$ is intuitively non-concave such that problem (OP) is a complicated non-convex problem which difficult to analyse. However, we are capable to leverage the convexity characteristics inherent in both the nonlinear EH model and the cosine modified antenna pattern, thereby facilitating the analytical process. Leveraging these convexities, in the subsequent section, we will focus on formulating a convex subproblem centered around any local point. By tackling this convex subproblem, we ensure a tangible enhancement in energy efficiency, thereby enabling an iterative algorithm to refine the solution progressively.

8.4.3 Convex Subproblem Construction

As all constraints in problem (OP) hold convexity, given that a concave objective function is vital for ensuring convexity in a maximization problem, we are prompted to construct a convex subproblem derived from (OP) through transforming the original objective into a concave form. Additionally, maximizing the objective of this subproblem ought to enhance energy efficiency, paving the way for an iterative approach to solve the problem. Denoting the iteration index by m, for the local point $\left(\boldsymbol{q}^{(m)}, \boldsymbol{P}^{(m)}, \mathrm{U}^{(m)}\right)$ in mth iteration, the corresponding energy efficiency is $\eta^{(m)} = \hat{E}_H\left(\boldsymbol{q}^{(m)}, \boldsymbol{P}^{(m)}, \mathrm{U}^{(m)}\right)E_{ex}(\boldsymbol{q}^{(m)}, \boldsymbol{P}^{(m)}, \mathrm{U}^{(m)})^{-1}$ which helps in constructing the following objective function:

$$\Delta E(\mathbf{q}, \mathbf{P}, \mathrm{U}) \triangleq E_H(\mathbf{q}, \mathbf{P}, \mathrm{U}) - \eta^{(m)} E_{ex}(\mathbf{q}, \mathbf{P}) \tag{8.8}$$

where $\Delta E(\mathbf{q}, \mathbf{P}, \mathrm{U}) = 0$ at the local point $\left(\mathbf{q},^{(m)}, \mathbf{P}^{(m)}, \mathrm{U}^{(m)}\right)$, implying that a non-negative value of $\Delta E(\mathbf{q}, \mathbf{P}, \mathrm{U})$ should be obtained through maximizing the constructed objective, while a positive value can lead to a larger energy efficiency. Consequently, $\Delta E(\mathbf{q}, \mathbf{P}, \mathrm{U})$ can be treated as the objective iteratively in the constructed subproblem in each iteration. Note that $\Delta E(\mathbf{q}, \mathbf{P}, \mathrm{U})$ is still not concave, to construct a convex subproblem in mth iteration, we need to further build up a lower bound concave approximation $\Delta E^{(m)}(\mathbf{q}, \mathbf{P}, \mathrm{U})$ of objective $\Delta E(\mathbf{q}, \mathbf{P}, \mathrm{U})$ where $\Delta E^{(m)}(\mathbf{q}, \mathbf{P}, \mathrm{U}) \leq \Delta E(\mathbf{q}, \mathbf{P}, \mathrm{U})$ and the equality $\Delta E^{(m)}(\mathbf{q}, \mathbf{P}, \mathrm{U}) = \Delta E(\mathbf{q}, \mathbf{P}, \mathrm{U}) = 0$ at the local point $\left(\mathbf{q}^{(m)}, \mathbf{P}^{(m)}, \mathrm{U}^{(m)}\right)$.

As $\Delta E(\mathbf{q}, \mathbf{P}, \mathrm{U})$ consists of the minimal harvested energy $E_H(\mathbf{q}, \mathbf{P}, \mathrm{U})$ and extra energy consumption $E_{ex}(\mathbf{q}, \mathbf{P}, \mathrm{U})$ related to the optimization variable, we will propose a lower bound concave approximation for $E_H(\mathbf{q}, \mathbf{P}, \mathrm{U})$, represented by $E_H^{(m)}(\mathbf{q}, \mathbf{P}, \mathrm{U})$ and upper bound convex approximation for $E_{ex}(\mathbf{q}, \mathbf{P})$, represented by $E_{ex}^{(m)}(\mathbf{q}, \mathbf{P})$.

Approximation for Harvested Energy

As the minimal harvested energy $E_H(\mathbf{q}, \mathbf{P}, \mathrm{U}) = \underset{k}{min} \sum_{n=1}^{N} \mathscr{F}\left(\frac{\beta_0 P[n]\tilde{G}_{A,k}[n]}{d_k[n]^2}\right)\delta_t$, the approximation can be promoted through finding a lower bound concave appromation of $\mathcal{F}\left(\frac{\beta_0 P[n]\tilde{G}_{A,k}[n]}{d_k[n]^2}\right)$ for SN k in nth slot. To start with the approximation, the convexity of $\mathcal{F}\left(\frac{\beta_0 P[n]\tilde{G}_{A,k}[n]}{d_k[n]^2}\right)$ in $\frac{d_k[n]^2}{\beta_0 P[n]\tilde{G}_{A,k}[n]}$ proving in [11] is first utilized which results in the following inequality:

$$\mathcal{F}\left(\frac{\beta_0 P[n]\tilde{G}_{A,k}[n]}{d_k[n]^2}\right) \geq -B_{1,k,n}^{(m)} \frac{d_k[n]^2}{P[n]\tilde{G}_{A,k}[n]} + B_{2,k,n}^{(m)} \tag{8.9}$$

where positive constants $B_{1,k,n}^{(m)} = \mathscr{F}'\left(\frac{\beta_0 P^{(m)}[n]\tilde{G}_{A,k}^{(m)}[n]}{d_k^{(m)}[n]^2}\right)\frac{\beta_0 P^{(m)}[n]^2\tilde{G}_{A,k}^{(m)}[n]^2}{d_k^{(m)}[n]^4}$ and $B_{2,k,n}^{(m)} = \mathcal{F}\left(\frac{\beta_0 P^{(m)}[n]\tilde{G}_{A,k}^{(m)}[n]}{d_k^{(m)}[n]^2}\right) + B_{1,k,n}^{(m)}\frac{d_k^{(m)}[n]^2}{P^{(m)}[n]\tilde{G}_{A,k}^{(m)}[n]}$. Then, to deal with $\frac{d_k[n]^2}{P[n]\tilde{G}_{A,k}[n]}$, the mean inequality is employed where $\forall b_1, \ldots, b_R \in R^+, b_1 b_2 \ldots b_R \geq (\frac{1}{R}(b_1+, \ldots, +b_R))^R$ and the equality holds at $b_1 =, \ldots = b_\mathrm{R}$, which leads that:

$$
\begin{aligned}
&- B_{1,k,n}^{(m)}\frac{d_k[n]^2}{P[n]\tilde{G}_{A,k}[n]} + B_{2,k,n}^{(m)} \\
&= -\frac{B_{1,k,n}^{(m)}\Pi_{i=1}^4 C_{i,k,n}^{(m)}}{5^5 N_t \Pi_{i=1}^4 C_{i,k,n}^{(m)}}\frac{d_k[n]^2}{P[n]\tilde{G}_x(\varphi_{x,k}[n])\tilde{G}_y(\varphi_{y,k}[n])\tilde{G}_z(\varphi_{z,k}[n])} + B_{2,k,n}^{(m)} \\
&\geq -\frac{B_{1,k,n}^{(m)}}{5^5 N_t \Pi_{i=1}^4 C_{i,k,n}^{(m)}}\left(\frac{1}{P[n]} + C_{1,k,n}^{(m)} d_k[n]^2 + \frac{C_{2,k,n}^{(m)}}{\tilde{G}_x(\varphi_{x,k}[n])} + \frac{C_{3,k,n}^{(m)}}{\tilde{G}_y(\varphi_{y,k}[n])}\right. \\
&\left. + \frac{C_{4,k,n}^{(m)}}{\tilde{G}_z(\varphi_{z,k}[n])}\right)^5 + B_{2,k,n}^{(m)}
\end{aligned}
\tag{8.10}
$$

where the positive constants $C_{1,k,n}^{(m)} = \frac{1}{P^{(m)}[n]d_k^{(m)}[n]^2}$, $C_{2,k,n}^{(m)} = \frac{\tilde{G}_x(\varphi_{x,k}^{(m)}[n])}{P^{(m)}[n]}$, $C_{3,k,n}^{(m)} = \frac{\tilde{G}_y(\varphi_{y,k}^{(m)}[n])}{P^{(m)}[n]}$, $C_{4,k,n}^{(m)} = \frac{\tilde{G}_z(\varphi_{z,k}^{(m)}[n])}{P^{(m)}[n]}$ are given based on the equal condition of mean equality to ensure that there is no approximation error at the local point, and $d_k^{(m)}[n]$, $\varphi_{x,k}^{(m)}[n]$, $\varphi_{y,k}^{(m)}[n]$, $\varphi_{z,k}^{(m)}[n]$ is value of $d_k[n]$, $\varphi_{x,k}[n]$, $\varphi_{y,k}[n]$, $\varphi_{z,k}[n]$ at the local point $(\mathbf{q}^{(m)}, \mathbf{P}^{(m)}, \mathrm{U}^{(m)})$.

Then, as long as $\frac{1}{\tilde{G}_i(\varphi_{i,k}[n])}, i \in \{x, y, z\}$ can be approximated into a upper bound convex form, we can finally obtain a lower bound concave approximation for $E_H(\boldsymbol{q}, \mathbf{P}, \mathrm{U})$. Recall that the increasing monotonicity and convexity of $\frac{1}{\tilde{G}_i(\varphi_{i,k}[n])}, i \in \{x, y, z\}$ in $\varphi_{i,k}[n]^2, i \in \{x, y, z\}$ has been proved in the last section, which imply that the convex approximation can be achieved through constructing upper bound convex approximation of $\varphi_{i,k}[n]^2, i \in \{x, y, z\}$, denoting by $\hat{\varphi}_{i,k}^{(m)}[n]^2, i \in \{x, y, z\}$ where $\varphi_{i,k}[n]^2 \leq \hat{\varphi}_{i,k}^{(m)}[n]^2, i \in \{x, y, z\}$.

We first start with the approximation for $\varphi_{x,k}[n]^2$:

$$\begin{aligned}\varphi_{x,k}[n]^2 &= \frac{\pi^2}{4}\left(\frac{w_{x,k}-x[n]}{d_k[n]}+u_x[n]\right)^2\\ &= \frac{\pi^2}{4}\left(\frac{(s_{x,k}-x[n])^2}{d_k[n]^2}-\frac{2\big(s_{x,k}-x[n]\big)}{d_k[n]}+2(1+u_x[n])\left(\frac{s_{x,k}-x[n]}{d_k[n]}+1\right)\right.\\ &\quad\left.-2u_x[n]-2+u_x[n]^2\right)\\ &\le \frac{\pi^2}{4}\left(\left(1+\frac{1}{C^{(m)}_{5,k,n}}\right)\frac{(s_{x,k}-x[n])^2}{d_k[n]^2}+\left(\frac{2}{C^{(m)}_{5,k,n}}-2\right)\frac{s_{x,k}-x[n]}{d_k[n]}\right.\\ &\quad\left.+C^{(m)}_{5,k,n}(1+u_x[n])^2+(u_x[n]-1)^2-3+\frac{1}{C^{(m)}_{5,k,n}}\right)\end{aligned}\tag{8.11}$$

where $C^{(m)}_{5,k,n}=\left(\frac{s_{x,k}-x^{(m)}[n]}{d_k^{(m)}[n]}+1\right)\frac{1}{1+u_x^{(m)}[n]}$ is to ensure the equality at local point. Define $\lambda^{(m)}_{k,n}(u_x[n])=C^{(m)}_{5,k,n}(1+u_x[n])^2+(u_x[n]-1)^2-3+\frac{1}{C^{(m)}_{5,k,n}}$ which is convex in $u_x[n]$, such that we just approximate the remaining part as:

$$\begin{aligned}\left(1+\frac{1}{C^{(m)}_{5,k,n}}\right)\frac{(s_{x,k}-x[n])^2}{d_k[n]^2} &= \left(1+\frac{1}{C^{(m)}_{5,k,n}}\right)\left(\frac{d_k[n]^2+(s_{x,k}-x[n])^2}{d_k[n]^2}-1\right)\\ &\le \left(1+\frac{1}{C^{(m)}_{5,k,n}}\right)\left(\frac{1}{2C^{(m)}_{6,k,n}}(d_k[n]^2+(s_{x,k}-x[n])^2\right)^2+\frac{C^{(m)}_{6,k,n}}{2d_k[n]^4}-1\end{aligned}\tag{8.12}$$

$$\begin{aligned}\left(\frac{2}{C^{(m)}_{5,k,n}}-2\right)\frac{s_{x,k}-x[n]}{d_k[n]} &= \frac{\left|\frac{2}{C^{(m)}_{5,k,n}}-2\right|d_k[n]+\left(\frac{2}{C^{(m)}_{5,k,n}}-2\right)(s_{x,k}-x[n])}{d_k[n]}-\left|\frac{2}{C^{(m)}_{5,k,n}}-2\right|\\ &\le \frac{1}{2C^{(m)}_{7,k,n}}\left(\left|\frac{2}{C^{(m)}_{7,k,n}}-2\right|d_k[n]+\left(\frac{2}{C^{(m)}_{5,k,n}}-2\right)(s_{x,k}-x[n])\right)^2\\ &\quad+\frac{C^{(m)}_{7,k,n}}{2}\frac{1}{d_k[n]^2}-\left|\frac{2}{C^{(m)}_{5,k,n}}-2\right|\end{aligned}\tag{8.13}$$

where positive constants $C^{(m)}_{6,k,n}=d_k^{(m)}[n]^2(d_k^{(m)}[n]^2+\big(s_{x,k}-x^{(m)}[n])^2\big)$ and $C^{(m)}_{7,k,n}=d_k^{(m)}[n]\left(\left|\frac{2}{C^{(m)}_{5,k,n}}-2\right|d_k^{(m)}[n]+\left(\frac{2}{C^{(m)}_{5,k,n}}-2\right)\left(s_{x,k}-x^{(m)}[n]\right)\right)$ are to ensure the equality at the local point. Noticing the expression of (8.12) and (8.13), $\mathrm{d_k[n]}$ is convex in q and $\mathrm{d_k[n]} > \left|\mathrm{s_{x,k}-x[n]}\right|$ such that we can complete the upper bound convex approximation of $\varphi_{x,k}[n]^2$ via build up a convex approximation for $\frac{1}{d_k[n]^2}$. For $\frac{1}{d_k[n]^2}$, the following inequality holds at the local point:

$$\frac{1}{d_k[n]^2} \leq \frac{2}{H^6}d_k[n]^4 - \left(\frac{1}{d_k^{(m)}[n]^4} + \frac{4d_k^{(m)}[n]^2}{H^6}\right)d_k[n]^2 + \frac{2}{d_k^{(m)}[n]^2} + \frac{2d_k^{(m)}[n]^4}{H^6}$$
$$\leq \frac{2}{H^6}d_k[n]^4 - \left(\frac{1}{d_k^{(m)}[n]^4} + \frac{4d_k^{(m)}[n]^2}{H^6}\right)\left(\left(\mathbf{q}^{(m)} - \mathbf{s}_k\right)^T(\mathbf{q} - \mathbf{s}_k) - d_k^{(m)}[n]^2\right)$$
$$+ \frac{2}{d_k^{(m)}[n]^2} + \frac{2d_k^{(m)}[n]^4}{H^6} \triangleq \chi_{k,n}^{(m)}(\mathbf{q}) \tag{8.14}$$

which can be proved through derivative comparison. Replacing $\frac{1}{d_k[n]^2}$ by $\boldsymbol{\chi}_{\boldsymbol{k},\boldsymbol{n}}^{(\boldsymbol{m})}(\mathbf{q})$ in (8.12) and (8.13), we have the upper bound convex approximation of the remaining two parts as $\left(1 + \frac{1}{C_{5,k,n}^{(m)}}\right)\frac{(s_{x,k}-x[n])^2}{d_k[n]^2} \geq \mu_{\boldsymbol{k},\boldsymbol{n}}^{(\boldsymbol{m})}(\mathbf{q})$ and $\left(\frac{2}{C_{5,k,n}^{(m)}} - 2\right)\frac{s_{x,k}-x[n]}{d_k[n]} \geq \vartheta_{\boldsymbol{k},\boldsymbol{n}}^{(\boldsymbol{m})}(\mathbf{q})$. Combining the approximation $\lambda_{k,n}^{(m)}(u_x[n])$, the convex approximation of $\varphi_{x,k}[n]^2$ is obstructed as:

$$\varphi_{x,k}[n]^2 \leq \frac{\pi^2}{4}\left(\mu_{k,n}^{(m)}(\mathbf{q}) + \vartheta_{k,n}^{(m)}(\mathbf{q}) + \lambda_{k,n}^{(m)}(u_x[n])\right) \triangleq \hat{\varphi}_{x,k}^{(m)}[n]^2 \tag{8.15}$$

where the equality hold at the local point $\left(\mathbf{q}^{(m)}, \mathbf{P}^{(m)}, \mathrm{U}^{(m)}\right)$. Recall that expressions for $\varphi_{x,k}[n]^2$ and $\varphi_{y,k}[n]^2$ are consistent in form, with the only difference being the denotations, i.e., x[n] comparing with y[n], $\mathrm{s}_{x,k}$ comparing with $\mathrm{s}_{y,k}$. Therefore, we can construct the upper bound convex approximation of $\varphi_{y,k}[n]^2$ through the same way like $\varphi_{x,k}[n]^2$ which is denoted as $\hat{\varphi}_{y,k}^{(m)}[n]^2$.

For the approximation of $\varphi_{z,k}[n]^2$, we have:

$$\varphi_{z,k}[n]^2 = \frac{\pi^2}{4}\left(\frac{H}{d_k[n]} + u_z[n]\right)^2$$
$$= \frac{\pi^2}{4}\left(\frac{H^2}{d_k[n]^2} - \frac{2H}{d_k[n]} + 2(1 + u_z[n])\frac{H}{d_k[n]} + u_z[n]^2\right)$$
$$\leq \frac{\pi^2}{4}\left(\left(1 + \frac{1}{C_{8,k,n}^{(m)}}\right)\frac{H^2}{d_k[n]^2} - \frac{2H}{d_k[n]} + C_{8,k,n}^{(m)}(1 + u_z[n])^2 + u_z[n]^2\right)$$
$$\leq \frac{\pi^2}{4}\left(\left(1 + \frac{1}{C_{8,k,n}^{(m)}}\right)H^2\chi_{k,n}^{(m)}(\mathbf{q}) + \frac{2H}{d_k^m[n]^2}d_k[n] - \frac{4H}{d_k^m[n]}\right.$$
$$\left. + C_{8,k,n}^{(m)}(1 + u_z[n])^2 + u_z[n]^2\right) \triangleq \hat{\varphi}_{z,k}^{(m)}[n]^2 \tag{8.16}$$

The lower bound concave approximation for the minimal harvested energy is

$$E_H(\boldsymbol{q}, \boldsymbol{P}, \boldsymbol{U}) = \min_k \sum_{n=1}^{N} \mathscr{F}\left(\frac{\beta_0 P[n]\tilde{G}_{A,k}[n]}{d_k[n]^2}\right)\delta_t$$

$$\geq \min_k \delta_t \sum_{n=1}^{N} -\frac{B_{1,k,n}^{(m)}}{5^5 N_t \Pi_{i=1}^4 C_{i,k,n}^{(m)}}\left(\frac{1}{P[n]} + C_{1,k,n}^{(m)} d_k[n]^2 + \sum_{i\in\{x,y,z\}} \frac{C_{i,k,n}^{(m)}}{\tilde{G}_i\left(\sqrt{\hat{\varphi}_{i,k}^{(m)}[n]^2}\right)}\right)^5$$

$$+ B_{2,k,n}^{(m)} \triangleq E_H^{(m)}(\boldsymbol{q}, \boldsymbol{P}, \boldsymbol{U}) \tag{8.17}$$

where $C_{x,k,n}^{(m)} = C_{2,k,n}^{(m)}, C_{y,k,n}^{(m)} = C_{3,k,n}^{(m)}, C_{z,k,n}^{(m)} = C_{4,k,n}^{(m)}$ and the equality hold at the local point $\left(\mathbf{q}^{(m)}, \mathbf{P}^{(m)}, \mathbf{U}^{(m)}\right)$.

Approximation for Extra Energy Consumption

Recall that we need construct an upper bound convex approximation for $E_{ex}(\boldsymbol{q}, \boldsymbol{P}) = \sum_{n=1}^{N} (P_c(v[n]) + P[n])\delta_t - E_0$ while $P_c(v[n] = \gamma_1\left(1 + \frac{3v[n]^2}{V_{tip}^2}\right) + \gamma_2\left(\sqrt{1 + \frac{v[n]^4}{4v_0^4}} - \frac{v[n]^2}{2v_0^2}\right)^{1/2} + \frac{1}{2}d_0\rho_0 s_0 A_0 v[n]^3$. Besides, $E_{ex}(\boldsymbol{q}, \boldsymbol{P})$ is affine function of $\mathrm{P_c}$ and P, $\gamma_1\left(1 + \frac{3v[n]^2}{V_{tip}^2}\right)$ and $\frac{1}{2}d_0\rho_0 s_0 A_0 v[n]^3$ is convex in q while $\gamma_2\left(\sqrt{1 + \frac{v[n]^4}{4v_0^4}} - \frac{v[n]^2}{2v_0^2}\right)^{1/2}$ is monotonically decreasing and convex in $\mathrm{v}^2[\mathrm{n}]$. Therefore, a lower bound concave approximation of $v^2[n]$ in $\gamma_2\left(\sqrt{1 + \frac{v[n]^4}{4v_0^4}} - \frac{v[n]^2}{2v_0^2}\right)^{1/2}$ will finalize the approximation for extra energy consumption which gives:

$$v[n]^2 \geq \frac{2\left(q^{(m)}[n+1] - q^{(m)}[n]\right)^T (q[n+1] - q[n])}{\delta_t^2} - v^{(m)}[n]^2 \triangleq v_{lo}^{(m)}[n]^2 \tag{8.18}$$

and accordingly, the upper bound convex approximation of extra energy consumption is given as follows:

$$E_{ex}(\boldsymbol{q}, \boldsymbol{P}) \leq E_0 + \sum_{n=1}^{N}\left(\gamma_1\left(1 + \frac{3v[n]^2}{V_{\text{tip}}^2}\right) + \frac{1}{2}d_0\rho_0 s_0 A_0 v[n]^3 + P[n]\right)\delta_t$$

$$+ \sum_{n=1}^{N}\left(\gamma_2\left(\left(1 + \frac{v_{lo}^{(m)}[n]^4}{4v_0^4}\right)^{1/2} - \frac{v_{lo}^{(m)}[n]^2}{2v_0^2}\right)^{1/2}\right)\delta_t \triangleq E_{ex}^{(m)}(\boldsymbol{q}, \boldsymbol{P}). \tag{8.19}$$

8.4.4 Iterative Algorithm

Combining the above approximation for harvested energy (8.17) and extra energy consumption (8.19), at the local point $\left(\mathbf{q}^{(m)}, \mathbf{P}^{(m)}, \mathbf{U}^{(m)}\right)$ in mth iteration, the lower bound concave approximation for the objective $\Delta E(\mathbf{q}, \mathbf{P}, \mathbf{U})$ can be eventually constructed as:

$$\Delta E(\mathbf{q}, \mathbf{P}, \mathbf{U}) \geq E_H^{(m)}(\mathbf{q}, \mathbf{P}, \mathbf{U}) - \eta^{(m)} E_{ex}^{(m)}(\mathbf{q}, \mathbf{P}) \triangleq \Delta E^{(m)}(\mathbf{q}, \mathbf{P}, \mathbf{U}) \tag{8.20}$$

Thereafter, a convex subproblem for the original problem to ensure stable improvement of objective in mth iteration is formulated as:

$$(\text{P1}) : \max_{q,P,\text{U}} \Delta E^{(m)}(\mathbf{q}, \mathbf{P}, \mathbf{U}) \tag{8.21}$$

$$\|\boldsymbol{q}[n+1] - \boldsymbol{q}[n]\| \leq V\delta_t, \forall n \in \mathcal{N} \tag{8.21a}$$

$$\mathbf{q}[1] = (x_1, y_1), \mathbf{q}[N+1] = (x_2, y_2) \tag{8.21b}$$

$$0 \leq P[n] \leq P_{max}, \forall n \in \mathcal{N} \tag{8.21c}$$

$$u_x[n], u_y[n], u_z[n] \in [-1, 1], \forall n \in \mathcal{N} \tag{8.21d}$$

which can be efficiently solved via convex optimization method due to the convexity. As pre-discussed, for a given local point $\left(\mathbf{q}^{(m)}, \mathbf{P}^{(m)}, \mathbf{U}^{(m)}\right)$, solving the formulated convex subproblem (P1) can at least lead to a non-negative objective. The non-negative objective of subproblem corresponds to a stable improvement for the objective function $\eta(\mathbf{q}, \mathbf{P}, \mathbf{U})$ of the original problem (OP) which motivate us to propose an iterative algorithm to tackling problem (OP).

Before introducing the iterative algorithm, note that starting with a satisfying initial point $\left(\boldsymbol{q}^{(0)}, \boldsymbol{P}^{(0)}, \boldsymbol{U}^{(0)}\right)$ can conspicuously improve the convergence speed. We first construct the UAV initial path along the straight line from (x_1, y_1) to (x_2, y_2) where the initial trajectory $\mathbf{q}^{(0)}$ is initialized as the position of N-equal points of this path which is obviously feasible for problem (OP). Then, the transmit power and analog beamforming variable are initialized as $\mathbf{P}^{(0)} = 0.5P_{max}, \mathbf{U}^{(0)} = 0.5$. After initializing the variable, an efficient iterative algorithm is provided to solve problem (OP). Specifically, in mth iteration, we first calculate the energy efficiency $\eta^{(m)}$ and then construct the lower-bound concave approximation of $\Delta E(\mathbf{q}, \mathbf{P}, \mathbf{U})$ which subsequently helps in formulating the convex subproblem (P1). Then problem (P1) is solved efficiently to get a current optimal UAV trajectory, transmit power and analog beamforming variables $(\mathbf{q}^*, \mathbf{P}^*, \mathbf{U}^*)$ which is then set as the local point $\left(\mathbf{q}^{(m+1)}, \mathbf{P}^{(m+1)}, \mathbf{U}^{(m+1)}\right)$ for the next iteration and repeat the above steps until the result converges to threshold ξ.

Following the complexity analysis in [30] and assuming that the ellipsoid method is applied to address convex subproblem (P1), we then provide the overall design complexity of the iterative algorithm. In the subproblem (P1), it contains $6N$ variables and $5N$ constraints such that in each iteration, $\mathcal{O}(\xi^{-1}(6N)^2)$ is cost for ellipsoid generation which in each ellipsoid generation, calculating the objective function and constraints will cost $\mathcal{O}((6N)^2)$ and $\mathcal{O}(6N \times 5N)$. Denoting the iteration number by Φ, the total complexity of proposed algorithm is $\mathcal{O}(\phi\xi^{-1}(6N)^2 \times ((6N)^2 + 6N \times 5N)) = \mathcal{O}(2376\phi\xi^{-1}N^4)$.

8.5 Simulation

We first evaluate the algorithm convergence in Fig. 8.3. As the iteration index m increases, the energy efficiency η consistently demonstrates improvement across various antenna array configurations N_t, exhibiting a convergence pattern within a limited iteration number. Furthermore, an observation of the energy efficiency under diverse antenna arrays reveals that a larger number of antenna elements results in a greater converged energy efficiency while it accompanies a slower convergence rate and costs more iteration until convergence. This stems from the substantial augmentation of antenna gains within the focused beam achieved by augmenting the number of antenna elements, which concurrently narrows the beamwidth. Consequently, a greater number of iterations are necessitated to optimize the directional alignment of these narrower beams.

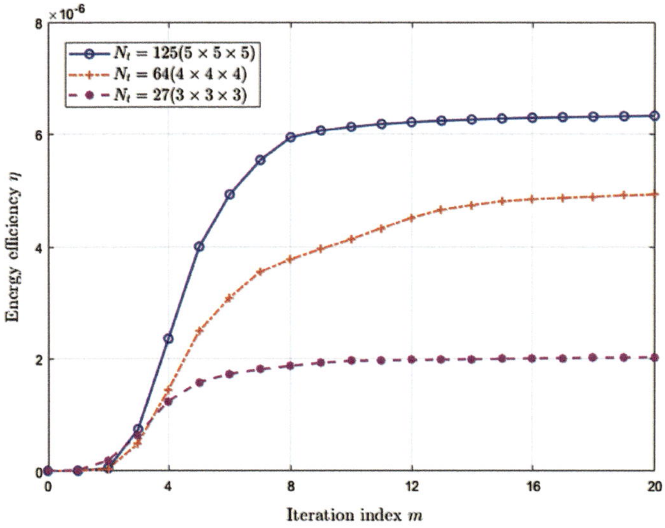

Fig. 8.3 Convergence evaluation of proposed algorithm

Next, in Fig. 8.4, we compare the energy efficiency achieved by our proposed solution with solution without employing directional antennas, i.e., with conventional omni-directional antennas. This comparison aims to demonstrate the benefits of analog beamforming in maximizing energy efficiency. This figure reveals that, across various operation time T, a longer operation time consistently yields greater flexibility in enhancing efficiency, resulting in a higher energy efficiency for WPT tasks. Notably, the solution with directional antennas demonstrates marked improvements in energy efficiency over the omni-directional. As the operation time *T* increases, the performance gain stemming from analog beamforming becomes more pronounced. This trend is attributed to the wider range of channel state variations experienced by the UAV during longer operations, underscoring the importance of analog beamforming in enhancing energy efficiency and highlighting its necessity in UAV-enabled WPT networks. Additionally, we observe that an increased UAV speed limit V contributes to enhanced flexibility, allowing for further optimization of WPT energy efficiency.

Then, we turn on to investigate the impact of UAV maximum transmit power P_{max} on the energy efficiency in Fig. 8.5. As demonstrated in this figure, adding more antenna elements N_t consistently boosts the energy efficiency of WPT by significantly enhancing the antenna gain within the focused beam, thereby elevating overall energy efficiency. Furthermore, as the maximum transmit power Pmax increases, reflecting a higher power transmission capability of the UAV, the combined solution illustrated in Fig. 8.5 exhibits a greater degree of flexibility in power allocation, enabling a superior energy efficiency for WPT performance. However, it is noteworthy that when Pmax reaches a sufficiently high level, further increments yield diminishing returns in terms of energy efficiency enhancement. This phenomenon stems from the nonlinear nature of the EH process, where conversion efficiency diminishes at both low and excessively high received power levels. Consequently, once the received

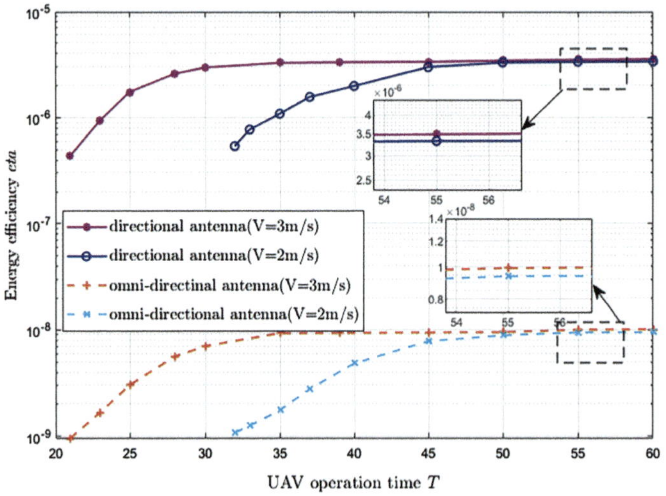

Fig. 8.4 Energy efficiency comparison between directional antenna with omni-directional antenna

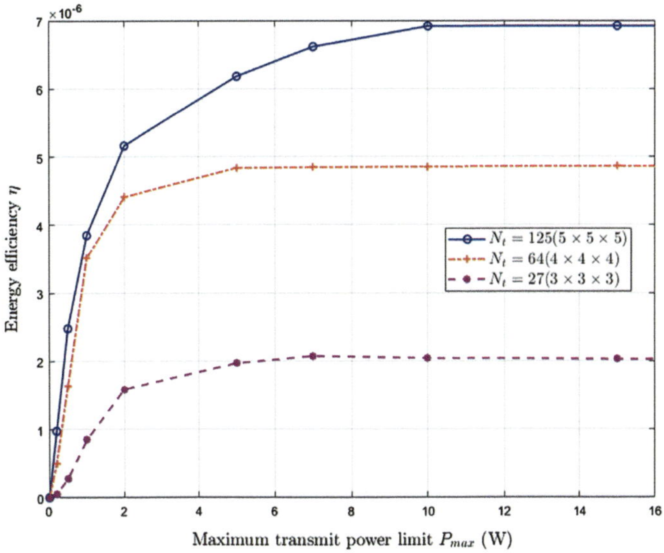

Fig. 8.5 Energy efficiency comparison over different maximum transmit power

power reaches a relatively high threshold, increasing transmit power does not necessarily translate to higher conversion efficiency, limiting the impact of relaxed transmit power constraints on overall energy efficiency.

After validating the iterative algorithm and assessing its optimized performance, we continue to observe the attained UAV trajectory and analog beamforming solution to gain additional insights into the joint design. Firstly, Fig. 8.6 illustrates the optimized UAV trajectories for varying maximum speed V and WPT operation times T. Notably, all trajectories initiate from $q[0] = (0\text{ m}, 0\text{ m})$ and terminate at $q[N] = (40\text{ m}, 40\text{ m})$ with a constant flying height H, adhering to the mobility constraints. Upon further observation, it is obvious that with a larger value of maximum speed V or WPT operation times T, the UAV performs greater flexibility in approaching SNs closer, thereby fostering the establishment of wireless channels with superior quality. Consequently, this proximity enhancement translates into an anticipated boost in WPT energy efficiency, and the above finding can be corroborated by Fig. 8.4.

Finally, in Fig. 8.7, we plot the harvested power distribution for $N_t = 27$ and $N_t = 125$ in $n = 10$th and $n = 25$th slot which is related to the antenna gain and distance, to gain a insight on the impact of analog beamforming. It is shown that through analog beamforming, the harvested power can be mainly focused on the nearest SN which is the most important factor in improving WPT energy efficiency. Then, by comparing with the beam when $N_t = 27$ with beam when $N_t = 125$, we can see that a larger value of antenna elements number leads to a narrower and stronger beam at the SN position which performs better in energy efficiency improvement.

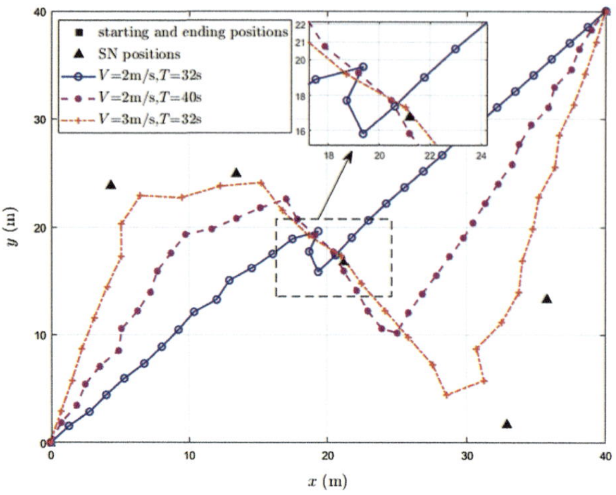

Fig. 8.6 Optimized trajectory with different maximum speeds and task times

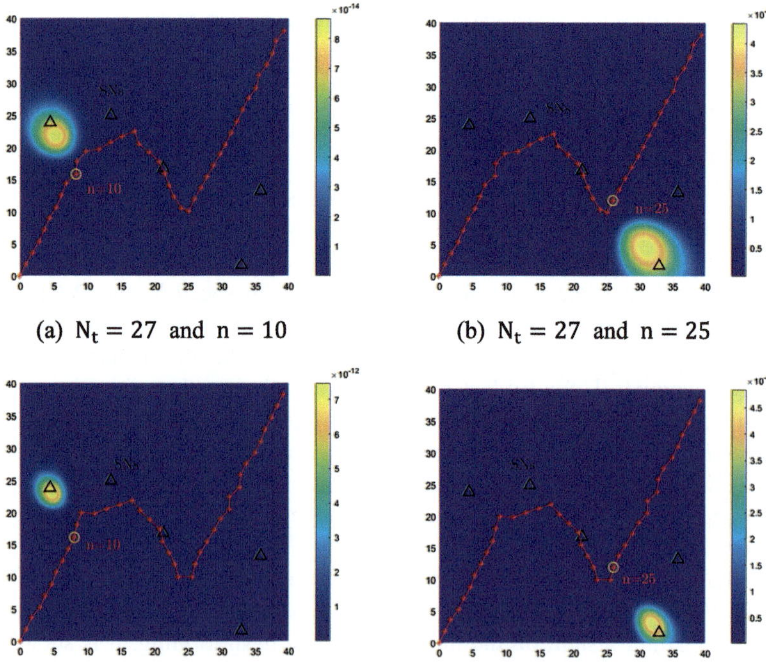

(a) $N_t = 27$ and $n = 10$ (b) $N_t = 27$ and $n = 25$

(c) $N_t = 125$ and $n = 10$ (d) $N_t = 125$ and $n = 25$

Fig. 8.7 Harvest power distribution with different antenna elements N_t and time slot n

8.6 Conclusion

In this chapter, we explored a UAV-enabled WPT network including multiple ground sensors, where the UAV is equipped with a 3D ULA antenna array for energy transmission. Considering the UAV energy consumption, we introduced a task-aware energy efficiency metric. For optimizing this WPT task, we formulated an energy efficiency maximization problem, through jointly designing the analog beamforming, UAV trajectory, and transmit power allocation. Due to the complicated expression of 3D ULA antenna gain, we design a cosine antenna pattern approximation that simplifies analysis and possesses convexity, which is different from conventional modified cosine patterns. This enabled us to construct convex subproblems for given local point, upon which we developed an iterative algorithm that iteratively enhances energy efficiency until convergence is achieved. Our numerical results validated the algorithm convergence and evaluate its effectiveness in maximizing energy efficiency. Furthermore, by analyzing iterative solutions under varying configurations, we demonstrated the adaptability of our approach in leveraging the extensive beam coverage and high gain advantages of 3D ULA-based analog beamforming.

References

1. Zeng, Y., Lu, C., Liu, R., He, X., Rong, C., & Liu, M. (2023). Wireless power and data transfer system using multidirectional magnetic coupler for swarm AUVs. *IEEE Transactions on Power Electronics, 38*, 1440–1444.
2. Wu, J., Li, Y., Dai, X., Gao, R., & He, M. (2024). A dynamic power transfer route construction and optimization method considering random node distribution for wireless power transfer network. *IEEE Transactions on Power Electronics, 39*, 4858–4869.
3. Xu, J., Zeng, Y., & Zhang, R. (2018). UAV-enabled wireless power transfer: Trajectory design and energy optimization. *IEEE Transactions on Wireless Communications, 17*, 5092–5106.
4. Xie, L., Cao, X., Xu, J., & Zhang, R. (2021). UAV-enabled wireless power transfer: A tutorial overview. *IEEE Transactions on Green Communications and Networking, 5*, 2042–2064.
5. Liu, Y., Xiong, K., Lu, Y., Ni, Q., Fan, P., & Letaief, K. B. (2021). UAV-aided wireless power transfer and data collection in Rician fading. *IEEE Journal on Selected Areas in Communications, 39*, 3097–3113.
6. Li, M., Li, H., Ma, P., & Wang, H. (2023). Energy maximization for ground nodes in UAV-enabled wireless power transfer systems. *IEEE Internet of Things Journal, 10*, 17096–171091.
7. Wu, S., Cai, C., Liu, X., Chai, W., & Yang, S. (2022). Compact and free-positioning omnidirectional wireless power transfer system for unmanned aerial vehicle charging applications. *IEEE Transactions on Power Electronics, 37*, 8790–8794.
8. Pan, H., Liu, Y., Sun, G., Fan, J., Liang, S., & Yuen, C. (2023). Joint power and 3D trajectory optimization for UAV-enabled wireless powered communication networks with obstacles. *IEEE Transactions on Communications, 71*, 2364–2380.
9. Liu, L., Xiong, K., Cao, J., Lu, Y., Fan, P., & Letaief, K. B. (2022). Average AoI minimization in UAV-assisted data collection with RF wireless power transfer: A deep reinforcement learning scheme. *IEEE Internet of Things Journal, 9*, 5216–5228.
10. Hu, Y., Yuan, X., Xu, J., & Schmeink, A. (2019). Optimal 1D trajectory design for UAV-enabled multiuser wireless power transfer. *IEEE Transactions on Communications, 67*, 5674–5688.

11. Hu, Y., Yuan, X., Yang, T., Clerckx, B., & Schmeink, A. (2020). On the convex properties of wireless power transfer with nonlinear energy harvesting. *IEEE Transactions on Vehicular Technology, 69*, 5672–5676.
12. Valenta, C. R., & Durgin, G. D. (2014). Harvesting wireless power: survey of energy-harvester conversion efficiency in far-field, wireless power transfer systems. *IEEE Micro Magazine, 15*, 108–120.
13. Boshkovska, E., Ng, D. W. K., Zlatanov, N., & Schober, R. (2015). Practical non-linear energy harvesting model and resource allocation for SWIPT systems. *IEEE Communications Letters, 19*, 2082–2085.
14. Yuan, X., Hu, Y., Schober, R., & Schmeink, A. (2022). Convexity analysis of nonlinear wireless power transfer with multiple RF sources. *IEEE Transactions on Vehicular Technology, 71*, 11311–11316.
15. Yan, H., Chen, Y., & Yang, S.-H. (2022). Time allocation and optimization in UAV-enabled wireless powered communication networks. *IEEE Transactions on Green Communications and Networking, 6*, 951–964.
16. Hu, Y., Yuan, X., Zhang, G., & Schmeink, A. (2021). Sustainable wireless sensor networks with UAV-enabled wireless power transfer. *IEEE Transactions on Vehicular Technology, 70*, 8050–8064.
17. Zargari, S., Hakimi, A., Tellambura, C., & Herath, S. (2023). User scheduling and trajectory optimization for energy-efficient IRS-UAV networks with SWIPT. *IEEE Transactions on Vehicular Technology, 72*, 1815–1830.
18. Yuan, X., Yang, T., Hu, Y., Xu, J., & Schemink, A. (2021). UAV trajectory design for UAV-enabled multiuser nonlinear wireless power transfer. *IEEE Transactions on Wireless Communications, 20*, 1105–1121.
19. Dai, H.-N., Ng, K.-W., Wong, R. C.-W., & Wu, M.-Y. (2008). On the capacity of multi-channel wireless networks using directional antennas. In *Proceedings of 27th International Conference on Computer Communication (INFOCOM)* (pp. 628–636). Phoenix.
20. Spoof, K., Tenhunen, M., Unnikrishnan, V., Stadius, K., Kosunen, M., & Ryynänen, J. (2022). True-time-delay receiver IC with reconfigurable analog and digital beamforming. *IEEE Access, 10*, 116375–116383.
21. Nasir, A. A., Tuan, H. D., Dutkiewicz, E., & Hanzo, L. (2022). Finite-resolution digital beamforming for multi-user millimeter-wave networks. *IEEE Transactions on Vehicular Technology, 71*, 9647–9662.
22. Kutty, S., & Sen, D. (2016). Beamforming for millimeter wave communications: An inclusive survey. *IEEE Communications Surveys & Tutorials, 18*, 949–973.
23. Ahmed, I., Khammari, H., Shahid, A., Musa, A., Kim, K. S., Poorter, E. D., & Moerman, I. (2018). A survey on hybrid beamforming techniques in 5G: Architecture and system model perspectives. *IEEE Communications Surveys & Tutorials, 20*, 3060–3097.
24. Wu, Y., Qiu, L., & Xu, J. (2018). UAV-enabled wireless power transfer with directional antenna: A two-user case. In *Proceedings of 15th International Symposium on Wireless Communication Systems (ISWCS)* (pp. 1–6). Lisbon.
25. Mozaffari, M., Saad, W., Bennis, M., & Debbah, M. (2016). Effificient deployment of multiple unmanned aerial vehicles for optimal wireless coverage. *IEEE Communications Letters, 20*, 1647–1650.
26. Jiang, R., Xiong, K., Yang, H.-C., Cao, J., Zhong, Z., & Ai, B. (2022). Coverage performance of UAV-assisted SWIPT networks with directional antennas. *IEEE Internet of Things Journal, 9*, 10600–10609.
27. Yuan, X., Hu, Y., & Schmeink, A. (2021). Joint design of UAV trajectory and directional antenna orientation in UAV-enabled wireless power transfer networks. *IEEE Journal on Selected Areas in Communications, 39*, 3081–3096.
28. Vesa, A., Alexa, F., & Balta, H. (2015). Comparisons between 2D and 3D uniform array antennas. In *2015 Federated Conference on Computer Science and Information Systems (FedCSIS)* (pp. 1285–1290).

29. Zeng, Y., Xu, J., & Zhang, R. (2019). Energy minimization for wireless communication with rotary-wing UAV. *IEEE Transactions on Wireless Communications, 18*, 2329–2345.
30. Wu, P., Yuan, X., Hu, Y., & Schmeink, A. (2024). Joint power allocation and trajectory design for UAV-enabled covert communication. *IEEE Transactions on Wireless Communications, 23*, 683–698.

Chapter 9
Joint Directional Antenna Orientation and UAV Trajectory Design in UAV-Enabled Networks

Yuxi Huang, Yulin Hu, and Xiaopeng Yuan

Abstract With the development of wireless communication systems, the integration of unmanned aerial vehicles (UAVs) for wireless power transfer (WPT) has emerged as a promising solution to extend the operational range and longevity of ground-based sensor networks. This chapter delves into the joint design of such a UAV-assisted WPT network, where the UAV is equipped with a uniform linear array (ULA) directional antenna. Our objective is to enhance the minimum energy collected by all sensor nodes (SNs) by jointly optimizing the UAV trajectory and the directional antenna orientation. To accurately capture real-world dynamics, we employ a non-linear model for energy harvesting (EH). The difficulties of the original problem, arising from the complex patterns of analog beamforming and the non-linearity of the energy harvesting model, are addressed by introducing an approximate model for the antenna's radiation pattern and discretizing the UAV trajectory. Capitalizing on the convex properties of the energy harvesting model, we develop a robust convex approximation for the modified problem and devise an efficient computational method to solve it. Simulation results validate the algorithm's convergence and its capability to significantly boost the system's performance.

Keywords Unmanned aerial vehicle · Nonlinear energy harvesting · Directional antenna · Trajectory design · Joint optimization

---

Y. Huang · Y. Hu
School of Electronic Information, Wuhan University, Wuhan, China
e-mail: hyxxii@whu.edu.cn

Y. Hu
e-mail: yulin.hu@whu.edu.cn

X. Yuan (✉)
Chair of Information Theory and Data Analytics, RWTH Aachen University, Aachen, Germany
e-mail: yuan@inda.rwth-aachen.de

Y. Hu et al. (eds.), *Reliable and Efficient Trajectory Design in UAV-Aided Wireless Networks*, https://doi.org/10.1007/978-981-96-5988-3_9

Abbreviations

2D	Two-Dimensional
3D	Three-Dimensional
DC	Direct-Current
EH	Energy Harvesting
LoS	Line of Sight
RF	Radio-Frequency
SN	Sensor Node
TSP	Travelling Salesman Problem
UAV	Unmanned Aerial Vehicle
ULA	Uniform Linear Array
WPT	Wireless Power Transfer

9.1 Introduction

9.1.1 Background

Unmanned aerial vehicles (UAVs) has emerged as pivotal components in various wireless communication systems, including logistics transportation, agricultural monitoring and security protection, thanks to their remarkable flexibility and mobility [1, 2]. By optimizing the UAV time-varying location, i.e., the UAV continuous trajectory, the performance of wireless network can be significantly enhanced for the following two reasons [3–5]. On the one hand, UAVs have a high probability of establishing line-of-sight (LoS) dominant air-ground channels, which offer substantial channel benefits over traditional terrestrial cellular networks [6]. On the other hand, UAVs can dynamically adjust their flight paths in real-time, minimizing the transmission distance to ground devices (GDs) and effectively circumventing obstacles in their environment to improve channel gain. Given this potential, numerous research has focused on integrating UAVs into wireless networks, especially the wireless power transfer (WPT) networks [7–9]. Hence, a joint optimization of the UAV trajectory and resource allocation is highly favorable UAV-enabled WPT networks, with directional antennas playing a critical role in enhancing WPT gain [10, 11]. Nevertheless, challenges remain due to the non-linear energy harvesting (EH) process and inaccuracies in directional antenna models, which result in significant estimation errors in the received power at ground devices [12, 13]. These errors can severely hinder the performance of WPT networks in energy harvesting. Thus, it is necessary to develop more accurate and analytically tractable characterizations of non-linear energy harvesting process and the directional antenna model for UAV-assisted WPT network.

9.1.2 Motivation

To address these difficulties in UAV-enabled networks, we explore a UAV-enabled multi-user WPT network, where a UAV equipped with a uniform linear array (ULA) directional antenna performs analog beamforming. In particular, we adopt an analytical non-linear EH model [14] to accurately characterize the WPT performance. We also consider a joint design aiming to maximize the minimum harvested energy among all sensor nodes (SNs) by optimizing the UAV trajectory and the directional antenna orientation. Subsequently, we construct a concave approximation for the objective function and successfully address the joint optimization problem with an iterative algorithm, in which the WPT performance is ensured to progressively improve with each iteration, ultimately converging to an efficient suboptimal solution. This approach ensures a significant enhancement in the energy harvesting performance of the UAV-enabled network, demonstrating the potential of integrating advanced trajectory design and beamforming strategies in practical WPT applications.

9.1.3 Main Contributions

Main contributions to this chapter are as follows:

(1) Modified Cosine Antenna Pattern: To address the complex analysis associated with existing models, we introduced a modified cosine antenna pattern. This model preserves accuracy while featuring a modified parameter that can indefinitely approach the standard cosine antenna pattern, thereby enhancing tractability. This modification benefits not only our current research but also other designs utilizing analog directional beamforming;
(2) Joint Design on UAV Trajectory and Directional Antenna Orientation: We focus on a joint optimization design, where both directional beamforming and UAV trajectory are optimized together in a single convex optimization problem during each iteration.
(3) Simulative Evaluation: Simulations are conducted to validate the convergence and WPT Performance of the proposed algorithm, as well as the corresponding UAV trajectory and antenna behavior. It is worth noting that our design with directional antennas always outperforms the omni-directional antennas.

9.1.4 Structural Organization

Relevant works are introduced in Sect. 9.2, including UAV-enabled WPT technology and directional beamforming strategy. Section 9.3 shows the system model, including antenna model, channel model and non-linear WPT model. In addition, Sects. 9.4 and 9.5 present the optimization joint problem and the proposed iterative algorithm. The numerical results of the simulation and the conclusions are provided and discussed in Sects. 9.6 and 9.7.

9.2 Related Works

9.2.1 UAV-Enabled WPT Technology

UAVs are increasingly viewed as promising mobile power station to enhance the performance of UAV-enabled WPT networks. Their high flight altitude facilitates the establishment of LoS links with ground devices, significantly reducing path loss and improving WPT performance. In the literature, numerous studies have focused on UAV trajectory design in pure WPT networks [12, 13, 15, 16]. Specifically, the works in [12, 13] employed a linear energy harvesting process. However, as noted in [17, 18], this approach is merely an approximation and lacks practicality, given the considerable variation in EH efficiency as received power fluctuates. To address this, two models have been widely adopted to capture the non-linearity in EH: a statistical model [18] based on curve fitting of empirical data, and an analytical model [14] derived from real-world conversion circuits. Utilizing the analytical non-linear model, our previous research [16] demonstrated that the optimal UAV trajectory structure in WPT networks using an omni-directional antenna is a successive-hover-fly pattern. Nevertheless, as confirmed by [16], the substantial signal attenuation encountered in practical EH setups significantly limit the amount of harvested power, highlighting the need for more effective strategies in optimizing UAV trajectories.

The studies mentioned above have improved the efficiency of energy transfer process to a certain extent, proving that UAV-enabled WPT technology can extend network device operational lifetime and ensure network sustainability.

9.2.2 Directional Beamforming Strategy

Directional beamforming is a technique that focuses the transmission of signals in specific directions, enhancing the overall performance of wireless communication systems, particularly in WPT networks [19–21]. Operating with a directional antenna has been shown to significantly enhance antenna gain and improving energy transfer efficiency, as highlighted in [22]. Note that there are two primary strategies to directional beamforming: digital and analog beamforming. Compared to digital beamforming, analog beamforming operates with a simpler structure, using fewer components and resulting in lower operational complexity and cost [23, 24]. While it may offer less flexibility compared to digital methods, it is particularly beneficial in scenarios where resources are limited or where immediate feedback from the channel is not available. For instance, the authors in [25] adopt an analog directional antenna without channel state feedback for a UAV-enabled WPT network. In addition, the authors in [26] analyze the coverage area of a UAV based on its location, beam direction, and beamwidth. However, they did not focus on finding their optimal values. Recognizing that optimal beam direction and beamwidth are location-dependent, the author in [27] jointly optimize the beam direction and beamwidth of a directional antenna along with the UAV location in the UAV-enabled network.

Although these efforts have contributed a lot to enhance the performance of UAV-enabled WPT systems, an efficient design for the UAV trajectory and antenna orientation is still missing, particularly with an accurate directional antenna model. By optimizing the beam direction and bandwidth based on UAV time-varying location, directional beamforming can enhance energy harvesting and mitigate signal loss. Additionally, controlling beam direction allows for a narrower beamwidth, further increasing antenna gain. Hence, we focus on a UAV-enabled multi-user WPT network that employs a directional antenna with a ULA to perform analog beamforming.

9.3 System Model

In this section, we first introduce the considered UAV-enabled WPT networks and relevant parameters. Subsequently, we provide detailed statements to the directional antenna model, channel model, and nonlinear WPT model.

We consider a UAV-enabled WPT network including multiple SNs and a single UAV as an energy supplier, which is illustrated in Fig. 9.1. We assume that in total K SNs on the ground and denote the corresponding horizontal position of each SN k by $(s_{x,k}, s_{y,k}), k \in \mathcal{K} \triangleq \{1, 2, \ldots, K\}$. The UAV is always at a fixed height H and has a fixed task duration for charging, i.e., $\mathcal{T} \triangleq [0, T]$ with length of $T > 0$. The task duration is pre-designed, which may be determined based on the UAV energy capacity or external scheduling policies. In addition, the positions of SNs are assumed to be a prior knowledge by the UAV. During task time, the horizontal location of UAV is denoted by $(x(t), y(t)),\ t \in [0, T]$ which continuously varies over time and location. Given that we adopt a rotary-wing UAV, it benefits from high mobility and adjustable speed, allowing it to efficiently maneuver within the network [11]. The maximum UAV speed is denoted as V, and the corresponding limit is expressed as $\sqrt{(\dot{x}(t))^2 + (\dot{y}(t))^2} \leq V, \forall t \in \mathcal{T}$ where $\dot{x}(t)$ and $\dot{y}(t)$ are the first-derivatives of $x(t)$ and $y(t)$ in t, respectively.

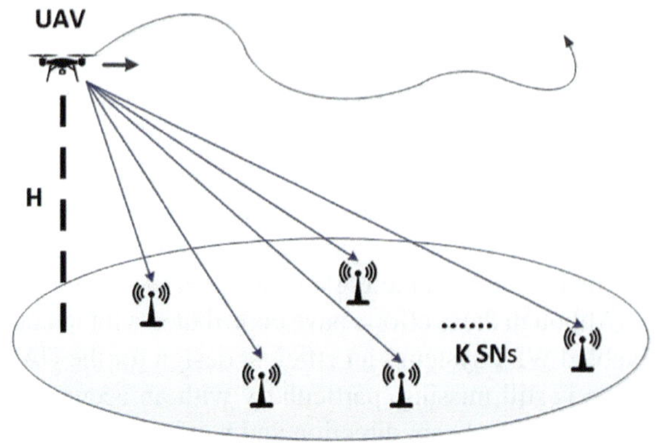

Fig. 9.1 An example of UAV-enabled WPT networks

9.3.1 *Uniform Linear Array of Directional Antenna*

Moreover, the WPT is performed using a directional antenna mounted on the UAV, which significantly enhances the channel gain at the orientation direction. The directional antenna we adopted is configured as a uniform linear array (ULA), where N_t isotropic antenna elements are uniformly placed along a vertical straight line with equal spacing Δ, as illustrated in Fig. 9.2. As widely adopted in ULA [29], the spacing between antenna elements is set to half the wavelength, i.e., $\Delta = \lambda/2$ where λ is the wavelength of the centre carrier frequency for RF signals.

We focus on analog directional beamforming due to its advantages of low cost and low power consumption, making it an attractive option for energy-efficient wireless systems. We denote by θ_0 the elevation angle at which the directional beamforming of the ULA concentrates power. The elevation angle θ_0 represents the direction of maximum power gain, where the main lobe of the antenna pattern is

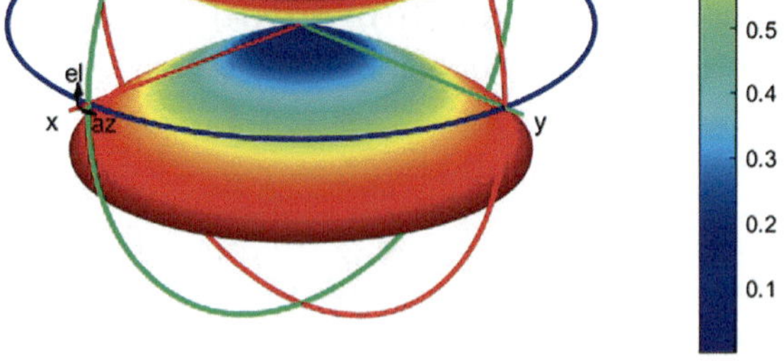

Fig. 9.2 An example of ULA antenna pattern ($N_t = 2$)

aligned. According to [30], the resulting radiation intensity of the optimal analog beamforming satisfies

$$U(\theta,\phi) \propto \left(\frac{1}{N_t}\frac{\sin(N_t\pi\varphi)}{\sin(\pi\varphi)}\right)^2 \triangleq g_{AP}(\theta,\phi), \tag{9.1}$$

where θ and ϕ represent the elevation angle and azimuth angle respectively, and $\varphi = (\cos\theta - \cos\theta_0)/2$ is the auxiliary variable. $g_{AP}(\theta,\phi)$ is the corresponding antenna pattern, as provided in Fig. 9.2. According to [29], when the antenna spacing $\Delta = \lambda/2$ and the optimal analog beamforming are applied, the directional antenna gain $g_A(\theta,\phi)$ is expressed by

$$g_A(\theta,\phi) = \frac{U(\theta,\phi)}{\frac{1}{4\pi}\int_0^{2\pi}\int_0^{\pi} U(\theta,\phi)\sin\theta d\theta d\phi} = \frac{1}{N_t}\left(\frac{\sin(N_t\pi\varphi)}{\sin(\pi\varphi)}\right)^2 = N_t g_{AP}(\theta,\phi). \tag{9.2}$$

For a ULA oriented along a straight line, the antenna gain is unaffected by the azimuth angle ϕ, therefore, simplifying the antenna gain in (9.2) to $g_A(\theta,\phi) = g_A(\theta) = N_t g_{AP}(\theta)$. In our considered network, to make the most of the directional antenna, the elevation angle θ_0 and the centre of the main lobe are optimized dynamically over time, i.e., $\theta_0 = \theta_0(t)$. Therefore, at any given time t, the antenna gain from the UAV for any elevation angle θ is represented by

$$g_A(\theta,\theta_0(t)) = N_t\left(\frac{\sin(N_t\pi\varphi(\theta,\theta_0(t)))}{N_t\sin(\pi\varphi(\theta,\theta_0(t)))}\right)^2 \triangleq N_t G_{AP}(\theta,\theta_0(t)), \tag{9.3}$$

where the auxiliary variable $\varphi(\theta,\theta_0(t))$ also becomes a function in t, i.e., $\varphi(\theta,\theta_0(t)) = (\cos\theta - \cos\theta_0(t))/2$ and $G_{AP}(\theta,\theta_0(t))$ denotes the antenna pattern when the directional antenna is designed for orientation in direction $\theta_0(t)$.

9.3.2 Channel Model

In UAV-enabled networks, air-to-ground links are more likely to maintain LoS paths, typically dominated by LoS components [3]. Accordingly, we use a free-space channel model to represent these characteristics accurately. At time point $t \in \mathcal{T}$ the channel gain from UAV to an SN k is represented by

$$g_{ch,k}(t) = \frac{\beta_0 N_t G_{AP}(\theta_k(t),\theta_0(t))}{(d_k(t))^2}, \tag{9.4}$$

where β_0 denotes the unit channel power gain with an isotropic antenna, $d_k(t)$ is the distance from UAV to SN k at time t, i.e.,

$$d_k(t) = \sqrt{(x(t) - w_{x,k})^2 + (y(t) - w_{y,k})^2 + H^2}, \tag{9.5}$$

and $\theta_k(t) \in [0, \frac{\pi}{2})$ is the elevation angle of the SN k with respect to the UAV at time t. In particular, we have

$$\cos(\theta_k(t)) = \frac{H}{d_k(t)}, \text{ i.e.,} \theta_k(t) = \arccos\left(\frac{H}{d_k(t)}\right). \tag{9.6}$$

Then, the received RF power of SN k at time t is obtained by

$$Q_k(t) = g_{\mathrm{ch},k}(t)P = \frac{\beta_0 P N_t G_{\mathrm{AP}}(\theta_k(t), \theta_0(t))}{(d_k(t))^2}, \tag{9.7}$$

where P represents the UAV transmission power.

9.3.3 Nonlinear Energy Harvesting

During the energy harvesting process, the received RF signal is first converted into a direct-current (DC) signal for the battery charging at SNs, which is inherently a non-linear process [24]. To characterize this process and make it analyzable, several works [20–22] consider simplified linear EH models. However, these models are often inaccurate and unrealistic. Notably, signal transmission experiences significant attenuation, and the large distances between UAVs and SNs result in lower received RF power levels, which leads us to employ a small signal model [31] to better characterize the diode non-linearity. According to [26], the output current I_{out} for harvesting can be characterized by a non-linear implicit function $I_{out}(Q)$ with respect to the received RF power Q. More specifically, as displayed in Eq. (6) in [26], we have

$$e^{\frac{R_L I_{out}(Q)}{n v_t}}(I_{out}(Q) + I_s) \approx \sum_{j=0}^{n_{o'}} \alpha_j Q^j, \tag{9.8}$$

where R_L, n, v_t represent the load resistance, ideality factor, and the reverse bias saturation current, respectively. Although $I_{out}(Q)$ is an implicit function of received RF power Q, an explicit expression for $I_{out}(Q)$ can still be obtained with the aid of the Lambert W function as follows

$$I_{out}(Q) \approx \frac{n v_t}{R_L} W_0\left(\frac{R_L}{n v_t} e^{\frac{R_L I_s}{n v_t}} \sum_{j=0}^{n_{o'}} \alpha_j Q^j\right) - I_s, \tag{9.9}$$

where W_0 is the principal branch of Lambert W function, which is the inverse relation of function $f(x) = xe^x$. Afterwards, the charged (harvested) power P_{ch} can be directly represented by the output current $I_{out}(Q)$, i.e., $P_{ch} = (I_{out}(Q))^2 R_L$, which implies that the charged power P_{ch} is also an implicit nonlinear function of received RF power Q, denoted by $\mathcal{F}_{nl}(Q)$. Therefore, at time t, the harvested power $P_{ch,k}(t)$ at SN k is denoted by

$$P_{ch,k}(t) = \mathcal{F}_{nl}(Q_k(t)) \triangleq (I_{out}(Q_k(t)))^2 R_L. \tag{9.10}$$

Convexity in nonlinear EH model: Although the charged power $P_{ch,k}(t)$ cannot be derived out in terms of elementary functions, we can apply the convexity of this non-linear process in $\mathcal{F}_{nl}$, which has been already proved in [26]. In particular, let $u(t) = \frac{\beta_0 P N_t}{Q_k(t)} = \frac{(d_k(t))^2}{G_{AP}(\theta_k(t), \theta_0(t))}$, and $Q_k(t)$ described in (9.7) can be reformulated as $Q_k(t) = Q_k^*(u) = \frac{\beta_0 P N_t}{u}$. Then, the first-order and second-order derivatives of $Q_k^*(u)$ are given by $\dot{Q}_k^*(u) = -\frac{\beta_0 P N_t}{u^2}$ and $\ddot{Q}_k^*(u) = \frac{2\beta_0 P N_t}{u^3}$, respectively. Therefore, the following inequality can be proved to hold

$$\ddot{Q}_k^*(u) Q_k^*(u) - \left(\dot{Q}_k^*(u)\right)^2 = \frac{2\beta_0 P N_t}{u^3} \cdot \frac{\beta_0 P N_t}{u} - \left(-\frac{\beta_0 P N_t}{u^2}\right)^2 = \frac{\beta_0^2 P^2 N_t^2}{u^4} > 0. \tag{9.11}$$

According to [26], satisfying the condition $\ddot{Q}_k^*(u) Q_k^*(u) - \left(\dot{Q}_k^*(u)\right)^2 \geq 0$ definitely makes the function $\mathcal{F}_{nl}(Q_k(t)) = \mathcal{F}_{nl}(Q_k^*(u))$ be convex with respect to u. In other words, for our considered network, $P_{ch,k}(t) = \mathcal{F}_{nl}(Q_k(t))$ is convex in u, namely convex in $\frac{(d_k(t))^2}{G_{AP}(\theta_k(t), \theta_0(t))}$.

9.4 Problem Statement and Reformulation

In this section, the original joint optimization problem is presented and reformulated as problem (P1) with assistance of the auxiliary variable. Subsequently, problem (P1) can be transformed into problem (P2) via trajectory discretization and approximating the antenna gain model, where problem (P2) will be addressed via an iterative algorithm in the next section.

9.4.1 Problem Statement

We assume that all SNs can continuously receive the wireless energy from the UAV. Therefore, the total harvested energy at SN k during the task time is given by

$$E_k(\{x(t), y(t), \theta(t)\}) = \int_0^T P_{ch,k}(t)dt = \int_0^T \mathcal{F}_{nl}(Q_k(t))dt, \tag{9.12}$$

where $Q_k(t)$ is given in (9.7) and is actually a function of $x(t)$, $y(t)$ and $\theta_0(t)$. To address the fairness issue, we aim at maximizing the minimal harvested energy among all SNs by designing the UAV trajectory $\{x(t), y(t)\}$ and the directional antenna orientation $\theta_0(t)$. Thus, the considered optimization problem can be formulated as

$$\begin{aligned} \text{(OP):} \quad & \max_{\{x(t),y(t),\theta_0(t)\}} \left\{ \min_{k\in\mathcal{K}} E_k(\{x(t), y(t), \theta_0(t)\}) \right\} \\ \text{s.t.} \quad & \text{C1: } \sqrt{(\dot{x}(t))^2 + (\dot{y}(t))^2} \le V, \forall t \in \mathcal{T}; \\ & \text{C2: } 0 \le \theta_0(t) < \frac{\pi}{2}, \forall t \in \mathcal{T}; \end{aligned} \tag{9.13}$$

Next, to address the max–min problem, we have introduced relaxation variables E to represent the $\min_{k\in\mathcal{K}} E_k(\{x(t), y(t), \theta_0(t)\},,$, therefore converting the problem (OP) to problem (P1) as:

$$\begin{aligned} \text{(P1):} \quad & \max_{\{x(t),y(t),\theta_0(t)\},E} E \\ \text{s.t.} \quad & C1\text{: } E_k(\{x(t), y(t), \theta_0(t)\}) \ge E, \forall k \in \mathcal{K}; \\ & \text{C2: } \sqrt{(\dot{x}(t))^2 + (\dot{y}(t))^2} \le V, \forall t \in \mathcal{T}; \\ & \text{C3: } 0 \le \theta_0(t) < \frac{\pi}{2}, \forall t \in \mathcal{T}; \end{aligned} \tag{9.14}$$

However, problem (P1) still involves an infinite number of variables to address due to the continuous time and space scale, which complicates the analysis significantly. Additionally, the implicit complexity in $\mathcal{F}_{nl}$ and multimodal behavior of the antenna gain function in (9.3) make it difficult to identify any concavity properties $E_k(\{x(t), y(t), \theta_0(t)\})$, rendering problem (P1) non-convex. As a result, efficient solutions through convex optimization techniques are not feasible. To overcome these challenges, we will apply appropriate approximations and reformulate the problem in the following subsection.

9.4.2 Problem Reformulation

Time Quantization on UAV Trajectory

To address the issue of optimizing a continuous UAV trajectory with infinitely many variables, we divide the trajectory interval into N time slots of sufficiently small lengths $dT = T/N$. Thus, in each time slot, it allows us to treat UAV's position as static and the directional antenna orientation as constant. However, a smaller resolution dT increases the number of time slots N, creating a trade-off between approximation accuracy and problem complexity.

Furthermore, during time slot $n \in \mathcal{N} \triangleq \{1, ..., N\}$, let $(x[n], y[n])$ represent the fixed position of the UAV and $\theta_0[n]$ represent the fixed orientation of the directional antenna. Consequently, the UAV speed constraint can be expressed as

$$(x[n+1] - x[n])^2 + (y[n+1] - y[n])^2 \leq V^2 dT^2, \forall n \in \mathcal{N}, n \neq N, \tag{9.15}$$

where the maximum movement permitted in a single time slot is VdT.

Antenna Pattern Approximation

Since the antenna pattern presented in features (9.3) sine functions in the numerator and the denominator, it is multimodal and challenging for performance analysis [28]. Therefore, the authors in [28] propose a cosine approximation, shown in Fig. 9.3, where the main lobe of the antenna pattern is modelled by the square of a cosine function, ignoring the side lobes. However, in our current analysis, it has been established that the charged power $P_{ch,k}(t)$ exhibits convex behavior in relation to $\frac{(d_k(t))^2}{G_A(\theta_k(t), \theta_0(t))}$, which approaches infinity when the gain is zero. This characteristic indicates that the approximation method proposed in [28] cannot be directly utilized in our context. To harness the benefits of convexity and consider both the main lobe and the side lobes, a modified cosine antenna pattern approximation is provided by us as following:

$$G_{AP-mc}(\theta_k(t), \theta_0(t)) = \mathcal{F}_A(\varphi(\theta_k(t), \theta_0(t)))$$
$$= \begin{cases} \cos^2\left(\frac{N_t \pi}{2} \varphi(\theta_k(t), \theta_0(t))\right), & \text{when } |\varphi(\theta_k(t), \theta_0(t))| < \frac{\delta}{N_t}; \\ \frac{\cos^2(\frac{\delta\pi}{2})}{\sqrt{2N_t \pi \tan(\frac{\delta\pi}{2})(|\varphi(\theta_k(t) - \theta_0(t))| - \frac{\delta}{N_t}) + 1}}, & \text{when } |\varphi(\theta_k(t), \theta_0(t))| \geq \frac{\delta}{N_t}, \end{cases} \tag{9.16}$$

where $\delta \in (0, 1)$ is the modification factor. Note that function $\mathcal{F}_{\rm A}$ is a continuous function. Combining (9.3) and (9.16), we can update the antenna gain as

$$G_{A-mc}(\theta_k[n], \theta_0[n]) = N_t G_{AP-mc}(\theta_k[n], \theta_0[n]) = N_t \mathcal{F}_A(\varphi_k[n]), \tag{9.17}$$

where $\theta_k[n]$ represents the elevation angle of SN k to the UAV in time slot n and is given by

$$\theta_k[n] = \arccos\left(\frac{H}{d_k[n]}\right), \tag{9.18}$$

$$d_k[n] = \sqrt{(x[n] - w_{x,k})^2 + (y[n] - w_{y,k})^2 + H^2}. \tag{9.19}$$

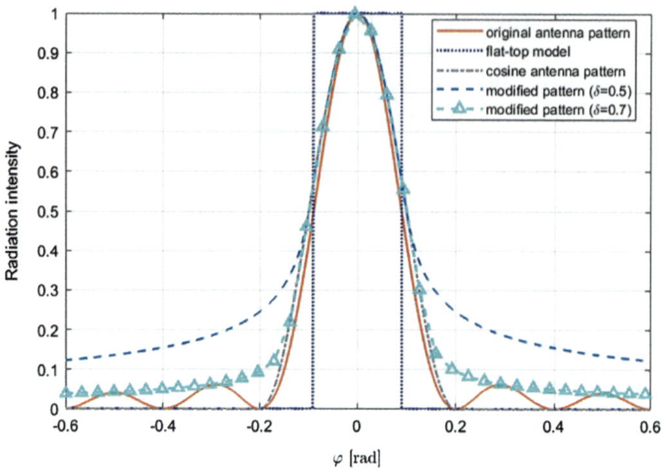

Fig. 9.3 Comparison of different antenna pattern approximation ($N_t = 5$)

Accordingly, we denote $\varphi_k[n] = \varphi(\theta_k[n], \theta_0[n]) = \frac{1}{2}(\frac{H}{d_k[n]} - \cos\theta_0[n])$ as the auxiliary variable. Since the variable $\theta_0[n]$ for directional antenna orientation only appears in $\varphi_k[n]$ within a term of $\cos\theta_0[n]$, we substitute the variables $\theta_0[n]$ with a new variable $\mu[n] = \cos\theta_0[n] \in (0, 1]$ for simplification and reformulate $\varphi_k[n]$ as

$$\varphi_k[n] = \frac{1}{2}\left(\frac{H}{d_k[n]} - \mu[n]\right). \tag{9.20}$$

As a result, the accumulated harvested energy at SN k within all N time slots are denoted by

$$E_k(\{x[n], y[n], \mu[n]\}) = \sum_{n=1}^{N} \mathcal{F}_{nl}\left(\frac{\beta_0 P N_t \mathcal{F}_A(\varphi_k[n])}{d_k[n]^2}\right) dT. \tag{9.21}$$

Reformulated Problem

With the above efforts of trajectory discretization and modified antenna pattern, we finally reformulate problem (P1) as

$$
\begin{aligned}
\text{(P2):} \quad & \max_{\{x(t),y(t),\mu[n]\},E} \quad E \\
\text{s.t.} \quad & C1: \sum_{n=1}^{N} \mathcal{F}_{nl}\left(\frac{\beta_0 P N_t \mathcal{F}_A(\varphi_k[n])}{d_k[n]^2}\right) dT \geq E, \forall k \in \mathcal{K}; \\
& \text{C2}: \quad (x[n+1]-x[n])^2 + (y[n+1]-y[n])^2 \leq V^2 dT^2, \forall n \in \mathcal{N}, n \neq N; \\
& \text{C3}: \quad 0 < \mu[n] \leq 1, \forall n \in \mathcal{N};
\end{aligned} \tag{9.22}
$$

Since the constraint in C1 in (9.22) is non-convex, problem (P2) also remains non-convex and cannot be solved optimally. In the next section, we will apply a tight convex approximation to enable efficient resolution through an iterative approach.

9.5 Proposed Joint Optimization Algorithm

In this section, we first construct a tight convex approximation for the reformulated problem (P2) and then propose an iterative algorithm, along with an analysis of its complexity.

9.5.1 Convex Approximation

As we mentioned above, to create a convex approximation of problem (P2), we need to approximate the constraint C1 in (9.22) to a convex form, specifically by identifying a concave function $P_{k,n}^{(r)}$ for the harvested power $\mathcal{F}_{nl}\left(\frac{\beta_0 P N_t \mathcal{F}_A(\varphi_k[n])}{d_k[n]^2}\right)$, so that $\mathcal{F}_{nl}\left(\frac{\beta_0 P N_t \mathcal{F}_A(\varphi_k[n])}{d_k[n]^2}\right) \geq P_{k,n}^{(r)}(x[n], y[n], \mu[n])$ and the equality holds at a local point $(x^{(r)}[n], y^{(r)}[n], \mu^{(r)}[n])$.

On the other hand, the function $\mathcal{F}_{nl}\left(\frac{\beta_0 P N_t \mathcal{F}_A(\varphi_k[n])}{d_k[n]^2}\right)$ is convex with respect to $\frac{d_k[n]^2}{\mathcal{F}_A(\varphi_k[n])}$, proved in Sect. 9.3.3. Therefore, for any pair of k and n, we can obtain an inequality

$$
\mathcal{F}_{nl}\left(\frac{\beta_0 P N_t \mathcal{F}_A(\varphi_k[n])}{d_k[n]^2}\right) \geq -A_{1,k,n}^{(r)}\left(\frac{d_k[n]^2}{\mathcal{F}_A(\varphi_k[n])}\right) + A_{2,k,n}^{(r)}, \tag{9.23}
$$

where the positive constants $A_{1,k,n}^{(r)}$ and $A_{2,k,n}^{(r)}$ are given by

$$
A_{1,k,n}^{(r)} = \mathcal{F}_{nl'}\left(\frac{\beta_0 P N_t \mathcal{F}_A(\varphi_k^{(r)}[n])}{d_k^{(r)}[n]^2}\right) \frac{\beta_0 P N_t d_k^{(r)}[n]^4}{\left(\mathcal{F}_A(\varphi_k^{(r)}[n])\right)^2}, \tag{9.24}
$$

$$A_{2,k,n}^{(r)} = \mathcal{F}_{nl}\left(\frac{\beta_0 P N_t \mathcal{F}_A(\varphi_k^{(r)}[n])}{d_k^{(r)}[n]^2}\right) + A_{1,k,n}^{(r)}\left(\frac{d_k^{(r)}[n]^2}{\mathcal{F}_A(\varphi_k^{(r)}[n])}\right), \tag{9.25}$$

and the terms $d_k^{(r)}[n]$ and $\varphi_k^{(r)}[n]$ are respectively the corresponding values of $d_k[n]$ and $\varphi_k[n]$ at local point $(x^{(r)}[n], y^{(r)}[n], \mu^{(r)}[n])$. Note that $-A_{1,k,n}^{(r)}$ is the first derivative of $\mathcal{F}_{\text{nl}}\left(\frac{\beta_0 P N_t \mathcal{F}_A(\varphi_k[n])}{d_k[n]^2}\right)$ to $\frac{d_k[n]^2}{\mathcal{F}_A(\varphi_k[n])}$ at $(x^{(r)}[n], y^{(r)}[n], \mu^{(r)}[n])$.

Given the non-convexity of the function $\frac{d_k[n]^2}{\mathcal{F}_A(\varphi_k[n])}$, we are led to perform an additional convex approximation to (9.23). According to the inequality of arithmetic and geometric means, we can derive from (9.23) the following expression as

$$\begin{aligned}
&\mathcal{F}_{\text{nl}}\left(\frac{\beta_0 P N_t \mathcal{F}_A(\varphi_k[n])}{d_k[n]^2}\right) \geq -A_{1,k,n}^{(r)}\left(\frac{d_k[n]^2}{\mathcal{F}_A(\varphi_k[n])}\right) + A_{2,k,n}^{(r)} \\
&= -A_{1,k,n}^{(r)}\frac{1}{F_{1,k,n}^{(r)}}\underline{F_{1,k,n}^{(r)} d_k[n]^2} \cdot \underline{\frac{1}{\mathcal{F}_A(\varphi_k[n])}} + A_{2,k,n}^{(r)} \\
&\geq -A_{1,k,n}^{(r)}\frac{1}{F_{1,k,n}^{(r)}}\underline{\frac{1}{2}\left(F_{1,k,n}^{(r)} d_k[n]^2\right)^2} - A_{1,k,n}^{(r)}\frac{1}{F_{1,k,n}^{(r)}}\underline{\frac{1}{2(\mathcal{F}_A(\varphi_k[n]))^2}} + A_{2,k,n}^{(r)} \\
&= -A_{1,k,n}^{(r)}\frac{F_{1,k,n}^{(r)} d_k[n]^4}{2} - A_{1,k,n}^{(r)}\frac{1}{2F_{1,k,n}^{(r)}(\mathcal{F}_A(\varphi_k[n]))^2} + A_{2,k,n}^{(r)},
\end{aligned} \tag{9.26}$$

where the underlined terms correspond to the terms in the applied inequality, and the positive constant $F_{1,k,n}^{(r)}$ is defined as

$$F_{1,k,n}^{(r)} = \frac{1}{\mathcal{F}_A(\varphi_k^{(r)}[n])}\frac{1}{d_k^{(r)}[n]^2}, \tag{9.27}$$

which guarantees the equality in (9.26) at the local point $(x^{(r)}[n], y^{(r)}[n], \mu^{(r)}[n])$. Subsequently, we find that the concave approximation $P_{k,n}^{(r)}$ can be completed when a tight upper-bound convex approximation of $\frac{1}{(\mathcal{F}_A(\varphi_k[n]))^2}$ can be obtained via Lemma 1 in our previous work [19]. Then, we focus on the function $\frac{1}{(\mathcal{F}_A(\varphi_k[n]))^2}$. According to the definition in (9.16), we have

$$\frac{1}{(\mathcal{F}_A(\varphi_k[n]))^2} = \begin{cases} \frac{1}{\cos^4(\frac{N_t\pi}{2}\varphi_k[n])}, & |\varphi_k[n]| < \frac{\delta}{N_t}, \\ \frac{2N_t\pi\tan(\frac{\delta\pi}{2})(|\varphi_k[n]| - \frac{\delta}{N_t}) + 1}{\cos^4(\frac{\delta\pi}{2})}, & |\varphi_k[n]| \geq \frac{\delta}{N_t}. \end{cases} \tag{9.28}$$

Function $\frac{1}{(\mathcal{F}_A(\varphi_k[n]))^2}$ is clearly continuous, and its first derivative to $\varphi_k[n]$ is given by

$$\left(\frac{1}{(\mathcal{F}_A(\varphi_k[n]))^2}\right)'_{\varphi_k[n]} = \frac{-2\mathcal{F}_{A'}(\varphi_k[n])}{(\mathcal{F}_A(\varphi_k[n]))^3} = \begin{cases} \frac{2N_t\pi\sin(\frac{N_t\pi}{2}\varphi_k[n])}{\cos^5(\frac{N_t\pi}{2}\varphi_k[n])}, & |\varphi_k[n]| < \frac{\delta}{N_t}, \\ \frac{2N_t\pi\sin(\frac{\delta\pi}{2})}{\cos^5(\frac{\delta\pi}{2})}, & \varphi_k[n] \geq \frac{\delta}{N_t}, \\ -\frac{2N_t\pi\sin(\frac{\delta\pi}{2})}{\cos^5(\frac{\delta\pi}{2})}, & \varphi_k[n] \leq -\frac{\delta}{N_t}, \end{cases} \tag{9.29}$$

which is apparently continuous. The second-order derivative of $\frac{1}{(\mathcal{F}_A(\varphi_k[n]))^2}$ to $\varphi_k[n]$ is

$$\left(\frac{1}{(\mathcal{F}_A(\varphi_k[n]))^2}\right)''_{\varphi_k[n]} = \begin{cases} \frac{N_t^2\pi^2 + 4N_t^2\pi^2\sin^2(\frac{N_t\pi}{2}\varphi_k[n])}{\cos^4(\frac{N_t\pi}{2}\varphi_k[n])}, & |\varphi_k[n]| < \frac{\delta}{N_t}, \\ 0, & |\varphi_k[n]| > \frac{\delta}{N_t}. \end{cases} \tag{9.30}$$

Thus, we have $\left(\frac{1}{(\mathcal{F}_A(\varphi_k[n]))^2}\right)''_{\varphi_k[n]} < \frac{5N_t^2\pi^2}{\cos^4(\frac{\delta\pi}{2})}$. According to Lemma 1 in our previous work [19], we can obtain that

$$\frac{1}{(\mathcal{F}_A(\varphi_k[n]))^2} \leq B_{1,k,n}^{(r)}(\varphi_k[n])^2 + B_{2,k,n}^{(r)}(\varphi_k[n]) + B_{3,k,n}^{(r)}, \tag{9.31}$$

where the constants $B_{1,k,n}^{(r)}$, $B_{2,k,n}^{(r)}$ and $B_{3,k,n}^{(r)}$ are formulated as

$$B_{1,k,n}^{(r)} = \frac{5N_t^2\pi^2}{2\cos^4(\frac{\delta\pi}{2})} > 0, \tag{9.32}$$

$$B_{2,k,n}^{(r)} = \frac{-2\mathcal{F}_{A'}(\varphi_k^{(r)}[n])}{\left(\mathcal{F}_A(\varphi_k^{(r)}[n])\right)^3} - \frac{5N_t^2\pi^2}{\cos^4(\frac{\delta\pi}{2})}\varphi_k^{(r)}[n], \tag{9.33}$$

$$B_{3,k,n}^{(r)} = \frac{1}{\left(\mathcal{F}_A(\varphi_k^{(r)}[n])\right)^2} + \frac{2\mathcal{F}_{A'}(\varphi_k^{(r)}[n])\varphi_k^{(r)}[n]}{\left(\mathcal{F}_A(\varphi_k^{(r)}[n])\right)^3} + \frac{5N_t^2\pi^2}{2\cos^4(\frac{\delta\pi}{2})}(\varphi_k^{(r)}[n])^2. \tag{9.34}$$

By combining (9.20) with the approximation in (9.31), we can obtain

$$\begin{aligned} \frac{1}{(\mathcal{F}_A(\varphi_k[n]))^2} &\leq \frac{B_{1,k,n}^{(r)}}{4}\left(\frac{H}{d_k[n]} - \mu[n]\right)^2 + \frac{B_{2,k,n}^{(r)}}{2}\left(\frac{H}{d_k[n]} - \mu[n]\right) + B_{3,k,n}^{(r)} \\ &= \frac{B_{1,k,n}^{(r)}}{4}\frac{H^2}{d_k[n]^2} + \frac{B_{1,k,n}^{(r)}}{4}\mu[n]^2 + \frac{B_{1,k,n}^{(r)}}{4}(3 - 2\mu[n])\frac{H}{d_k[n]} \\ &\quad + \left(\frac{B_{2,k,n}^{(r)}}{2} - \frac{3B_{1,k,n}^{(r)}}{4}\right)\frac{H}{d_k[n]} - \frac{B_{2,k,n}^{(r)}}{2}\mu[n] + B_{3,k,n}^{(r)}. \end{aligned} \tag{9.35}$$

Note that $B^{(r)}_{1,k,n} > 0$ and that $\mu[n] \in (0, 1]$ makes $3 - 2\mu[n] > 0$ hold. Therefore, by defining a positive constant $F^{(r)}_{2,k,n} = \frac{H}{d_k^{(r)}[n]} \cdot \frac{1}{3-2\mu^{(r)}[n]} > 0$, based on the inequality of arithmetic and geometric means, we have the approximation as (9.36)

$$\begin{aligned}
\frac{1}{(\mathcal{F}_A(\varphi_k[n]))^2} &\le \frac{B^{(r)}_{1,k,n}}{4}\frac{H^2}{d_k[n]^2} + \left(\frac{B^{(r)}_{2,k,n}}{2} - \frac{3B^{(r)}_{1,k,n}}{4}\right)\frac{H}{d_k[n]} \\
&+ \frac{B^{(r)}_{1,k,n}}{4F^{(r)}_{2,k,n}} \underline{\underline{F^{(r)}_{2,k,n}(3-2\mu[n]) \cdot \frac{H}{d_k[n]}}} + \frac{B^{(r)}_{1,k,n}}{4}\mu[n]^2 - \frac{B^{(r)}_{2,k,n}}{2}\mu[n] + B^{(r)}_{3,k,n} \\
&\le \underbrace{\frac{B^{(r)}_{1,k,n}}{4}\frac{H^2}{d_k[n]^2} + \left(\frac{B^{(r)}_{2,k,n}}{2} - \frac{3B^{(r)}_{1,k,n}}{4}\right)\frac{H}{d_k[n]} + \frac{B^{(r)}_{1,k,n}}{4F^{(r)}_{2,k,n}}\underline{\underline{\frac{H^2}{2d_k[n]^2}}}}_{h^{(r)}_{1,k,n}(d_k[n])} \\
&+ \underbrace{\frac{B^{(r)}_{1,k,n}}{4F^{(r)}_{2,k,n}}\underline{\underline{\frac{\left(F^{(r)}_{2,k,n}(3-2\mu[n])\right)^2}{2}}} + \frac{B^{(r)}_{1,k,n}}{4}\mu[n]^2 - \frac{B^{(r)}_{2,k,n}}{2}\mu[n] + B^{(r)}_{3,k,n}}_{h^{(r)}_{2,k,n}(\mu[n])} \\
&\triangleq h^{(r)}_{1,k,n}(d_k[n]) + h^{(r)}_{2,k,n}(\mu[n]).
\end{aligned} \tag{9.36}$$

The equality in (9.36) holds at the local point $(x^{(r)}[n], y^{(r)}[n], \mu^{(r)}[n])$ with the aid of $F^{(r)}_{2,k,n}$. We can observe that the function $h^{(r)}_{1,k,n}(d_k[n])$ and $h^{(r)}_{2,k,n}(\mu[n])$ exhibits convexity in $d_k[n]$ and $\mu[n]$, respectively. However, $h^{(r)}_{1,k,n}(d_k[n])$ does not exhibit joint convexity in both $x[n]$ and $y[n]$, as shown in (9.19). Thus, our next step is to focus on the function $h^{(r)}_{1,k,n}(d_k[n])$, reformulated as

$$h^{(r)}_{1,k,n}(d_k[n]) = \frac{B^{(r)}_{1,k,n}}{4}\left(1 + \frac{1}{2F^{(r)}_{2,k,n}}\right)\frac{H^2}{d_k[n]^2} + \left(\frac{B^{(r)}_{2,k,n}}{2} - \frac{3B^{(r)}_{1,k,n}}{4}\right)\frac{H}{d_k[n]}. \tag{9.37}$$

We consider the term $(x[n] - w_{x,k})^2 + (y[n] - w_{y,k})^2$ in $h^{(r)}_{1,k,n}$ as a variable γ and define a function $h_0(\gamma) = \frac{B^{(r)}_{1,k,n}}{4}\left(1 + \frac{1}{2F^{(r)}_{2,k,n}}\right)\frac{H^2}{\gamma+H^2} + \left(\frac{B^{(r)}_{2,k,n}}{2} - \frac{3B^{(r)}_{1,k,n}}{4}\right)\frac{H}{\sqrt{\gamma+H^2}}$ $(\gamma \ge 0)$. With a single variable γ, we can readily prove that its second-order derivative meets the following condition

$$\begin{aligned}
h_{0''}(\gamma) &= \frac{B^{(r)}_{1,k,n}}{4}\left(1 + \frac{1}{2F^{(r)}_{2,k,n}}\right)\frac{2H^2}{(\gamma+H^2)^3} + \left(\frac{B^{(r)}_{2,k,n}}{2} - \frac{3B^{(r)}_{1,k,n}}{4}\right)\frac{3H}{4(\gamma+H^2)^{\frac{5}{2}}} \\
&\le \frac{B^{(r)}_{1,k,n}}{4}\left(1 + \frac{1}{2F^{(r)}_{2,k,n}}\right)\frac{2}{H^4} + \left|\frac{B^{(r)}_{2,k,n}}{2} - \frac{3B^{(r)}_{1,k,n}}{4}\right|\frac{3}{4H^4} \triangleq C^{(r)}_{1,k,n}, \ \forall \gamma \ge 0.
\end{aligned} \tag{9.38}$$

Since $C_{1,k,n}^{(r)} > 0$, according to Lemma 1 in work [19], we can get

$$\begin{aligned} h_{1,k,n}^{(r)}(d_k[n]) &= h_0((x[n]-w_{x,k})^2+(y[n]-w_{y,k})^2) \\ &\leq \frac{C_{1,k,n}^{(r)}}{2}((x[n]-w_{x,k})^2+(y[n]-w_{y,k})^2)^2 \\ &+ C_{2,k,n}^{(r)}((x[n]-w_{x,k})^2+(y[n]-w_{y,k})^2)+C_{3,k,n}^{(r)} \\ &\triangleq h_{3,k,n}^{(r)}(x[n],y[n]), \end{aligned} \tag{9.39}$$

where $C_{2,k,n}^{(r)}$ and $C_{3,k,n}^{(r)}$ are obtained as

$$C_{2,k,n}^{(r)} = h_{0'}(\gamma_0^{(r)}) - C_{1,k,n}^{(r)}\gamma_0^{(r)}, \tag{9.40}$$

$$C_{3,k,n}^{(r)} = h_0(\gamma_0^{(r)}) - \frac{C_{1,k,n}^{(r)}}{2}(\gamma_0^{(r)})^2 - C_{2,k,n}^{(r)}\gamma_0^{(r)}, \tag{9.41}$$

$$\gamma_0^{(r)} = (x^{(r)}[n]-w_{x,k})^2+(y^{(r)}[n]-w_{y,k})^2. \tag{9.42}$$

The constant $C_{2,k,n}^{(r)}$ in the function $h_{3,k,n}^{(r)}(x[n],y[n])$ will affect its convexity based on its sign. Thus, Therefore, we examine two cases, i.e., case $C_{2,k,n}^{(r)} \geq 0$ and case $C_{2,k,n}^{(r)} < 0$, respectively.

(1) **Case** $C_{2,k,n}^{(r)} \geq 0$: the function $h_{3,k,n}^{(r)}(x[n],y[n])$ will be convex with respect to $x[n]$ and $y[n]$. In this case, we define the approximation for $h_{1,k,n}^{(r)}(d_k[n])$ as $h_{4,k,n}^{(r)}(x[n],y[n]) \triangleq h_{3,k,n}^{(r)}(x[n],y[n])$.
(2) **Case** $C_{2,k,n}^{(r)} < 0$: the term $C_{2,k,n}^{(r)}((x[n]-w_{x,k})^2+(y[n]-w_{y,k})^2)$ will be concave in $x[n]$ and $y[n]$. Thus, we have

$$\begin{aligned} h_{1,k,n}^{(r)}(d_k[n]) \leq h_{3,k,n}^{(r)}(x[n],y[n]) &\leq \frac{C_{1,k,n}^{(r)}}{2}((x[n]-w_{x,k})^2+(y[n]-w_{y,k})^2)^2 \\ &+ C_{2,k,n}^{(r)}(x^{(r)}[n]-w_{x,k})(2x[n]-x^{(r)}[n]-w_{x,k}) \\ &+ C_{2,k,n}^{(r)}(y^{(r)}[n]-w_{y,k})(2y[n]-y^{(r)}[n]-w_{y,k})+C_{3,k,n}^{(r)} \\ &\triangleq h_{4,k,n}^{(r)}(x[n],y[n]), \end{aligned} \tag{9.43}$$

where $h_{4,k,n}^{(r)}(x[n],y[n])$ is clearly a convex function in $x[n]$ and $y[n]$. Combining (9.26) and (9.43), we can obtain a convex approximation for function $\frac{1}{(\mathcal{F}_A(\varphi_k[n]))^2}$

$$\frac{1}{(\mathcal{F}_A(\varphi_k[n]))^2} \leq h_{1,k,n}^{(r)}(d_k[n]) + h_{2,k,n}^{(r)}(\mu[n]) \leq h_{4,k,n}^{(r)}(x[n],y[n]) + h_{2,k,n}^{(r)}(\mu[n]). \tag{9.44}$$

By combining (9.44) with the inequality in (9.26), we have the final concave approximation for harvested power as

$$\begin{aligned}
\mathcal{F}_{\text{nl}}\left(\frac{\beta_0 PN_t f_A(\varphi_k[n])}{d_k[n]^2}\right) &\geq -A_{1,k,n}^{(r)}\frac{F_{1,k,n}^{(r)} d_k[n]^4}{2} - A_{1,k,n}^{(r)}\frac{1}{2F_{1,k,n}^{(r)}(\mathcal{F}_A(\varphi_k[n]))^2} + A_{2,k,n}^{(r)} \\
&\geq -A_{1,k,n}^{(r)}\frac{F_{1,k,n}^{(r)} d_k[n]^4}{2} - A_{1,k,n}^{(r)}\frac{h_{4,k,n}^{(r)}(x[n], y[n]) + h_{2,k,n}^{(r)}(\mu[n])}{2F_{1,k,n}^{(r)}} + A_{2,k,n}^{(r)} \\
&\triangleq P_{k,n}^{(r)}(x[n], y[n], \mu[n]).
\end{aligned} \tag{9.45}$$

Since $P_{k,n}^{(r)}(x[n], y[n], \mu[n])$ is concave and the inequality holds for any feasible point $(x[n], y[n], \mu[n])$, we can conclude that equality in (9.45) must hold at the local point $(x^{(r)}[n], y^{(r)}[n], \mu^{(r)}[n])$ with the above approximations.

Therefore, we have established a tight concave approximation for harvested power. By substituting the harvested power with its concave approximation, we can derive a convex approximation (P3) for problem (P2).

$$\begin{aligned}
\text{(P3)}: \quad & \max_{\{x[n], y[n], \mu[n]\}, E} \quad E \\
\text{s.t.}\ \ & \text{C1}: \sum_{n=1}^{N} P_{k,n}^{(r)}(x[n], y[n], \mu[n]) dT \geq E, \forall k \in \mathcal{K}; \\
& \text{C2, C3 in (P2);}
\end{aligned} \tag{9.46}$$

Since the problem (P3) is convex, an optimal solution can be efficiently obtained using convex optimization tools like CVX and the ellipsoid method. With the convex approximation of (P3), problem (P2) is then enabled to be solved iteratively. In the following subsection, we will outline our proposed iterative algorithm for addressing problem (P2).

9.5.2 Proposed Iterative Solution

In the initialization step, we establish a feasible local point $(x^{(0)}[n], y^{(0)}[n], \mu^{(0)}[n])$. Then, in each iteration r, we construct the concave approximation $P_{k,n}^{(r)}(x[n], y[n], \mu[n])$ for the harvested power using the local point $(x^{(r)}[n], y^{(r)}[n], \mu^{(r)}[n])$, according to (9.45), which allows us to solve the convex problem (P3). Once we optimally solve problem (P3), the resulting solution serves as the local point for the next iteration. By continuing this iterative process, we will consistently enhance the objective for problem (P2).

Note that the tight approximation in (9.45), which ensures zero approximation error at the local point, guarantees that the objective value for problem (P2) improves with each iteration. Specifically, during iteration r, the relation holds at the local point:

$$\begin{aligned}\min_k\left\{\sum_{n=1}^{N}\mathcal{F}_{nl}\left(\frac{\beta_0 PN_t\mathcal{F}_A(\varphi_k^{(r)}[n])}{d_k^{(r)}[n]^2}\right)dT\right\} &= \min_k\left\{\sum_{n=1}^{N}P_{k,n}^{(r)}(x^{(r)}[n],y^{(r)}[n],\mu^{(r)}[n])dT\right\}\\ &\le \min_k\left\{\sum_{n=1}^{N}P_{k,n}^{(r)}(x^{(r+1)}[n],y^{(r+1)}[n],\mu^{(r+1)}[n])dT\right\}\\ &\le \min_k\left\{\sum_{n=1}^{N}\mathcal{F}_{nl}\left(\frac{\beta_0 PN_t\mathcal{F}_A(\varphi_k^{(r+1)}[n])}{d_k^{(r+1)}[n]^2}\right)dT\right\}. \end{aligned} \quad (9.47)$$

Therefore, through repeated iterations, the minimum harvested energy among all SNs steadily increases. Since this minimum energy is upper-bounded, the solution cannot improve indefinitely and will eventually converge to a suboptimal point. Ultimately, from this solution, we can derive the jointly optimized UAV trajectory $\{x[n], y[n]\}$ and directional antenna orientation scheme $\{\theta_0[n]\}$ according to $\theta_0[n] = \arccos \mu[n], \forall n \in \mathcal{N}$.

9.6 Simulation Results

This section shows the simulation results of our proposed iterative algorithm aimed at addressing the fairness issue in energy harvesting among SNs. The minimum average harvested power during the charging period serves as a key performance indicator for the system, namely WPT performance.

We begin by evaluating the convergence of our proposed algorithm, which utilizes a modified cosine antenna pattern. Next, we compare the WPT performance against a benchmark design that employs an omnidirectional antenna, where only the UAV trajectory needs to be optimized, allowing for a similar iterative algorithm to be developed. Finally, we provide insights into the UAV trajectory and antenna behaviors across different algorithms in both low-density and high-density SN cases.

9.6.1 Algorithm Convergence Validation

To assist in choosing δ, we first investigate the convergence performance with different modification factors δ in Fig. 9.4a and relationship between estimated WPT performance and δ under different patterns in Fig. 9.4b. In Fig. 9.4a, we can observe that a smaller δ results in faster convergence. This is due to the fact that a small δ indicates a less accurate approximation of the directional antenna pattern compared

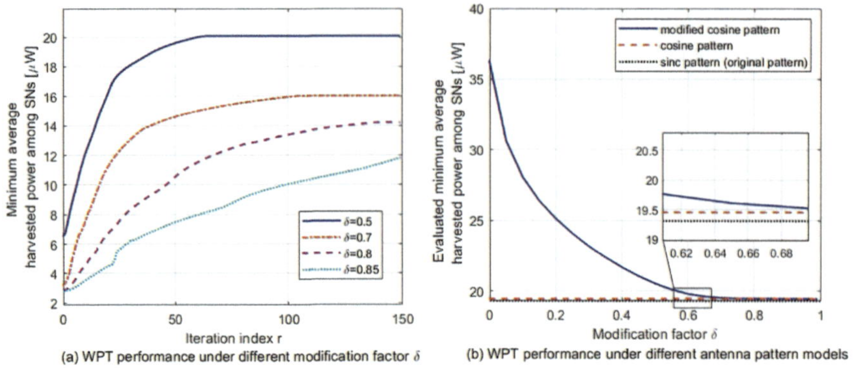

Fig. 9.4 Convergence performance and evaluated WPT performance under different setups and patterns

to the cosine antenna pattern, as we noted in Fig. 9.3. On the other hand, from the results in Fig. 9.4b, both the cosine antenna pattern [28] and our proposed modified cosine pattern exhibit small approximation errors (around 1%) compared to the accurate sinc pattern. Notably, the error for the cosine antenna pattern remains constant across different δ values, while the modified pattern's error decreases as δ increases, becoming significant at lower δ values. We conclude that when $\delta \geq 0.8$, the approximation error of the modified cosine pattern matches that of the original pattern. Thus, δ introduces a trade-off between convergence speed and model accuracy, requiring careful selection. It is important to note that a small δ is not without value. It allows for efficient initial optimization, providing a high-quality starting point. For the main iterations, a larger δ is preferred for accuracy. In the subsequent subsection evaluating our proposed joint design, we set $\delta = 0.5$ for initialization and $\delta = 0.9$ for the main iterations.

9.6.2 WPT Performance Comparison Between Different Algorithms

Then we compared the WPT performance with different charging duration T under different algorithms in Fig. 9.5, where the UAV maximum speed V is set differently. For both directional and omni-directional antennas, a large value of T can lead to a better WPT performance. The phenomenon of larger V is also similar. The improvement is attributed to the fact that a larger T allows the UAV to spend more time in positions with relatively higher WPT performance, potentially closer to the SNs. Additionally, a greater V enhances the UAV's mobility, resulting in less time spent flying and more time available for hovering at optimal locations. Furthermore, it is evident that the directional antenna case consistently performs better than the omni-directional antenna case. This advantage is primarily due to the increased

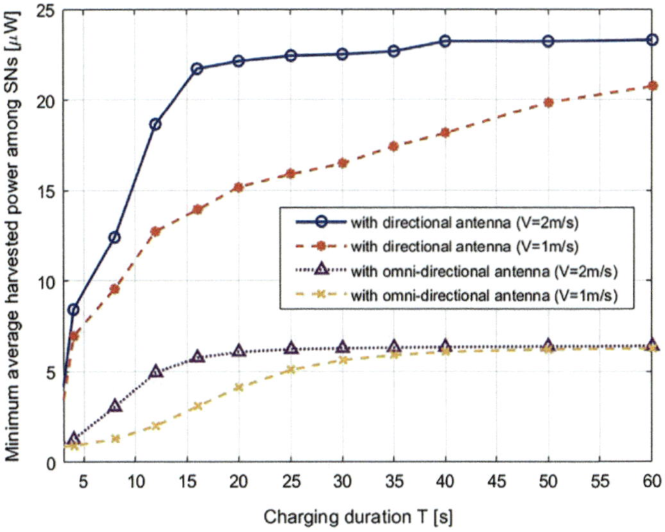

Fig. 9.5 WPT performance comparison with different charging duration

antenna gain from the directional antenna, which significantly offsets path loss during transmission, enabling more power to be wirelessly transferred through optimized directional antenna orientation. In summary, increasing either T or V can enhance WPT performance. Significantly, the enhancement in performance when increasing the UAV's speed or the charging period is more marked in scenarios utilizing directional antennas as opposed to omni-directional ones. This finding underscores the advantages of employing a more agile UAV and extending the charging cycle for optimizing the performance of a UAV-enabled WPT network that employs directional antennas.

Now, we investigate how the UAV's altitude H affects the WPT performance under different algorithms, as shown in Fig. 9.6. First, Fig. 9.6a focuses on a relatively lower SN density, while Fig. 9.6b addresses a relatively higher density. In both cases, an increased altitude leads to a decrease in the average power harvested by the SNs because of the increased path loss. To enhance the robustness in our simulations, when employing a larger number of K, we expand the scenario by adding additional random SNs based on the setup with fewer SNs. It is important to recognize that both the extent of coverage and the intensity of the signal significantly impact WPT performance. The directional antenna exhibits a narrower beamwidth with more antenna elements, leading to reduced coverage compared to the omni-directional antenna. This creates a dichotomy: the omni-directional antenna offers broader coverage, while the directional antenna provides stronger signal strength. Interestingly, as seen in Fig. 9.6a and b, the system with a directional antenna consistently achieves better performance than the omni-directional case for both scenarios. In the low-density case, the directional antenna can be directed toward different SNs in each time slot, resulting in superior WPT performance. In the high-density

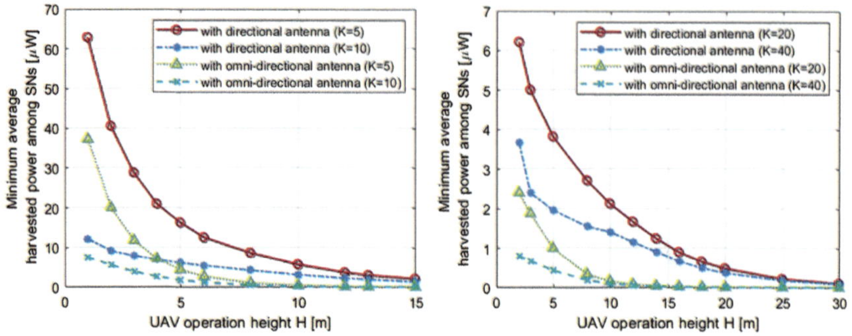

(a) WPT performance comparison in relatively lower SN density (b) WPT performance comparison in relatively higher SN density

Fig. 9.6 WPT performance comparison in different SN density cases

case, further investigation is required to understand why the directional antenna still surpasses the omni-directional antenna in performance, despite its limited coverage.

9.6.3 UAV Trajectory and Antenna Behaviors

As discussed above, the observations in Fig. 9.6 prompt us to explore the corresponding UAV trajectory and directional antenna behaviors across different SN density. This analysis aims to provide fundamental insights into the improvements in WPT performance compared to the benchmark. In the following, we will focus on the differences in UAV trajectories and directional antenna behaviors in two scenarios: a lower SN density (with $K = 5$) and a higher SN density (with $K = 40$).

Low-Density SNs

We start with the low-density SN case and focus on the UAV trajectory difference in Fig. 9.7 and directional antenna orientation observation in Fig. 9.8, where the charging durations are set as 20 and 50s. First, as shown in Fig. 9.7, we can observe the UAV with a directional antenna generally operates at a closer proximity to the SNs compared to a UAV with an omni-directional antenna. This distinction is attributed to the broadcasting characteristic of the omni-directional antenna, which disperses the signal across all directions, eliminating the need for the UAV to be in close proximity to certain SNs for performance enhancement. However, the directional antenna concentrates its energy in a specific direction, allowing for more efficient energy transfer when UAV flies near specific SNs at different time slot. This observation is further supported by Fig. 9.8a and b. With the increasing of charging period, the UAV equipped with a directional antenna gains more flexibility to fly closer to the SNs. When positioned directly above an SN, the optimized elevation angle

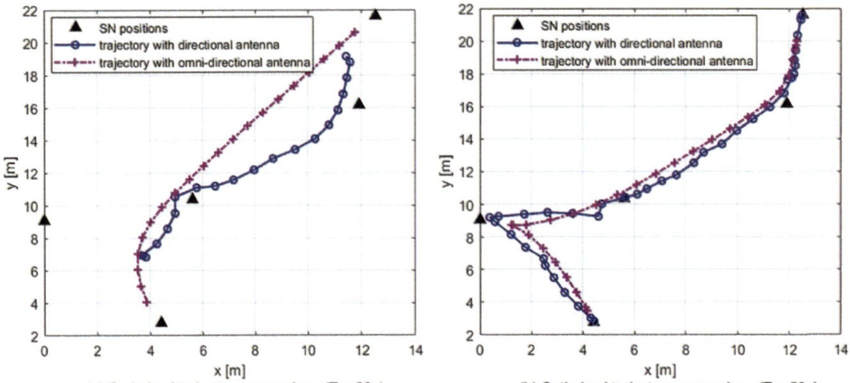

(a) Optimized trajectory comparison (T = 20s)

(b) Optimized trajectory comparison (T = 50s)

Fig. 9.7 UAV trajectory comparison in relatively lower SN density

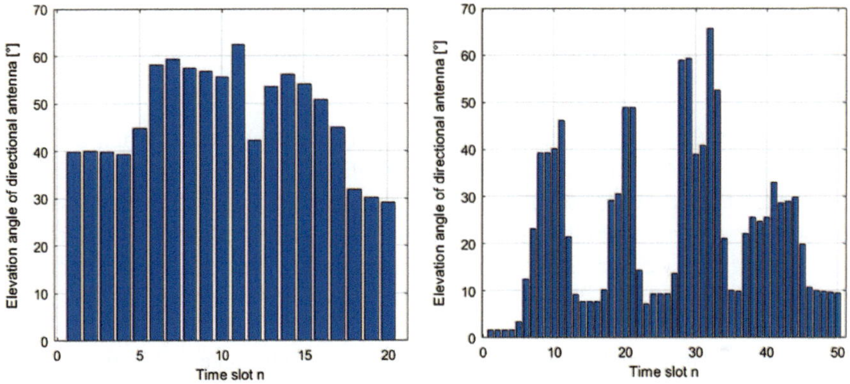

(a) Optimized elevation angle θ_0 (T=20s)

(b) Optimized elevation angle θ_0 (T=50s)

Fig. 9.8 Directional antenna orientation in relatively lower SN density

approaches zero, enabling efficient charging of that SN. However, as the distance between the UAV and SNs grows, the optimized elevation angle increases to provide broader coverage.

High-Density SNs

Then we move on to investigate the high-density SN case and focus on the UAV trajectory difference in Fig. 9.9 and directional antenna orientation observation in Fig. 9.10, where the UAV flying heights are set as 2 and 30 m. First, we can observe that the UAV trajectories vary only slightly with different heights in Fig. 9.9. Since we adopt convex optimization methods, the optimized UAV trajectory is primarily

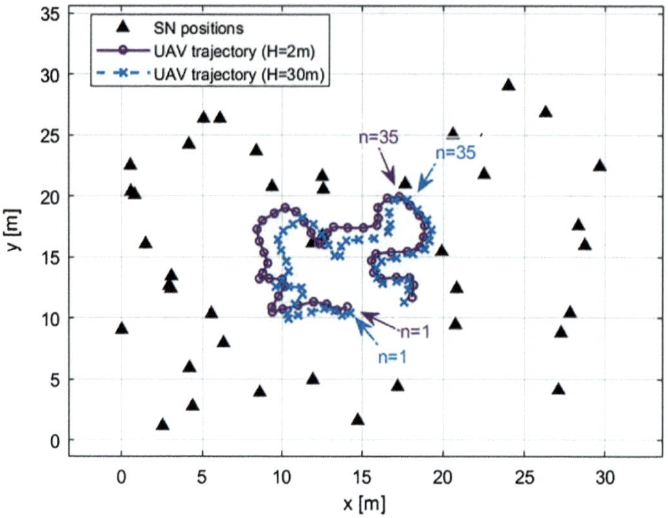

Fig. 9.9 UAV trajectory comparison in relatively higher SN density

influenced by the layout of the SNs and the initial feasible point. When the UAV operates at higher heights, the locations on its flight path tend to cluster more densely. Operating at a greater height allows the UAV to provide broader coverage with its directional antenna, making slight variations in position sufficient to ensure that the UAV trajectory covers all SNs. Additionally, we pay attention to the directional antenna behaviors in Fig. 9.10. As shown in Fig. 9.10a, when the UAV flies at the height of $H = 2$ m, the optimized elevation angles for the directional antenna alignment tend toward 90°. This alignment suggests that the antenna's primary beam is oriented nearly horizontally, enabling it to encompass a broader region and connect with a greater number of SNs. Conversely, as depicted in Fig. 9.10b, at a greater height $H = 30$ m, the optimized elevation angle is relatively smaller but adequate for achieving extensive coverage.

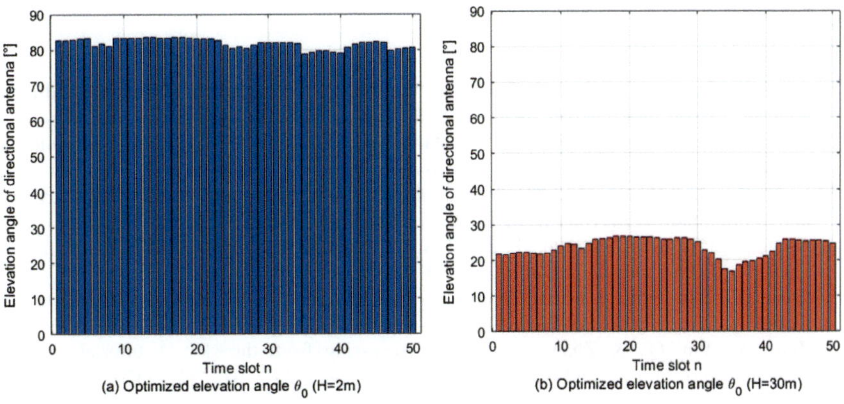

Fig. 9.10 Directional antenna orientation in relatively higher SN density

9.7 Conclusion

In this chapter, we have successfully addressed the optimization problem in UAV-supported WPT networks with complex antenna patterns and nonlinear energy harvesting models. Since the UAV employs a directional antenna array, we aim at maximizing the minimum harvested energy among all SNs by optimizing the UAV trajectory and the antenna elevation angle. By proposing an approximate antenna pattern model and quantizing the UAV trajectory, we transformed the original complex problem into a non-convex optimization problem. Subsequently, by constructing a concave lower-bound function for the harvested power, we can build a tight convex approximation for the reformulated problem that can be effectively tackled using convex optimization techniques. Our simulations confirmed the convergence and effectiveness of our proposed algorithm. The results demonstrated significant advantages of directional antennas over omni-directional ones in various SN densities and UAV heights.

We conclude by emphasizing the extensibility of our methodologies, highlighting the novel joint optimization of directional antenna orientation and UAV trajectory. This approach can be adapted to other joint design scenarios, including those with unknown SN positions or varying resource allocations. Additionally, while we focused on a 1D ULA, future work could expand to 2D and 3D antenna arrays, where the modified cosine pattern remains applicable.

References

1. Zeng, Y., Zhang, R., & Lim, T. J. (2016). Wireless communications with unmanned aerial vehicles: Opportunities and challenges. *IEEE Communications Magazine, 54*, 36–42.
2. Wu, Q., Zeng, Y., & Zhang, R. (2018). Joint trajectory and communication design for multi-uav enabled wireless networks. *IEEE Transactions on Wireless Communications, 17*, 2109–2121.
3. Li, B., Fei, Z., & Zhang, Y. (2019). UAV communications for 5g and beyond: Recent advances and future trends. *IEEE Internet of Things Journal, 6*, 2241–2263.
4. Gupta, L., Jain, R., & Vaszkun, G. (2016). Survey of important issues in UAV communication networks. *IEEE Communications Surveys & Tutorials, 18*, 1123–1152.
5. Zhang, G., Wu, Q., Cui, M., & Zhang, R. (2019). Securing UAV communications via joint trajectory and power control. *IEEE Transactions on Wireless Communications, 18*, 1376–1389.
6. Zeng, Y., Wu, Q., & Zhang, R. (2019). Accessing from the sky: A tutorial on UAV communications for 5G and beyond. *Proceedings of the IEEE, 107*, 2327–2375.
7. Meng, K., et al. (2024). UAV-enabled integrated sensing and communication: Opportunities and challenges. *IEEE Wireless Communications, 31*, 97–104.
8. Li, P., & Xu, J. (2020). Fundamental rate limits of UAV-enabled multiple access channel with trajectory optimization. *IEEE Transactions on Wireless Communications, 19*, 458–474.
9. Li, M., Li, H., Ma, P., & Wang, H. (2023). Energy maximization for ground nodes in UAV-enabled wireless power transfer systems. *IEEE Internet of Things Journal, 10*, 17096–17109.
10. Zeng, Y., & Zhang, R. (2017). Energy-efficient UAV communication with trajectory optimization. *IEEE Transactions on Wireless Communications, 16*, 3747–3760.
11. Zeng, Y., Xu, J., & Zhang, R. (2019). Energy minimization for wireless communication with rotary-wing UAV. *IEEE Transactions on Wireless Communications, 18*, 2329–2345.
12. Xu, J., Zeng, Y., & Zhang, R. (2018). UAV-enabled wireless power transfer: Trajectory design and energy optimization. *IEEE Transactions on Wireless Communications, 17*, 5092–5106.
13. Hu, Y., Yuan, X., Xu, J., & Schmeink, A. (2019). Optimal 1D trajectory design for UAV-enabled multiuser wireless power transfer. *IEEE Transactions on Communications, 67*, 5674–5688.
14. Hu, Y., Yuan, X., Yang, T., Clerckx, B., & Schmeink, A. (2020). On the convex properties of wireless power transfer with nonlinear energy harvesting. *IEEE Transactions on Vehicular Technology, 69*, 5672–5676.
15. Mo, X., Huang, Y., & Xu, J. (2020). Radio-map-based robust positioning optimization for UAV-enabled wireless power transfer. *IEEE Wireless Communications Letters, 9*(2), 179–183.
16. Yuan, X., Yang, T., Hu, Y., Xu, J., & Schmeink, A. (2021). Trajectory design for UAV-enabled multiuser wireless power transfer with nonlinear energy harvesting. *IEEE Transactions on Wireless Communications, 20*(2), 1105–1121.
17. Clerckx, B., Zhang, R., Schober, R., Ng, D. W. K., Kim, D. I., & Poor, H. V. (2019). Fundamentals of wireless information and power transfer: From RF energy harvester models to signal and system designs. *IEEE Journal on Selected Areas in Communications, 37*(1), 4–33.
18. Boshkovska, E., Ng, D. W. K., Zlatanov, N., & Schober, R. (2015). Practical non-linear energy harvesting model and resource allocation for SWIPT systems. *IEEE Communications Letters, 19*(12), 2082–2085.
19. Yuan, X., Hu, Y., & Schmeink, A. (2021). Joint design of UAV trajectory and directional antenna orientation in UAV-enabled wireless power transfer networks. *IEEE Journal on Selected Areas in Communications, 39*(10), 3081–3096.
20. Matsuda, T., Kaneko, M., Hiraguri, T., Nishimori, K., Kimura, T., & Nakao, A. (2020). Adaptive direction control for UAV full-duplex relay networks using multiple directional antennas. *IEEE Access, 8*, 85083–85093.
21. Li, S., Duo, B., Yuan, X., Liang, Y.-C., & Di Renzo, M. (2020). Reconfigurable intelligent surface assisted UAV communication: Joint trajectory design and passive beamforming. *IEEE Wireless Communications Letters, 9*(5), 716–720.
22. Dai, H., Ng, K., Wong, R.C., & Wu, M. (2008). On the capacity of multi-channel wireless networks using directional antennas. In *IEEE INFOCOM 2008—The 27th Conference on Computer Communications* (pp. 628–636).

23. Ahmed, I., Khammari, H., Shahid, A., Musa, A., Kim, K. S., De Poorter, E., & Moerman, I. (2018). A survey on hybrid beamforming techniques in 5G: Architecture and system model perspectives. *IEEE Communications Surveys & Tutorials, 20*(4), 3060–3097.
24. Kutty, S., & Sen, D. (2016). Beamforming for millimeter wave communications: An inclusive survey. *IEEE Communications Surveys & Tutorials, 18*(2), 949–973.
25. Wu, Y., Qiu, L., & Xu, J. (2018) UAV-enabled wireless power transfer with directional antenna: A two-user case (Invited paper). In *2018 15th International Symposium on Wireless Communication Systems (ISWCS)* (pp. 1–6).
26. Qureshi, H. N., & Imran, A. (2019). On the tradeoffs between coverage radius, altitude, and beamwidth for practical UAV deployments. *IEEE Transactions on Aerospace and Electronic Systems, 55*(6), 2805–2821.
27. Won, J., Kim, D.-Y., & Lee, J.-W. (2024). Joint optimization of location, beam, and radio resource for an aerial base station with controllable directional antennas. *IEEE Internet of Things Journal, 11*(16), 27571–27583.
28. Yu, X., Zhang, J., Haenggi, M., & Letaief, K. B. (2017). Coverage analysis for millimeter wave networks: The impact of directional antenna arrays. *IEEE Journal on Selected Areas in Communications, 35*(7), 1498–1512.
29. Bettstetter, C., Hartmann, C., & Moser, C. (2005). How does randomized beamforming improve the connectivity of ad hoc networks? In *IEEE International Conference on Communications* (Vol. 5, pp. 3380–3385).
30. Wang, Q., Dai, H. N., Zheng, Z., Imran, M., & Vasilakos, A. V. (2017). On connectivity of wireless sensor networks with directional antennas. *Sensors, 17*(1), 134.
31. Clerckx, B., & Bayguzina, E. (2016). Waveform design for wireless power transfer. *IEEE Transactions on Signal Processing, 64*(23), 6313–6328.

Chapter 10
Trajectories and Resource Allocation Design for Multi-UAV-Assisted Wireless Power Transfer with Nonlinear Energy Harvesting

Ning Guo, Xinran Wang, Peng Wu, Xiaopeng Yuan, and Yulin Hu

Abstract The integration of Unmanned Aerial Vehicles (UAVs) with wireless power transfer (WPT) technology represents a transformative advancement in modern energy delivery systems, offering unprecedented flexibility and efficiency in powering remote or hard-to-reach devices. This chapter considers a multi-UAV-assisted WPT network, where multiple UAVs are deployed to extend the lifespan of multiple ground devices (GDs) by providing WPT services. To ensure fairness in energy distribution among multiple devices, this chapter aims to maximize the harvested energy of the GD with the worst conditions through the joint optimization of UAV trajectories and transmit power design. Key factors such as UAV trajectories, collision avoidance, and power constraints are carefully accounted for in the optimization process. Diverging from traditional linear models, this chapter adopts a practical sophisticated multi-source nonlinear energy harvesting (EH) model, allowing a more precise representation of the energy conversion dynamics. Due to the highly non-convex nature of the original problem, arising from the presence of coupled variables, this chapter introduces a convex approximation method, which facilitates the transformation of the problem into a series of tightly convex subproblems. This approach enables efficient iterative solutions, leading to a high-quality solution for the joint design optimization. Finally, numerical results are provided to demonstrate

N. Guo · X. Wang · P. Wu · Y. Hu (✉)
School of Electronic Information, Wuhan University, Wuhan, China
e-mail: yulin.hu@whu.edu.cn

N. Guo
e-mail: ning.guo@whu.edu.cn

X. Wang
e-mail: wxr0812@whu.edu.cn

P. Wu
e-mail: peng.wu@whu.edu.cn

X. Yuan
Chair of Information Theory and Data Analytics, RWTH Aachen University, Aachen, Germany
e-mail: yuan@inda.rwth-aachen.de

Y. Hu et al. (eds.), *Reliable and Efficient Trajectory Design in UAV-Aided Wireless Networks*, https://doi.org/10.1007/978-981-96-5988-3_10

the convergence of the proposed algorithm and to validate the performance benefits of the proposed multi-UAV WPT scheme with a nonlinear EH model compared to benchmark methods.

Keywords Unmanned aerial vehicle (UAV) · Wireless power transfer · Trajectory design · Resource allocation

Abbreviations

EH	Energy Harvesting
GDs	Ground Devices
IoT	Internet of Things
LoS	Line-of-Sight
MEC	Mobile Edge Computing
PLoS	Probabilistic Line-of-Sight
RF	Radio Frequency
SWIPT	Simultaneous Wireless Information and Power Transfer
UAV	Unmanned Aerial Vehicle
WPT	Wireless Power Transfer
WPCN	Wireless Powered Communication Networks
WSNs	Wireless Sensor Networks

10.1 Introduction

10.1.1 Background

Unmanned Aerial Vehicles (UAVs) have emerged as a pivotal technology in various applications, demonstrating exceptional versatility, mobility, and cost-effectiveness. Their adaptability has led to widespread use in fields such as cargo transportation [1], disaster management [2], and data collection [3]. This expanding application of UAVs has also catalyzed their integration into wireless communication networks, where they have been employed to enhance channel capacity [4–6], expand coverage [7, 8], and improve communication security [9–11]. Among these applications, the role of UAVs in wireless power transfer (WPT) networks has gained significant attention due to their potential to extend the operational life of energy-constrained ground devices (GDs). In such scenarios, UAVs serve as mobile energy transmitters, offering an innovative solution for wirelessly powering devices over wide geographical areas. For instance, authors in [12, 13] optimized a UAV trajectory the and transmission power for ground users throughput improvement. Additionally, when UAVs function as relay nodes, as explored in [14, 15], through the UAV trajectory design, the minimum information rate across the network is maximized. Moreover, the strategic planning of UAV trajectories has also been applied to various domains to improve overall system performance. In UAV-enabled data collection systems [16, 17] and UAV-assisted data offloading scenarios [18], trajectory optimization plays a crucial role in enhancing efficiency and reliability. When considering more general channel models, researchers in [19, 20] have employed the Rician fading model to represent the channel characteristics for UAVs flying at higher altitudes. They developed three-dimensional trajectory designs combined with user scheduling to boost performance in UAV-supported wireless communication networks. Conversely, for UAVs operating at lower altitudes, the probabilistic line-of-sight (PLoS) model has been utilized in [20] to craft flight paths that improve energy utilization efficiency.

Parallel to these developments, the rapid growth of wireless sensor networks (WSNs) and the Internet of Things (IoT) has spurred interest in integrating wireless power transfer (WPT) technology into UAV-assisted wireless networks. This integration serves as a promising approach to extend the operational lifespan of energy-limited nodes. Prior research, including works like [21–25], has investigated UAV trajectory design within WPT networks to enhance energy transfer efficiency. To further improve energy transfer outcomes, extensive studies have focused on the joint optimization of UAV trajectories and resource allocation strategies. Specifically, some researchers have examined the simultaneous design of flight paths and transmission strategy [26, 27]. Others have explored integrating UAV trajectory planning with user scheduling mechanisms [28], directional antenna adjustments [25], and the allocation of computing resources [29]. Moreover, the application of UAV trajectory design has been extended to additional fields such as mobile edge computing (MEC) systems [30, 31], wireless powered communication networks (WPCN) [32], and simultaneous wireless information and power transfer (SWIPT) systems [33]. Notably, the study in [33] delves into an effective joint design strategy for UAV trajectories and resource allocation, addressing both the communication needs and energy provisioning for ground nodes concurrently.

10.1.2 Motivation

Integrating UAVs into wireless communication networks has the potential to significantly enhance data transmission, improve network coverage, and increase overall system efficiency. Nevertheless, one of the most critical challenges in modern WSNs and the IoT lies in extending the operational lifespan of energy-constrained GDs. Conventional energy sources are often impractical or inefficient, especially for devices deployed in remote or inaccessible locations, necessitating an alternative energy transfer method.

WPT technology, when combined with the high mobility and flexibility of UAVs, offers a promising solution to address this challenge. By leveraging UAVs as mobile energy transmitters, it becomes feasible to provide efficient and targeted energy replenishment to GDs over extensive areas. However, to maximize the effectiveness of UAV-assisted WPT, the precise optimization of UAV trajectories and transmit power allocation is essential. This optimization becomes especially challenging when striving to balance energy distribution fairly among multiple GDs with diverse energy demands.

Existing studies have often employed simplified linear energy harvesting (EH) models to analyze the performance of UAV-assisted WPT systems. While these models are computationally manageable, they fail to capture the inherent nonlinearities present in practical energy harvesting circuits. As a result, these linear models often lead to suboptimal performance, especially in dynamic environments where UAVs operate at varying distances and orientations relative to GDs. Therefore, a more accurate representation of energy harvesting behavior, with a nonlinear EH model, is necessary to better understand and optimize the energy transfer process in multi-UAV systems.

Another challenge in UAV-assisted WPT networks is the need to ensure reliable and efficient resource allocation while addressing constraints such as UAV mobility, collision avoidance, and power budgets. These factors significantly impact the overall efficiency of energy transfer, especially in scenarios involving multiple UAVs and GDs. Therefore, there is a pressing need to develop a robust optimization framework that can handle these complexities and improve the overall energy harvesting performance of WPT systems.

The motivation for this chapter stems from the need to address these challenges by introducing a multi-UAV-assisted WPT network that utilizes a more realistic multi-source nonlinear EH model. This chapter develops a joint optimization strategy for UAV trajectories and transmit power, which will not only enhance the efficiency of WPT but also ensure fair energy distribution among GDs. This research seeks to bridge the gap between theoretical models and practical implementations, providing insights into the potential of UAV-assisted WPT networks in real-world applications and paving the way for more efficient energy transfer solutions in future IoT and WSN deployments.

10.1.3 Main Contributions

Main contributions to this chapter are as follows:

(1) Application of Multi-source Nonlinear Energy Harvesting Model: This chapter introduces the application of a practical multi-source nonlinear EH model within the framework of multi-UAV-assisted WPT systems, characterizing the energy harvested by each GD as a function of UAV trajectories and transmit power, allowing for a more accurate representation of energy transfer.
(2) Joint Optimization of UAV Trajectories and Power Allocation: This chapter formulates a joint optimization problem aimed at maximizing the harvested energy for multiple GDs while adhering to critical constraints such as collision avoidance, maximum speed of UAVs, and power transmission limits.

(3) Iteration Algorithm based on Convex Approximation: To tackle the formulated challenging optimization problem, this chapter proposes an efficient iterative algorithm based on convex approximation techniques. By exploiting the convex properties of the nonlinear EH model, this work develops concave lower-bound functions that relate to UAV distance and harvested energy, thereby simplifying the optimization process and ensuring convergence to a satisfying solution.
(4) Simulative Validation and Performance Evaluation: Through extensive simulations, the performance of the proposed algorithm against existing benchmark approaches are validated. The results demonstrate that the proposed method significantly outperforms scenarios using a single UAV and those relying on linear EH models, confirming the effectiveness of incorporating a multi-source nonlinear EH model in UAV-assisted WPT systems.

10.1.4 Structural Organization

Relevant WPT technologies and UAV trajectory designs are introduced in Sect. 10.2. Section 10.3 shows the system model and discusses the channel model and nonlinear EH model that this work is based on. Section 10.4 formulates the fairness-aware harvested energy max–min problem and provides an efficient iterative algorithm based on the convex approximation method. In Sect. 10.5, simulative results are provided. A comprehensive summary of the work has been conducted in Sect. 10.6.

10.2 Related Works

WPT and EH technologies have emerged as promising solutions to supply power to energy-constrained devices, especially in scenarios where battery replacement or wired charging is impractical. Traditional studies often employ simple linear EH models to characterize the energy harvesting process. For example, in [21], Xu et al. investigated UAV-enabled wireless power transfer system with two energy receivers, focusing on 1D UAV trajectory design and energy region characterization under the maximum speed constraints. In [22], the authors extended their work to optimize the 2D UAV trajectory for energy maximization, and achieves a complexity reduction. Nevertheless, single UAV charging systems face inherent limitations such as restricted coverage areas, limited energy supply, and vulnerability to obstacles or failures. To address these challenges, multi-UAV assisted WPT systems have been proposed to coordinate multiple UAVs for efficient energy delivery [34–36]. Authors in [34] provided a comprehensive tutorial overview of UAV-enabled wireless power transfer, highlighting the benefits of employing multiple UAVs to enhance coverage and reliability. In [34], the authors extended their design strategy from a single-UAV to a multiple-UAV scenario, and proposed a heuristic algorithm that combines the evolutionary algorithm to achieve balanced energy consumption. Authors in [36] characterized the optimal trajectory as being evenly distributed over the circumference in the scenario of a dual-UAV-assisted WPT for a single energy receiver (Table 10.1).

Despite these advancements, the linear EH model does not accurately capture the nonlinear characteristics of practical energy harvesting circuits, which can significantly impact system performance. Recognizing this limitation, recent works have incorporated nonlinear EH models into UAV-enabled wireless power transfer systems. Authors in [24] examined trajectory design for UAV-enabled multiuser wireless power transfer with a nonlinear EH model. They formulated an optimization problem to maximize the minimum harvested energy among all users by jointly optimizing the UAV's trajectory and transmit power in a single UAV scenario. In [25], authors extended their study to jointly design the UAV's trajectory and directional antenna orientation, aiming to enhance energy transfer efficiency. Xiong et al. [37] investigated joint optimization of trajectory, task offloading, and CPU control in UAV-assisted wireless powered fog computing networks, considering nonlinear EH behavior to improve overall system performance.

However, these studies are confined to single UAV scenarios, which may not fully exploit the potential benefits of multi-UAV systems in terms of coverage, scalability, and robustness. Integrating nonlinear EH models with multi-UAV assisted wireless power transfer remains an unexplored solution that could significantly improve energy transfer efficiency and network performance. Therefore, this section aims to fill this research gap by jointly designing the trajectories and resource allocation for multi-UAV assisted wireless power transfer systems with nonlinear energy harvesting. By considering the practical nonlinear characteristics of energy harvesting devices and leveraging the advantages of multi-UAV coordination, this chapter seeks to enhance the efficiency and reliability of wireless power transfer.

Table 10.1 Comparison of current related work

References	EH model	Number of UAVs	Optimization objectives
[21]	Linear	Single	1D UAV trajectory
[22]	Linear	Single	2D UAV trajectory
[23]	Linear	Single	1D UAV trajectory (globally optimal)
[34]	Linear	Multiple	2D UAV trajectory
[38]	Linear	Multiple	2D UAV trajectory
[24]	Nonlinear	Single	2D UAV trajectory
[25]	Nonlinear	Single	2D UAV trajectory and directional antenna orientation
[37]	Nonlinear	Single	2D UAV trajectory, task offloading, and computing resources
Ours	Nonlinear	Multiple	2D UAV trajectory and transmit power

10.3 System Model

10.3.1 UAV and GD Description

The system contains two main types of devices: UAVs and GDs. UAVs, characterized by high mobility, utilize wireless power transfer technology to transmit energy wirelessly to the GDs. These UAVs have the ability to adjust their flight trajectories and transmission power to optimize energy delivery to the GDs. Multiple UAVs can operate simultaneously, enhancing the efficiency and coverage of the WPT system. Unlike fixed ground stations, UAVs can move dynamically, which allows them to achieve better positioning for efficient energy transfer.

GDs, on the other hand, are typically stationary or have limited mobility and are the energy-constrained devices that rely on the UAVs for wireless power transfer. They receive energy from the UAVs through RF signals and convert this energy into DC power using an energy harvesting circuit. The efficiency of energy harvesting at the GDs depends on various factors, such as the distance from the UAVs and the UAVs' transmission power.

10.3.2 Multi-UAV Assisted Multi-User WPT Scenarios

The multi-UAV assisted multi-user WPT network is shown in Fig. 10.1, which is including M UAVs and k randomly distributed GDs, with their locations prior known by UAVs. Serving as mobile energy sources, UAVs wirelessly transmit power to the energy-constrained GDs during the flight to help extend the potential operational lifetime of GDs. It is assumed that all UAVs fly from their respective initial points to final points at a constant height with a maximum flying speed, which is determined by the available battery energy.

We consider a WPT system consisting of M UAVs and k randomly distributed GDs. In this network, each UAV flies from its initial position $q_{m,0}, m \in \mathcal{M} \triangleq \{1, ..., M\}$ to its final destination $q_{m,F}, m \in \mathcal{M} = \{1, ..., M\}$, maintaining a fixed

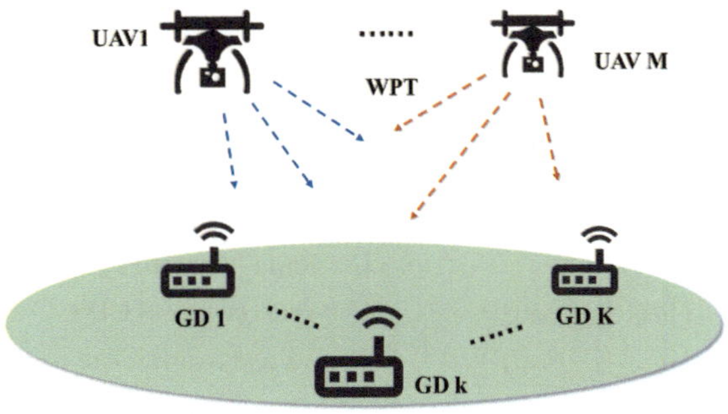

Fig. 10.1 An example of a multi-UAV-assisted multi-user WPT network

altitude H during the entire mission duration. The horizontal position and transmit power of UAV m at time t are represented as $q_m(t) = (x_m(t), y_m(t))$ and $P_m(t)$, respectively, where $m \in \mathcal{M} \triangleq \{1, ..., M\}$ denotes the UAV index.

To facilitate analysis, the total mission duration T is discretized into L equal time slots, denoted as $\mathcal{L} \triangleq \{1, \ldots, L\}$, with each time slot having a length of Δ. In the l th time slot, the trajectories and transmit power of UAV m are represented as $\boldsymbol{q}_m[l] = (x_{m,l}, y_{m,l}), l \in L$ and $P_m[l], l \in \mathcal{L}\ominus$, respectively (Table 10.2).

In order to ensure that the UAVs can complete their tasks within the designated time while moving efficiently and safely, it is necessary to establish the initial and final position constraint and the maximum speed constraint. First, each UAV m must start from a given initial position $\boldsymbol{q}_{m,0}$ and reach a final target position $\boldsymbol{q}_{m,F}$ by the end of the mission duration, which is represented by:

$$\boldsymbol{q}_m[0] = \boldsymbol{q}_{m,0}, \boldsymbol{q}_m[L] = \boldsymbol{q}_{m,F} \tag{10.1}$$

Second, to prevent any UAV from exceeding its speed limit, the distance travelled between consecutive time slots must be less than or equal to the maximum distance that can be travelled at the maximum speed V_{max} during a time slot of duration Δ:

$$\|\boldsymbol{q}_m[l] - \boldsymbol{q}_m[l-1]\|^2 \leq V_{max}^2 \Delta^2, \forall m \in \mathcal{M}, l \in \mathcal{L} \tag{10.2}$$

This constraint ensures that UAVs do not move faster than their physical capability, maintaining a realistic and achievable trajectory.

Besides the mobility constraints, UAVs also need to adhere to power-related limitations to ensure proper operation within hardware capabilities and energy budgets. In particular, over the entire mission duration, the average transmit power of UAV m should not exceed a given average power limit $\overline{P}$:

Table 10.2 Important notation description

Notation	Description
w_k	Location of GD k
$q_m[l]$	Location of UAV m at l th slot
$P_m[l]$	Transmit power of UAV m at l th slot
$d_k(\boldsymbol{q}_m[l])$	Distance between UAV m and GD k at l th slot
$h_{k,m}[l]$	Channel power gain between UAV m with GD k at lth slot
$Q_{k,m}[l]$	Received RF power at GD k from UAV m at l th slot
$I_{out,k}$	Output current of rectification circuit for GD k at l th slot
λ_{m,j_m}	Constant waveform factor for unit power
$P_{dc,k}(\mathbf{Q}_k[l])$	DC power harvested by GD k at l th slot
$E_k(\{\boldsymbol{q}_m[l]\}, \{P_m[l]\})$	Harvested energy of GD k
$\hat{E}_k(\{\boldsymbol{q}_m[l]\}, \{P_m[l]\})$	Convex approximation of harvested energy of GD k

$$\frac{1}{L}\sum_{l=1}^{L} P_m[L] \leq \overline{P}, \forall m \in \mathcal{M} \tag{10.3}$$

which prevents excessive energy consumption by the UAVs over the course of the mission. Moreover, limited by the UAVs hardware conditions, the transmit power at each time slot $P_m[l]$ must be within the allowable range of 0 to the maximum allowed peak transmit power P_{max}:

$$0 \leq P_m[l] \leq P_{\max}, \forall m \in \mathcal{M}. \tag{10.4}$$

Given the simultaneous operation of multiple UAVs, it's critical to ensure that they maintain a safe distance from each other to avoid collisions. For any pair of UAVs i and j, the Euclidean distance between them at any time slot l must be greater than or equal to a minimum safe distance d_{min}. The anti-collision constraint is formulated as

$$\|\boldsymbol{q}_i[l] - \boldsymbol{q}_j[l]\|^2 \geq d_{min}^2, i \neq j, \forall n \in \{1, 2, ..., L-1\}, \tag{10.5}$$

which is a crucial safety requirement, especially when multiple UAVs are working within a confined space.

10.3.3 Channel Model

The channels between the UAVs and GDs are assumed predominantly line-of-sight (LoS), which is practical in scenarios with few obstructions or a sufficiently high flying height. Thus, the channel power gain is directly affected by the distance, given by

$$h_{k,m}[l] = \frac{\beta_0}{d_k^2\left(\boldsymbol{q}_m[l]\right)}, \tag{10.6}$$

where β_0 denotes the gain at a distance of 1 m, and $d_k\left(\boldsymbol{q}_m[l]\right) = \sqrt{\boldsymbol{q}_m[l] - \boldsymbol{w}_k^2 + H^2}$ denotes the distance function between UAV m and GD k.

10.3.4 Nonlinear EH Model

During the EH process, the received RF signals at each GD from UAVs are converted into available DC signals in the operating circuit through internal nonlinear rectification circuit. The RF-DC conversion circuit consists of the receiving antenna, combined matching network, rectification circuit, and low-pass filter. The combined matching network is designed to filter the harmonics generated by the rectification elements and to match the impedance of the receiving antenna and rectification circuit correctly. Additionally, the low-pass filter is employed to suppress out-of-band fundamental and harmonics.

In contrast to most existing studies on multiple UAV WPT networks that utilize the simplified yet inaccurate linear EH model [35, 36], this chapter adopts a precise multi-source nonlinear EH model, as employed in the previous work [39]. This model effectively captures the non-linear characteristics of the rectifier diode during the DC conversion process in the GD's rectification circuit. Specifically, the rectification circuit operates as a half-wave rectifier, using a diode to achieve nonlinear RF-to-DC conversion. Based on the model, the output current for GD k is given by

$$I_{out,k} \approx \frac{1}{a} W_0\left(ae^{aI_s}\varphi\left(\boldsymbol{Q}_k[l]\right)\right) - I_s, \tag{10.7}$$

where $\boldsymbol{Q}_k[l] \triangleq \{Q_{k,1}[l], ..., Q_{k,M}[l]\}$ denotes the set of received RF power at GD k from all UAVs at slot l, $W_0(\cdot)$ refers to the principle branch of the Lambert W function, I_s is the reverse bias saturation current, and $a = \frac{R_L}{n_{ideal}v_t}$, where R_L denotes the resistance of the energy storage unit, n_{ideal} and v_t respectively denote the ideality factor and the thermal voltage of the diode respectively. Furthermore, the explicit expression for $\varphi(\boldsymbol{Q}_k[l])$ according to [39] is

$$\varphi\left(\boldsymbol{Q}_k[l]\right) = I_s + \sum_{i=1}^{n} \beta_i \sum_{\{j_{\hat{m}}\}_{[n]}} C_{l,\{j_{\hat{m}}\}} \prod_{m=1}^{M} Q_{k,m}\left[\frac{j_m}{2}\right], \tag{10.8}$$

where β_i is a positive constant and $C_{l,\{j_{\hat{m}}\}}$ is

$$C_{l,\{j_{\hat{m}}\}} = \frac{M!}{\prod_{m=1}^{M} j_m!} \prod_{m=1}^{M} \lambda_{m,j_m} \geq 0, \tag{10.9}$$

where $\{j_{\hat{m}}\}_{[l]}$ denotes $\sum_{\hat{m}=1}^{M} j_{\hat{m}} = l$ and λ_{m,j_m} is the constant waveform factor. Therefore, the DC power harvested by GD k from the superposition of multiple UAV signal sources is calculated by

$$P_{dc,k}\left(\boldsymbol{Q}_k[l]\right) = I_{out,k}^2 R_L \approx \left(\frac{1}{a} W_0\left(ae^{aI_s}\varphi\left(\boldsymbol{Q}_k[l]\right)\right) - I_s\right)^2 R_L. \tag{10.10}$$

Hence, the harvested energy of GD k during whole time is

$$E_k\left(\{\boldsymbol{q}_m[l]\}, \{P_m[l]\}\right) = \sum_{l=1}^{L} P_{dc,k}\left(\boldsymbol{Q}_k[l]\right)\Delta. \tag{10.11}$$

10.4 Joint Trajectories and Resource Allocation Algorithm

10.4.1 Problem Statement

This chapter focuses on a multi-UAV-assisted WPT network, where several UAVs are deployed to provide energy transfer services to multiple GDs. Our goal is to optimize the trajectories and transmit power of UAVs to maximize the minimum harvested energy across all GDs while considering UAV mobility, collision avoidance, and power limitations. The formulated optimization problem is given as:

$$
\begin{aligned}
\text{(P1):} \quad & \max_{\{\boldsymbol{q}_m[l]\},\{P_m[l]\}} \; \min_k E_k\left(\{\boldsymbol{q}_m[l]\}, \{P_m[l]\}\right), \\
\text{s.t.}: \; & \boldsymbol{q}_m[0] = \boldsymbol{q}_{m,0}, \boldsymbol{q}_m[L] = \boldsymbol{q}_{m,F}, \\
& \|\boldsymbol{q}_m[l] - \boldsymbol{q}_m[l-1]\|^2 \le V_{max}^2 \Delta^2, \forall m \in \mathcal{M}, l \in \mathcal{L}, \\
& \frac{1}{L}\sum_{l=1}^{L} P_m[L] \le \overline{P}, \forall m \in \mathcal{M} \\
& 0 \le P_m[l] \le P_{max} \\
& \|\boldsymbol{q}_i[l] - \boldsymbol{q}_j[l]\|^2 \ge d_{min}^2, i \ne j, \forall n \in \{1, 2, ..., L-1\}
\end{aligned} \tag{10.12}
$$

Nevertheless, both the harvested power function $E_k\left(\{\boldsymbol{q}_m[l]\}, \{P_m[l]\}\right)$ and the function $\|\boldsymbol{q}_i[l] - \boldsymbol{q}_j[l]\|^2$ in anti-collision constraint are nonconvex, resulting in the non-convexity of problem (P1) and complicating its efficient resolution. To address this challenge, in the following subsections, this chapter introduces a convex approximation method to construct a convex problem for (P1) at each local point, which facilitates an effective iterative algorithm for solving Problem (P1).

10.4.2 Anti-collision Constraint Approximation

Note that $\|\boldsymbol{q}_i[l] - \boldsymbol{q}_j[l]\|^2$ is a convex function, it can be approximated as:

$$
\begin{aligned}
\|\boldsymbol{q}_i[l] - \boldsymbol{q}_j[l]\|^2 \ge \; & \|\hat{\boldsymbol{q}}_i[l] - \hat{\boldsymbol{q}}_j[l]\|^2 + 2(\hat{\boldsymbol{q}}_i[l] - \hat{\boldsymbol{q}}_j[l])^T \\
& \times \left(q_i[l] - q_j[l] - \left(\hat{q}_i[l] - \hat{q}_j[l]\right)\right) \\
\triangleq \; & (\hat{d}_{i,j}[l])^2.
\end{aligned} \tag{10.13}
$$

where $(\hat{d}_{i,j}[l])^2$ is an affine term. Therefore, the anti-collision constraint can be reformulated into the following convex form:

$$
(\hat{d}_{i,j}[l])^2 \ge d_{\min}^2, l \in \{1, 2, ..., L-1\}. \tag{10.14}
$$

10.4.3 Harvested Power Convex Approximation

Although the harvested power is not convex in $P_m[l]$, the convexity between $P_{dc,k}(\boldsymbol{Q}_k[l])$ and $\frac{1}{Q_{k,m}[l]}, \forall m \in \mathcal{M}$ has been proved in [39]. Consequently, as indicated by the formulation (10.11), $E_k(\{\boldsymbol{q}_m[l]\}, \{P_m[l]\})$ is convex in $\frac{1}{Q_{k,m}[l]}$. Utilizing the properties of convex functions, $E_k(\{\boldsymbol{q}_m[l]\}, \{P_m[l]\})$ can be approximated as

$$\begin{aligned} E_k(\{\boldsymbol{q}_m[l]\}, \{P_m[l]\}) &\geq \sum_{l=1}^{L}\left(\sum_{m=1}^{M} -\hat{A}_{k,m,l}\frac{1}{Q_{k,m}[l]} + \hat{B}_{k,l}\right)\Delta \\ &= \sum_{l=1}^{L}\left(\sum_{m=1}^{M} -\hat{A}_{k,m,l}\frac{d_k^2(\boldsymbol{q}_m[l])}{\beta_0 P_m[l]} + \hat{B}_{k,l}\right)\Delta, \end{aligned} \tag{10.15}$$

where the equality holds when $Q_{k,m}[l] = \hat{Q}_{k,m}[l]$, implying $P_m[l] = \hat{P}_m[l]$ and $\boldsymbol{q}_m[l] = \hat{\boldsymbol{q}}_m[l]$. Parameters $\hat{A}_{k,m,l}$ and $\hat{B}_{k,l}$ are positive constants, defined as:

$$\begin{aligned} \hat{A}_{k,m,l} &= -\frac{\partial P_{dc,k}(\boldsymbol{Q}_k[l])}{\partial \frac{1}{Q_{k,m}[l]}}\Big|_{Q_{k,m}[l]=\hat{Q}_{k,m}[l]}, \\ \hat{B}_{k,l} &= P_{dc,k}(\hat{\boldsymbol{Q}}_k[l]) + \sum_{m=1}^{M}\hat{A}_{k,m,l}\frac{1}{\hat{Q}_{k,m}[l]}, \end{aligned} \tag{10.16}$$

where $\hat{A}_{k,m,l}$ is expressed in a partial derivative form since the harvested DC power $P_{dc,k}(Q_k[l])$ is an implicit function, whose specific value can be obtained through numerical calculation. To separate the multiplicative terms $d_k^2(\boldsymbol{q}_m[l])$ and $P_m[l]$, we employ the arithmetic–geometric mean inequality. For non-negative μ, ν, and F, we have $\mu\nu \leq \frac{\mu^2}{2F} + \frac{F\nu^2}{2}$, where equality is achieved when $\mu = F\nu$. Utilizing this property, Eq. (10.14) can be refined as follows:

$$\begin{aligned} &E_k(\{\boldsymbol{q}_m[l]\}, \{P_m[l]\}) \geq \sum_{l=1}^{L}\left(\sum_{m=1}^{M} -\hat{A}_{k,m,l}\frac{d_k^2(\boldsymbol{q}_m[l])}{\beta_0 P_m[l]} + \hat{B}_{k,l}\right)\Delta \\ &\geq \sum_{l=1}^{L}\left(\sum_{m=1}^{M}(-\frac{\hat{A}_{k,m,l}\hat{F}_{k,m,l}}{2}d_k^4(\boldsymbol{q}_m[l]) - \frac{\hat{A}_{k,m,l}}{2\hat{F}_{k,m,l}}\frac{1}{\beta_0^2 P_m^2[l]}\right) + \hat{B}_{k,l})\Delta \\ &\triangleq \hat{E}_k(\{\boldsymbol{q}_m[l]\}, \{P_m[l]\}), \end{aligned} \tag{10.17}$$

where equality holds at the local point $(\hat{\boldsymbol{q}}_m[l], \hat{P}_m[l])$ and $\hat{F}_{k,m,l}$ is defined as:

$$\hat{F}_{k,m,l} = \frac{1}{\beta_0 \hat{P}_m[l](\hat{d}_k(\boldsymbol{q}_m[l]))^2}. \tag{10.18}$$

Since $\hat{E}_k(q_m[l], P_m[l])$ is jointly concave in terms of $\{\boldsymbol{q}_m[l]\}$ and $\{P_m[l]\}$, we have derived a tight lower bound concave function for the harvested energy function $E_k(\boldsymbol{Q}_k[l])$ and the distance function $\|\boldsymbol{q}_i[l] - \boldsymbol{q}_j[l]\|^2$ at $(\hat{\boldsymbol{q}}_m[l], \hat{P}_m[l])$. Substituting $P_{dc,k}(\boldsymbol{Q}_k[l])$ and $\|\boldsymbol{q}_i[l] - \boldsymbol{q}_j[l]\|^2$ with $\hat{E}_k(\boldsymbol{Q}_k[l])$ and $(\hat{d}_{i,j}[l])^2$, the convex approximated problem (P1) is formulated as follows:

$$
\begin{aligned}
(p_2): \quad & \max_{\{\boldsymbol{q}_m[l]\},\{P_m[l]\}} \min_k \hat{E}_k(\{\boldsymbol{q}_m[l]\}, \{P_m[l]\}), \\
s.t.: \quad & \boldsymbol{q}_m[0] = \boldsymbol{q}_{m,0}, \boldsymbol{q}_m[L] = \boldsymbol{q}_{m,F}, \\
& \|\boldsymbol{q}_m[l] - \boldsymbol{q}_m[l-1]\|^2 \le V_{max}^2 \Delta^2, \forall m \in \mathcal{M}, l \in \mathcal{L}, \\
& \frac{1}{L}\sum_{l=1}^{L} P_m[L] \le \overline{P}, \forall m \in \mathcal{M} \\
& 0 \le P_m[l] \le P_{max} \\
& (\hat{d}_{i,j}[l])^2 \ge d_{min}^2
\end{aligned}
\tag{10.19}
$$

The convexity of problem (P2) allows it to be solved efficiently using methods such as the ellipsoid algorithm. Based on this property, an iterative optimization algorithm is proposed to address the original problem (P1) by leveraging the convex approximation (P2). In the following subsection, we will introduce the detailed algorithm and provide an in-depth analysis of its convergence properties and computational complexity.

10.4.4 Iterative Algorithm

The proposed iterative algorithm starts by initializing the iteration index $r = 0$ and setting up a feasible initial point $(\{\boldsymbol{q}_m^{(0)}[l]\}, \{P_m^{(0)}[l]\})$, where the initial values $(\hat{\boldsymbol{q}}_m^{(0)}[l], \hat{P}_m^{(0)}[l])$ are defined as $(\boldsymbol{q}_m^{(0)}[l], P_m^{(0)}[l])$. Specifically, the initial UAV trajectory is constructed such that each UAV flies along a straight-line path between its initial and final positions, with each path discretized into N trajectory points. If the trajectory satisfies the anti-collision constraints, it is directly accepted as the initial trajectory $\boldsymbol{q}_m^{(0)}[l]$. Otherwise, any points where collisions are detected are adjusted to avoid conflict, resulting in a collision-free initial trajectory $\boldsymbol{q}_m^{(0)}[l]$. For the initial transmit power, we set $P_m^{(0)}[l] = \overline{P}$ for each UAV.

Once the initialization is completed, the iterative process begins. At each iteration r, a convex approximation of the harvested energy $\hat{E}_k(\boldsymbol{q}_m[l], P_m[l])$ is constructed for each GD k at the current local point $(\{\hat{\boldsymbol{q}}_m[l]\}, \{\hat{P}_m[l]\})$ using the convex approximation method introduced in the previous subsection, as shown in Eq. (10.17). The objective of problem (P1) is then replaced with $\hat{E}_k(\{\boldsymbol{q}_m[l]\}, \{P_m[l]\})$, resulting in the tightly convex approximation problem (P2) at iteration r. This convex problem can be solved to yield the updated optimal solution $(\{\boldsymbol{q}_m^*[l]\}, \{P_m^*[l]\})$.

After obtaining the solution, we update $r = r + 1$ and set $\left(\{\hat{\boldsymbol{q}}_m[l]\}, \{\hat{P}_m[l]\}\right) = \left(\{\boldsymbol{q}_m^*[l]\}, \{P_m^*[l]\}\right)$. The algorithm continues to iterate until a convergence criterion is met. The complete iterative procedure is detailed in Algorithm 10.1.

Algorithm: Jointly Optimizing UAV Trajectory and Transmit Power

---

Initialization

Initialize $(\{\boldsymbol{q}_m^{(0)}[l]\}, \{P_m^{(0)}[l]\})$ and default setups of the iteration index $r = 0$ and convergence threshold ϵ. Set $(\{\hat{\boldsymbol{q}}_m^{(0)}[l]\}, \{\hat{P}_m[l]\}) = (\{\boldsymbol{q}_m^{(0)}[l]\}, \{P_m^{(0)}[l]\})$.

Iteration

a) Construct convex approximation $\tilde{E}_k(\{\boldsymbol{q}_m[l]\}, \{P_m[l]\})$ according to (10.17);

b) Address the convex problem (P2) and acquire current optimal * and optimal point$(\{\boldsymbol{q}_m^\star[l]\}, \{P_m^\star[l]\})$;

c) **If** the iterative difference of the objective is below ϵ

Obtain the final solution as$(\{\boldsymbol{q}_m^\star[l]\}, \{P_m^\star[l]\})$;

stop iteration;

Else

$r = r + 1$;

$(\{\hat{\boldsymbol{q}}_m[l]\}, \{\hat{P}_m[l]\}) = (\{\boldsymbol{q}_m^\star[l]\}, \{P_m^\star[l]\})$;

return to a).

10.5 Numerical Results

In this part, the effectiveness of the proposed iterative-based joint trajectories and transmit power optimization algorithm is evaluated via simulation.

To demonstrate the benefits of the proposed algorithm utilizing a multi-UAV system with a nonlinear EH model, we introduce the following two benchmark schemes:

- Single UAV with a nonlinear EH model: The trajectory design for a single UAV under the nonlinear EH model has been explored in previous studies [21–23]. These studies have demonstrated considerable performance gains when employing the accurate nonlinear EH model. However, the energy harvesting efficiency of a single UAV configuration remains extremely low.
- Multi-UAV with a linear EH model: Given the low EH efficiency of the single UAV scheme, multi-UAV deployments have been explored in several existing works [34, 36]. While these studies overlook nonlinear conversion in EH and rely on a simplified linear EH model, the joint design of trajectories and transmit power follows a comparable approach to ours. This approach can act as a benchmark to assess the benefits of incorporating a realistic nonlinear EH model.

Figure 10.2 illustrates the convergence behaviour of the proposed iterative algorithm for the joint trajectory design and resource allocation. The convergence rates indicate that the proposed iterative algorithm reaches a steady state within 25–30 iterations, demonstrating its efficiency in finding a near-optimal solution in a reasonable number of iterations. By comparing the results of '(P1) objective' and '(P2) objective', i.e., the results before and after the approximation, the objective value of problem (P1) consistently remains higher than that of problem (P2), particularly during the initial iterations. This is because the objective of (P2) serves as a lower bound approximation for (P1), maintaining equality at the beginning of each iteration but dropping below (P1) as iterations progress, as discussed in the convergence analysis. However, this does not imply that the proposed algorithm will result in a performance degradation. As iterations continue, the approximation gap between the two objectives gradually decreases and becomes almost negligible after the 15th iteration. This trend demonstrates a diminishing performance loss and confirms the effectiveness of the proposed convex approximation method. Moreover, to evaluate the advantage of the utilization of nonlinear EH model and multiple UAVs, benchmarks are provided for system performance comparisons, including the objective with the simplified linear EH model and the objective with the single UAV with both linear and nonlinear EH model. Note that the objective of the linear EH model is calculated by substituting the optimized trajectory and transmit power with the linear EH model into the nonlinear EH model, such that the harvested energies of the two schemes can be compared. Obviously, for the single-UAV scenarios, both nonlinear and linear models exhibit significantly lower performance in comparison to their multi-UAV counterparts. This limitation is due to the severe path loss when only a single UAV is used to cover all the GDs, making it difficult to supply adequate power across a large area. In contrast, deploying multiple UAVs contributes to better spatial distribution, providing a more efficient power delivery to GDs. Moreover, designs with nonlinear EH model outperform ones with linear EH model. The result underscores the impact of using a more accurate EH model, which considers the nonlinear characteristics of the rectification process, leading to a more efficient energy harvesting design. The nonlinear model captures the saturation effects in the rectification circuit that the linear model cannot, allowing for a more realistic and effective optimization. In conclusion, the simulation results in Fig. 10.2 validate the effectiveness of the proposed multi-UAV-assisted WPT system with nonlinear EH. It significantly improves the minimal harvested energy compared to the benchmark schemes, particularly demonstrating the importance of considering multiple UAVs and a nonlinear EH model to maximize the efficiency of wireless power transfer in UAV-assisted networks.

Figure 10.3 provides insights into the relationship between task time and minimum harvested energy under different configurations. From the results, it is evident that the minimum harvested energy for all schemes exhibits a linear increase as the task duration increases, and the multi-UAV nonlinear EH model consistently outperforms the other schemes across all task times. The gap between the multi-UAV linear and nonlinear EH models, labeled as the 'Nonlinear model gap' in the figure, gradually widens as the task duration increases, highlighting the limitations of using a simplified linear EH model. In addition, when the number of UAVs increases from 1 to 2, the system performance shows an improvement of more than twofold. This is because, within the same time period, the amount of energy harvested while UAVs fly around the GDs is significantly higher than the energy collected while flying towards the GDs. As a result, deploying multiple UAVs can considerably enhance the minimum

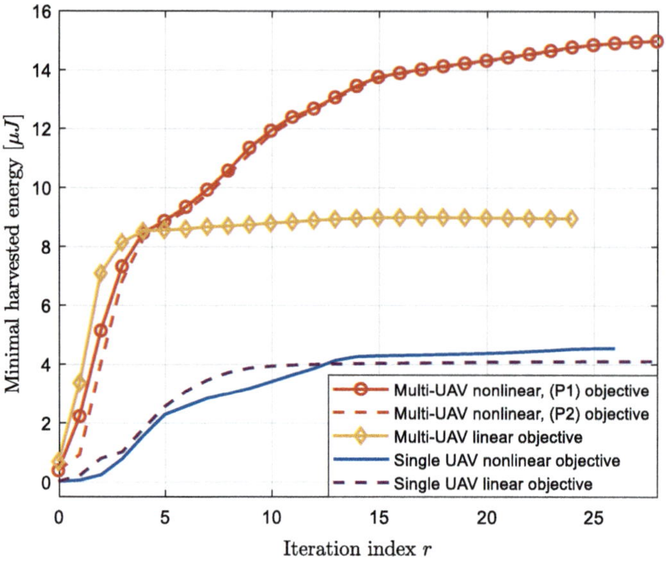

Fig. 10.2 Convergence behaviour illustration

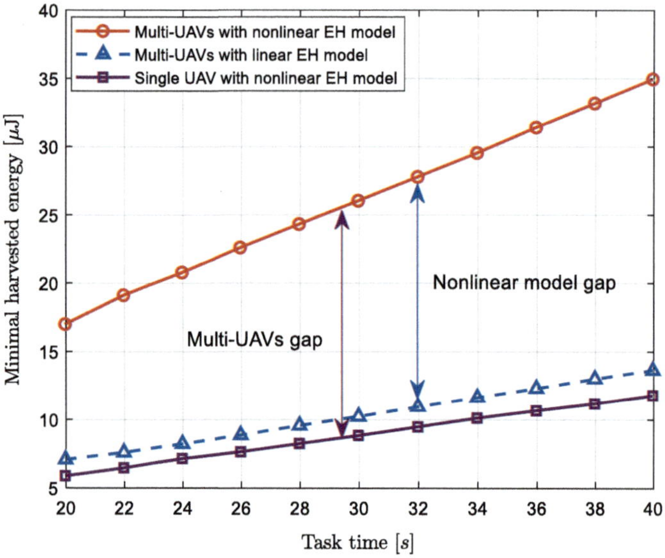

Fig. 10.3 Minimal harvested energy performance with different task times

harvested energy performance. By better distributing UAVs across the target area, energy harvesting efficiency is improved, leading to a substantial performance gain.

The bar graph Fig. 10.4 illustrates the harvested energy achieved by different GDs for three configurations. The results reveal that the multi-UAV configuration with a nonlinear EH model consistently achieves the highest harvested energy for each GD. More interestingly, the multi-UAV system with the nonlinear EH model achieves nearly identical harvested energy for all GDs, with values consistently around 18 μJ. The result indicates that the proposed optimization framework explicitly incorporates fairness considerations in the joint trajectory and power allocation design. By targeting a fairness-aware objective, the algorithm aims to maximize the minimum harvested energy among all GDs, thereby ensuring that even the GD experiencing the most adverse conditions receives adequate power. This approach yields a balanced energy distribution across all devices, as evidenced by the uniformity in harvested energy values for all GDs. Such a fairness-driven strategy is crucial in maintaining consistent system performance, particularly in scenarios with diverse environmental and channel conditions. In contrast, the linear model oversimplifies the EH process, leading to mismatches between the expected and actual energy harvested by the GDs, thereby contributing to unfair energy distribution. For example, even though the centrally located GD 3 generally has the best channel power gain due to favorable positioning, it ends up with the lowest harvested energy in the linear model scenario. This outcome can be attributed to the unrealistic assumptions of the linear energy harvesting model, which fails to capture the true dynamics of the energy conversion process. When the fairness-aware optimization, initially performed under the linear model, is applied to the actual nonlinear environment, a noticeable performance deviation occurs. GD 3, being centrally positioned with a strong channel gain, is particularly vulnerable to such deviations, as the linear model's simplification fails to accurately adjust for the nonlinear saturation effects present in practical scenarios. Consequently, the central GD, despite its advantageous position, suffers from insufficient harvested energy, highlighting the limitations of the linear model in achieving fair and energy distribution.

Subsequently, Fig. 10.5 provides a comparison of optimized UAV trajectories and corresponding transmit power for the proposed design and the benchmarks. The figure reveals several insightful behaviours and differences in system performance between these schemes. In both multi-UAV schemes, a notable aggregation of discrete trajectory points can be observed around the GDs. This behaviour indicates that the UAVs prefer to hover near the GDs for a longer duration to achieve better channel power gain when sufficient task time is available. Such a hovering strategy maximizes the harvested energy for the GDs by reducing the path loss between the UAVs and the devices. The UAV trajectories for the proposed multi-UAV nonlinear EH scheme (subfigure a) exhibit a more complex, tortuous, and space-coupled characteristic compared to the trajectories of the multi-UAV linear EH scheme (subfigure c), which display a smoother and space-decoupled behaviour. The more intertwined paths of the nonlinear model do not imply that the UAVs will collide; rather, they indicate a more tightly coordinated movement between the UAVs, which is achieved by decoupling their positions in the time domain. This temporal separation allows the UAVs to maintain close spatial paths without risk of collision. The coordinated, space-coupled trajectory results in improved system performance, as both UAVs

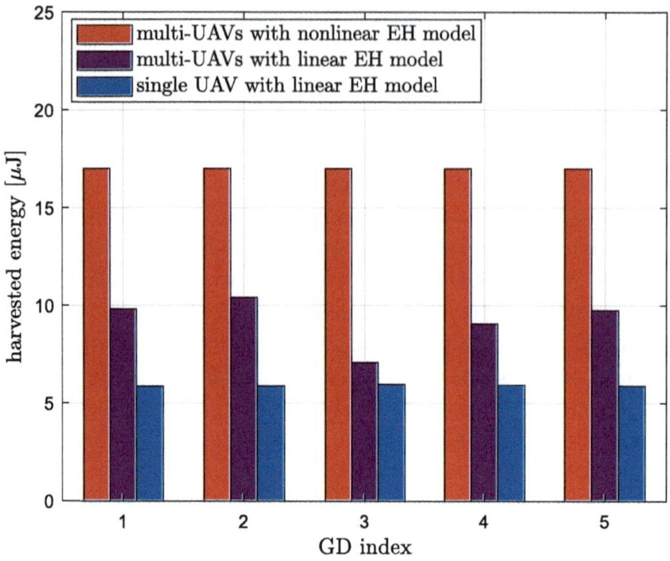

Fig. 10.4 Harvested energy of each GD

adjust their positions to collectively enhance the overall power delivery to the GDs. When comparing the transmit power of the two multi-UAV schemes (subfigures b and d), the proposed nonlinear EH model demonstrates a more efficient utilization of transmit power. In the nonlinear EH scheme, the UAVs dynamically adjust their power levels in response to their distance from the GDs: reducing power when they are far from the GDs to conserve energy and increasing power as much as possible when they are near the GDs to maximize energy transfer. This adaptive power allocation strategy is crucial for achieving high energy harvesting efficiency, as it ensures that power is used where it is most effective. Conversely, the linear EH scheme lacks such nuanced power control, leading to suboptimal power usage, which ultimately results in lower overall performance. The single UAV nonlinear EH scheme (subfigures e and f) further highlights the limitations of relying on a single UAV for energy transfer to multiple GDs. The UAV trajectory shows the necessity of covering a larger area and visiting each GD in sequence, resulting in greater path losses and reduced harvested energy. Additionally, while the UAV does adapt its power output, it cannot maintain the same level of efficiency as the coordinated multi-UAV approach, due to the increased distance and limited spatial distribution.

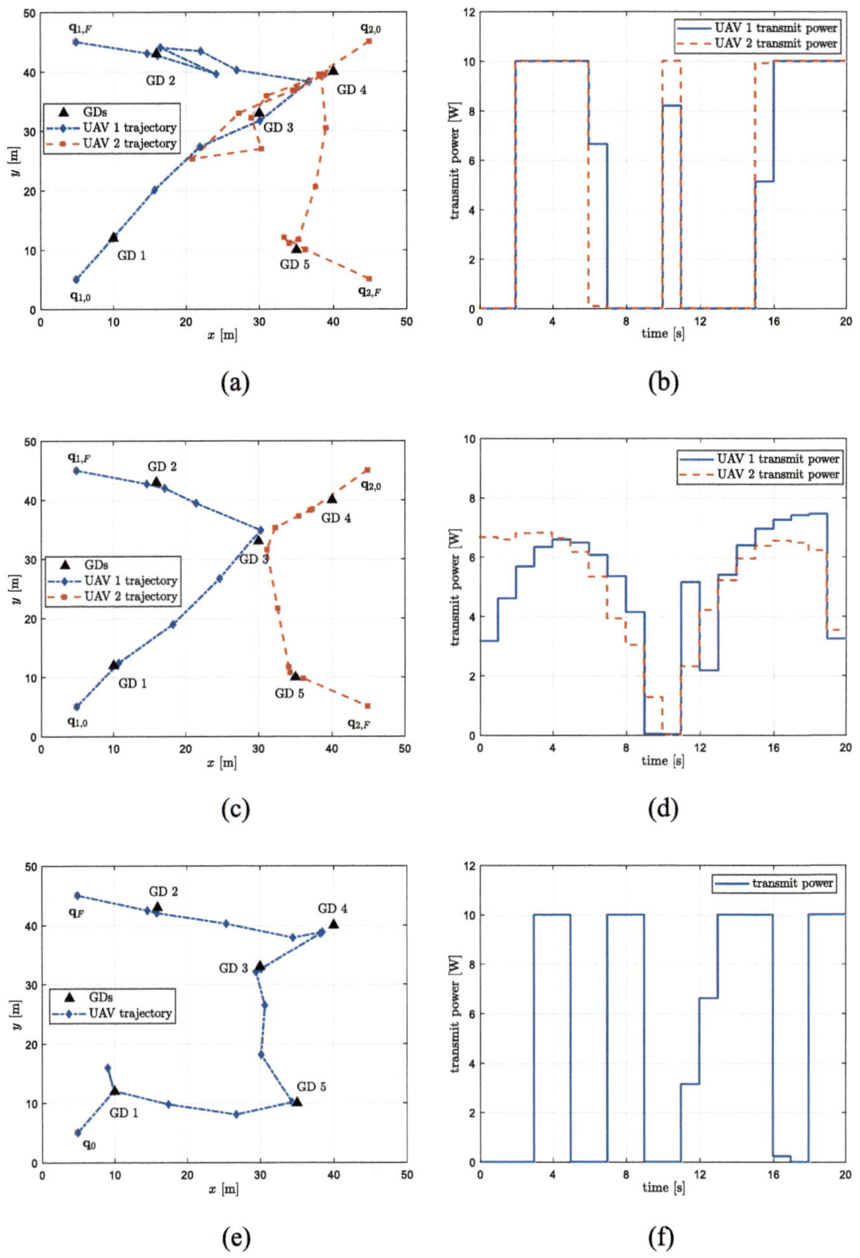

Fig. 10.5 Optimized trajectories and transmit power where (**a**) is UAV trajectories of proposed scheme, (**b**) is UAV transmit power of proposed scheme, (**c**) is UAV trajectories of multi-UAV with linear EH model scheme, (**d**) is UAV transmit power of multi-UAV with linear EH model scheme, (**e**) is UAV trajectory of single UAV with nonlinear EH model scheme and (**f**) is UAV transmit power of single UAV with nonlinear EH model scheme

10.6 Conclusion

This chapter considered a multi-UAV-assisted multi-user WPT network. To enhance the network's performance, this chapter concentrated on maximizing the minimum energy harvested by GDs. This was achieved by jointly optimizing the UAVs' trajectory and their transmit power, while ensuring proper UAV operation under constraints such as anti-collision protocols, maximum speed limits, and power budget restrictions. The joint design problem was formulated based on a practical multi-source nonlinear EH model. To tackle the non-convex problem, this chapter introduced a convex approximation method that transformed the original problem into local convex ones, which enables the effective iterative algorithm based near-optimal solutions. Through numerical simulations, the proposed scheme was compared with existing benchmarks. The results demonstrated that the proposed approach outperforms both single-UAV setups and multiple UAVs utilizing a linear EH model, highlighting the advantages of the proposed scheme.

References

1. Liang, X., Zhang, Z., Yu, H., Wang, Y., Fang, Y., & Han, J. (2022). Antiswing control for aerial transportation of the suspended cargo by dual quadrotor UAVs. *IEEE/ASME Transactions on Mechatronics, 27*, 5159–5172.
2. Mohsan, S. A. H., Khan, M. A., Noor, F., Ullah, I., & Alsharif, M. H. (2022). Towards the unmanned aerial vehicles (UAVs): A comprehensive review. *Drones, 6*, 147.
3. Khoufi, I., Laouiti, A., Adjih, C., & Hadded, M. (2021). UAVs trajectory optimization for data pick up and delivery with time window. *Drones, 5*, 27.
4. Wu, Y., Xu, J., Qiu, L., & Zhang, R. (2018). Capacity of UAV-enabled multicast channel: Joint trajectory design and power allocation. In *2018 IEEE International Conference on Communications (ICC)*, Kansas City, MO, USA (pp. 1–7). https://doi.org/10.1109/ICC.2018.8422581
5. Li, P., & Xu, J. (2019). Capacity characterization of UAV-enabled multiple access channel with trajectory optimization. In *2019 IEEE/CIC International Conference on Communications China (ICCC)*, Changchun, China (pp. 630–635). https://doi.org/10.1109/ICCChina.2019.8855837
6. Wu, Q., Xu, J., & Zhang, R. (2018). Capacity characterization of UAV-enabled two-user broadcast channel. *IEEE Journal on Selected Areas in Communications, 36*, 1955–1971.
7. Liu, C. H., Chen, Z., Tang, J., Xu, J., & Piao, C. (2018). Energy-efficient UAV control for effective and fair communication coverage: A deep reinforcement learning approach. *IEEE Journal on Selected Areas in Communications, 36*, 2059–2070.
8. Chowdhury, M. M. U., Maeng, S. J., Bulut, E., & Güvenç, I. (2020). 3-D trajectory optimization in UAV-assisted cellular networks considering antenna radiation pattern and backhaul constraint. *IEEE Transactions on Aerospace and Electronic Systems, 56*, 3735–3750.
9. Cui, M., Zhang, G., Wu, Q., & Ng, D. W. K. (2018). Robust trajectory and transmit power design for secure UAV communications. *IEEE Transactions on Vehicular Technology, 67*, 9042–9046.
10. Zhou, X., Wu, Q., Yan, S., Shu, F., & Li, J. (2019). UAV-enabled secure communications: Joint trajectory and transmit power optimization. *IEEE Transactions on Vehicular Technology, 68*, 4069–4073.

11. Lu, W., Ding, Y., Gao, Y., Hu, S., Wu, Y., Zhao, N., & Gong, Y. (2022). Resource and trajectory optimization for secure communications in dual unmanned aerial vehicle mobile edge computing systems. *IEEE Transactions on Industrial Informatics, 18*, 2704–2713.
12. Wu, Q., Zeng, Y., & Zhang, R. (2018). Joint trajectory and communication design for multi-UAV enabled wireless networks. *IEEE Transactions on Wireless Communications, 17*, 2109–2121.
13. Wang, H., Ren, G., Chen, J., Ding, G., & Yang, Y. (2018). Unmanned aerial vehicle-aided communications: Joint transmit power and trajectory optimization. *IEEE Wireless Communications Letters, 7*, 522–525.
14. Jiang, X., Wu, Z., Yin, Z., & Yang, Z. (2019). Joint power and trajectory design for UAV-relayed wireless systems. *IEEE Wireless Communications Letters, 8*, 697–700.
15. Jiang, X., Wu, Z., Yin, Z., Yang, W., & Yang, Z. (2019). Trajectory and communication design for UAV-relayed wireless networks. *IEEE Wireless Communications Letters, 8*, 1600–1603.
16. Ebrahimi, D., Sharafeddine, S., Ho, P. H., & Assi, C. (2019). UAV-aided projection-based compressive data gathering in wireless sensor networks. *IEEE Internet of Things Journal, 6*, 1893–1905.
17. Samir, M., Sharafeddine, S., Assi, C. M., Nguyen, T. M., & Ghrayeb, A. (2020). UAV trajectory planning for data collection from time-constrained IoT devices. *IEEE Transactions on Wireless Communications, 19*, 34–46.
18. Cheng, F., Zhang, S., Li, Z., Chen, Y., Zhao, N., Yu, F. R., & Leung, V. C. M. (2018). UAV trajectory optimization for data offloading at the edge of multiple cells. *IEEE Transactions on Vehicular Technology, 67*, 6732–6736.
19. You, C., & Zhang, R. (2019). 3D trajectory optimization in Rician fading for UAV-enabled data harvesting. *IEEE Transactions on Wireless Communications, 18*, 3192–3207.
20. Zeng, Y., Xu, J., & Zhang, R. (2019). Energy minimization for wireless communication with rotary-wing UAV. *IEEE Transactions on Wireless Communications, 18*, 2329–2345.
21. Xu, J., Zeng, Y., & Zhang, R. (2017). UAV-enabled wireless power transfer: Trajectory design and energy region characterization. In *2017 IEEE Globecom Workshops (GC Wkshps)*, Singapore (pp. 1–7). https://doi.org/10.1109/GLOCOMW.2017.8269097
22. Xu, J., Zeng, Y., & Zhang, R. (2018). UAV-enabled wireless power transfer: trajectory design and energy optimization. *IEEE Transactions on Wireless Communications, 17*, 5092–5106.
23. Hu, Y., Yuan, X., Xu, J., & Schmeink, A. (2019). Optimal 1D trajectory design for UAV-enabled multiuser wireless power transfer. *IEEE Transactions on Communications, 67*, 5674–5688.
24. Yuan, X., Yang, T., Hu, Y., Xu, J., & Schmeink, A. (2021). Trajectory design for UAV-enabled multiuser wireless power transfer with nonlinear energy harvesting. *IEEE Transactions on Wireless Communications, 20*, 1105–1121.
25. Yuan, X., Hu, Y., & Schmeink, A. (2021). Joint design of UAV trajectory and directional antenna orientation in UAV-enabled wireless power transfer networks. *IEEE Journal on Selected Areas in Communications, 39*, 3081–3096.
26. Wang, Z., Zhang, G., Wang, Q., Wang, K., & Yang, K. (2021). Completion time minimization in wireless-powered UAV-assisted data collection system. *IEEE Communications Letters, 25*, 1954–1958.
27. Wang, Z., Wen, M., Dang, S., Yu, L., & Wang, Y. (2021). Trajectory design and resource allocation for UAV energy minimization in a rotary-wing UAV-enabled WPCN. *Alexandria Engineering Journal, 60*, 1787–1796.
28. Zhang, D., Wang, Y., Hua, M., Ji, B., & Shah, S. A. (2019). UAV-based mobile wireless power transfer systems with joint optimization of user scheduling and trajectory. *Mobile Networks and Applications, 27*, 1813–1827.
29. Liu, J., Zhao, X., Qin, P., Geng, S., & Meng, S. (2022). Joint dynamic task offloading and resource scheduling for WPT-enabled space-air-ground power Internet of Things. *IEEE Transactions on Network Science and Engineering, 9*, 660–677.
30. Wang, Y., Ru, Z. Y., Wang, K., & Huang, P. Q. (2020). Joint deployment and task scheduling optimization for large-scale mobile users in multi-UAV-enabled mobile edge computing. *IEEE Transactions on Cybernetics, 50*, 3984–3997.

31. Hu, Q., Cai, Y., Yu, G., Qin, Z., Zhao, M., & Li, G. Y. (2019). Joint offloading and trajectory design for UAV-enabled mobile edge computing systems. *IEEE Internet of Things Journal, 6*, 1879–1892.
32. Park, J., Lee, H., Eom, S., & Lee, I. (2019). UAV-aided wireless powered communication networks: Trajectory optimization and resource allocation for minimum throughput maximization. *IEEE Access, 7*, 134978–134991.
33. Huang, F., Chen, J., Wang, H., Ding, G., Xue, Z., Yang, Y., & Song, F. (2019). UAV-assisted SWIPT in Internet of Things with power splitting: trajectory design and power allocation. *IEEE Access, 7*, 68260–68270.
34. Xie, L., Cao, X., Xu, J., & Zhang, R. (2021). UAV-enabled wireless power transfer: A tutorial overview. *IEEE Transactions on Green Communications and Networking, 5*, 2042–2064.
35. Wu, P., Xiao, F., Huang, H., & Wang, R. (2020). Load balance and trajectory design in multi-UAV aided large-scale wireless rechargeable networks. *IEEE Transactions on Vehicular Technology, 69*, 13756–13767.
36. Mu, J., & Sun, Z. (2022). Trajectory design for multi-UAV-aided wireless power transfer toward future wireless systems. *Sensors, 22*, 68510.
37. Xiong, K., Liu, Y., Zhang, L., Gao, B., Cao, J., Fan, P., & Letaief, K. B. (2022). Joint optimization of trajectory, task offloading, and CPU control in UAV-assisted wireless powered fog computing networks. *IEEE Trans. Green Commun. Netw., 6*, 1833–1845.
38. Wu, P., Xiao, F., Huang, H., Sha, C., & Yu, S. (2020). Adaptive and extensible energy supply mechanism for UAV-aided wireless-powered Internet of Things. *IEEE Internet of Things Journal, 7*, 9201–9213.
39. Yuan, X., Hu, Y., Schober, R., & Schmeink, A. (2022). Convexity analysis of nonlinear wireless power transfer with multiple RF sources. *IEEE Transactions on Vehicular Technology, 71*, 11311–11316.
40. Bubeck, S. (2015). *Convex optimization: Algorithms and complexity*. arXiv:1405.4980
41. Yan, H., Chen, Y., & Yang, S. H. (2020). UAV-enabled wireless power transfer with base station charging and uav power consumption. *IEEE Transactions on Vehicular Technology, 69*, 12883–12896.
42. Yu, Y., Tang, J., Huang, J., Zhang, X., & Ka Chun So, D. W. K. (2021). Multi-objective optimization for UAV-assisted wireless powered IoT networks based on extended DDPG algorithm. *IEEE Transactions on Communications, 69*, 6361–6374.

Chapter 11
Joint Power Allocation and Trajectory Design for UAV-Enabled Covert Communication

Peng Wu, Xiaopeng Yuan, and Yulin Hu

Abstract This chapter investigates covert communication in a UAV-assisted network, where a UAV covertly disseminates data to multiple ground users while evading detection by a concealed warden. To ensure transmission fairness, our objective is to maximize the minimum throughput among users through a joint optimization of the UAV's trajectory, transmit power, and power allocation variables, subject to UAV mobility and covertness constraints. Under the covertness constraint, we derive a closed-form expression for the maximum transmit power. Leveraging the expression and optimal successive hover-and-fly (SHF) structure, we construct a joint trajectory and power allocation design problem containing a reduced set of variables. This new problem is efficiently solved by the sequential convex programming (SCP) approach which involves iteratively constructing tight concave approximations of the original problem, enabling convergence to a high-quality solution.

Keywords Unmanned aerial vehicle · Covert communication · Trajectory optimization · Power allocation · Successive-hover-and-fly

Abbreviations

A2G	Aerial to Ground
DDPG	Deep Deterministic Policy Gradient
ISAC	Integrated Sensing and Communication
LoS	Line of Sight

P. Wu · Y. Hu (✉)
School of Electronic Information, Wuhan University, Wuhan, China
e-mail: yulin.hu@whu.edu.cn

P. Wu
e-mail: peng.wu@whu.edu.cn

X. Yuan
Chair of Information Theory and Data Analytics, RWTH Aachen University, Aachen, Germany
e-mail: yuan@inda.rwth-aachen.de

Y. Hu et al. (eds.), *Reliable and Efficient Trajectory Design in UAV-Aided Wireless Networks*, https://doi.org/10.1007/978-981-96-5988-3_11

PLS	Physical Layer Security
SCP	Sequential Convex Programming
SHF	Successive Hover-and-Fly
TD	Time Discretization
UAV	Unmanned Aerial Vehicle

11.1 Introduction

11.1.1 Background

Owing to their outstanding performance characteristics, including remarkable flexibility, extended endurance, expansive coverage, and especially the high quality line-of-sight (LoS) aerial-to-ground (A2G) channel [1], unmanned aerial vehicles (UAVs)-enabled wireless communication have been expected to be an emerging technology of future networks [2], where the system performance can be significantly improved through UAV trajectory design through utilizing its inherent performance in mobility [3]. However, the inherently open nature of communication channels in A2G links poses significant challenges for UAV-enabled communication systems which are vulnerable to malicious user called warden interventions [4]. The covert communication technology is key to ensuring transmission privacy and security in UAV-enabled communication networks. Through effective transmit power control and collaborative mechanism, the transmission exposure probability can be notably reduced [5]. In the meanwhile, the combination of effectively covert communication scheme with UAV trajectory design can provide more private service for users [6]. Nevertheless, due to the position uncertainty of warden and high design complexity of UAV trajectory design, efficient UAV trajectory design in covert communication scenario is still challenging. Thus, it is necessary to study characterization of covert performance for randomly distributed warden and low complexity trajectory design method.

11.1.2 Motivation

With the development of UAV covert communication which can slope the communication privacy via power control, collaborative jamming, and trajectory design technology, the performance including transmission covertness, throughput, latency, and so on can be highly enhanced. Nevertheless, almost all work on UAV trajectory design in covert communication scenario are under the assumption of perfect knowledge of warden and has extremely high design complexity based on the time discretization (TD) sequential convex programming (SCP) which is difficult to bear. It is presently very difficult to apply the covert communication technology in practical application. With the increasing emphasis on transmission covertness and algorithm complexity, it is significant to enhance the practical significance when investigate the covet communication scheme while considering the incompleteness of warden information, and to study efficient trajectory design algorithm for reducing the demand for hardware and cope with diverse applications.

11.1.3 Main Contributions

Main contributions to this chapter are as follows:

(1) Given the uncertainty in background noise, we deduce the average minimal detection error probability in UAV-enabled covert communication scenario considering the position uncertainty of warden, which enables us to evaluate the covert communication performance and further to derive a close form expression of optimal transmit power under covert communication constraint;
(2) Upon the optimal SHF structure with turning points, the joint trajectory and power allocation design problem with infinite variables is reformulated as a non-convex optimization problem with a finite set of variables. To tackle the intricacies of the non-concave throughput, an algorithm based on sequential convex programming (SCP) is leveraged to iteratively solve the non-convex optimization problem;
(3) Through simulation analyses, we evaluate the performance of proposed methodology is evaluated in comparison to benchmark where the convergence and advantages of proposed scheme is verified.

11.1.4 Structural Organization

Relevant UAV trajectory design and covert communication technologies are introduced in Sect. 11.2. Section 11.3 shows the system model and formulates the considered joint design problem. Section 11.4 presents the derivation on the covert communication performance metric. In Sect. 11.5, a low complexity design based on the SHF structure and convex approximation is provided. Section 11.6 provides numerical results to evaluate the proposed scheme and Sect. 11.7 is the conclusion.

11.2 Related Works

11.2.1 UAV Secure Communication

To improve the transmission security, one well-known and widely studied physical layer security (PLS) technology is secure communication technology. It mainly utilizes the differences between legitimate and illegitimate channels to protect the confidential transmission information through physical layer encoding and other methods [7–9], which is also applied in UAV communication network and combining with trajectory design to improve the security for UAV delivering critical data [1, 2, 10–13]. Authors in [10] developed a Deep Deterministic Policy Gradient (DDPG) based trajectory and transmit power design for dual-UAV aided secure communication system. Regarding the transmission security issues of UAV relay networks, the trajectory and resource allocation scheme have been studied in [11, 12] where the SCP-based algorithm was developed to maximize the secrecy rate. Reference [13] presented a method for real-time trajectory design in UAV secure integrated sensing and communication (ISAC) network. To further enhance the transmission security, the flexibility in channel design of Intelligent Reflecting Surface (IRS) is utilized where the UAV trajectory and resource allocation were jointly designed [1, 2].

The above studies have to some extent improved the security of UAV communication from the perspective of protecting privacy data, proving that UAV secure communication technology can be used in UAV communication application.

11.2.2 UAV Covert Communication

The aforementioned research pertaining to secure communication facilitated by UAV predominantly safeguarding the confidentiality transmitted information content. Nonetheless, in scenarios such as military applications, the necessity to hide the information transmission behaviour becomes paramount and possibly even more crucial. As a result, ensuring the covertness of UAV transmissions has emerged as a compelling concern and topic of interest [14–16]. Until now, current studies have demonstrated that it is feasible to enhance the level of covertness by optimizing the transmit power and trajectory of UAV. This approach offers a means to improve the stealth of information transmission [5, 6, 17–23]. Within the context of jammer-assisted UAV covert communication scenario, authors in [5, 17, 18] have analytically derived close form expressions for key covert performance metrics which help to draft an joint jamming power and trajectory design algorithm with the objective of maximizing the covert communication rate. In another way to achieve covert communication, the inherent uncertainty of background noise was explored where the UAV trajectory and resource allocation were jointly designed under the assumption of perfect knowledge of warden to boost system performance which offers a holistic approach to enhancing the transmission efficiency and covertness of the communication system [6, 19, 20]. The emerging IRS technology was also utilized in UAV covert communication networks and investigated by growing works which has been proved that the covertness performance be further strengthened through joint design the UAV trajectory, resource allocation and IRS phase shift matrix [21–23].

Although these researches have contributed a lot to UAV covert communication and transmission security, there is remaining for improvement in practical applications. For application scenarios, due to non-cooperative features for illegal users, the perfect information of warden is almost impossible to obtain. It is meaningful to investigate the covert communication scheme considering the uncertainty of warden. In the meanwhile, design complexity is a crucial concern in UAV trajectory design while the above works based on TD-SCP are generally leads to an extremely high complexity. Therefore, in this work, we focus on the scenario of covert communication facilitated by UAV, to investigate covert communication strategy with practical significance and efficient UAV trajectory and resource allocation algorithm for system performance enhancement.

11.3 System Model and Problem Statement

11.3.1 Considered Scenario

Here is the considered UAV-enabled covert communication network shown as Fig. 11.1 where K users and an illegal warden is randomly distributed on the ground. One UAV is starting from the initial position $\boldsymbol{q}_0 = (x_0, y_0)$ to the final position $\boldsymbol{q}_F = (x_F, y_F)$ at fixed height H limited by the maximum speed V, to transmit data to users covertly against the detection of warden. The horizontal position of user k is $\boldsymbol{u}_k = \left(u_{x,k}, u_{y,k}\right), \forall k \in \mathcal{K} \triangleq \{1, \ldots, K\}$ which is accurately known as collaborative user. The horizontal position of warden is $\boldsymbol{w} = \left(w_x, w_y\right)$ which is not perfectly known as a non-cooperative user, and following the work [24], the Gaussian error model is employed to represent the inaccuracy of the warden position such that its position can be modelled as $\boldsymbol{w} = \hat{\boldsymbol{w}} + \Delta\boldsymbol{w}$ where $\hat{\boldsymbol{w}} = \left(\hat{w}_x, \hat{w}_y\right)$ is the estimated position of warden and $\Delta\boldsymbol{w}_d = \left(\Delta w_{x,d}, \Delta w_{y,d}\right)$ is two-dimensional Gaussian random variables to denote position uncertainty which follows $N\left(0, \varepsilon_w^2\right)$ in each dimention. The whole task time T such that we have the timely position and transmit power of UAV as $\boldsymbol{q}(t) = (x(t), y(t))$ and $P(t),\ t \in \mathcal{T} \triangleq [0, T]$.

The A2G channels are generally LoS-dominant such that the channel power gain from UAV to user k is $h_k(t) = \beta_0 d_k(\boldsymbol{q}(t))^{-2}$ and from UAV to warden is $h_w(t) = \beta_0 d_w(\boldsymbol{q}(t))^{-2}$, where β_0 is the unit gain, $d_k(\boldsymbol{q}(t)) = \sqrt{\boldsymbol{q}(t) - \boldsymbol{u}_k^2 + H^2}$ and $d_w(\boldsymbol{q}(t)) = \sqrt{\boldsymbol{q}(t) - \boldsymbol{w}^2 + H^2}$ is the distance form UAV to user k and warden.

To against interference, the frequency-division multiple access (FDMA) communication model is adopted where the whole transmit power $P(t)$ is allocated to the

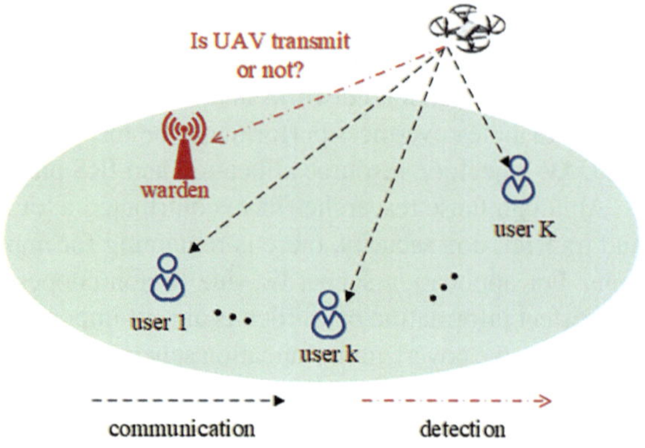

Fig. 11.1 UAV-enabled covert communication network

corresponding frequency band of each user and the allocated variables is $s_k(t)$ such that we have the signal power at user k is $Q_k(t) = s_k(t)h_k(t)P(t)$ and $Q_w(t) = h_w(t)P(t)$. Denoting the allocation coefficient set by $\mathbf{s}(t)$, the overall throughput of user k is:

$$U_k(\mathbf{q}(t), P(t), \mathbf{s}(t)) = \int_0^T log_2\left(1 + \frac{Q_k(t)}{\sigma^2}\right)dt \tag{11.1}$$

where σ^2 is the background noise power.

11.3.2 Problem Statement

For fairness-aware consideration, we aim at jointly optimizing UAV trajectory, transmit power and power allocation variable to maximize the minimum throughout of all users, where the original problem is formulated as:

$$\text{(OP)}: \quad \underset{\{q(t)\},\{P(t)\},\{s(t)\}}{max} \ \underset{k}{min}\, U_k(\boldsymbol{q}(t), P(t), \mathbf{s}(t)) \tag{11.2}$$

$$\boldsymbol{q}(0) = \boldsymbol{q}_0, \boldsymbol{q}(T) = \boldsymbol{q}_F \tag{11.2a}$$

$$\dot{\boldsymbol{q}}(t) \leq V, t \in \mathcal{T} \tag{11.2b}$$

$$P(t) \leq P_{max}, t \in \mathcal{T} \tag{11.2c}$$

$$0 \le s_k(t) \le 1, \sum_{k=1}^{K} s_k(t) = 1, k \in \mathcal{K}, t \in \mathcal{T} \tag{11.2d}$$

$$\xi(t) \ge 1 - \rho_w, t \in \mathcal{T} \tag{11.2e}$$

where constrain (11.2a) and (11.2b) are mobility constraints,Eq. (11.2c) is transmit power constraint, Eq. (11.2d) is power allocation constraint, and covert communication constraint means the detection error probability $\xi(t)$ must over than a threshold $1 - \rho_w$. The above problem is hard to solve for the following three reasons: (i) The throughput $U_k(\boldsymbol{q}(t), P(t), s(t))$ is highly non-concave integral form due to non-convex problem. (ii) The number of variables is infinite. (iii) The expression of $\xi(t)$ is hard to derive.

It is worthwhile to recognize that the primary objective of warden is to identify the optimal detection threshold $\tau^{\star}(t)$ that minimizes $\xi(t)$. Additionally, we must ensure that $\xi(t)$ remains above $1 - \rho_w$, regardless of detection threshold $\tau(t)$ selected by warden, while achieving the optimal solution. This necessitates contemplating the worst-case scenario, where the detector employs $\tau^{\star}(t)$ to attain the minimum $\xi^{\star}(t)$. Subsequently, we will derive in the subsequent section the expression for the minimal detection error probability $\xi^{\star}(t)$.

11.4 Covert Communication Performance

In this section, the optimal detection threshold $\tau^{\star}(t)$ and minimal detection error probability $\xi^{\star}(t)$ are derived firstly with a perfect knowledge of warden. Then, considering the distribution uncertainty of warden, the average of minimal detection error probability $\overline{\xi}^{\star}(t)$ is deduced to evaluate the convet communication performance.

11.4.1 Minimal Detection Probability

From the perspective of warden, it monitors whether UAV is transmitting or not through comparing the set detection threshold $\tau(t)$ with the overall signal power which is given by:

$$P_w(t) = \begin{cases} \sigma_w^2, & \mathcal{H}_0 \\ Q_w(t) + \sigma_w^2, & \mathcal{H}_1 \end{cases} \tag{11.3}$$

where $\mathcal{H}_0$ denotes the hypothesis that there is no transmission behaviour and $\mathcal{H}_1$ is its opposite hypothesis. σ_w^2 is the estimated background noise power at warden is uncertain for the warden as it has no does not have strong real-time channel estimation ability, we approximately assume it follows a uniform distribution within

$\left[\sigma^2 - \rho, \sigma^2 + \rho\right]$. The warden detection process can be expressed as:

$$\frac{f\left(P_w(t)|\mathscr{H}_0\right)}{f\left(P_w(t)|\mathscr{H}_1\right)} \underset{\mathcal{D}_1}{\overset{\mathcal{D}_0}{\gtrless}} 1 \tag{11.4}$$

where $f\left(P_w(t)|\mathscr{H}_0\right)$ and $f\left(P_w(t)|\mathscr{H}_1\right)$ are the probability density function of the overall signal power at warden when UAV is not or is transmitting respectively, $\mathcal{D}_0$ is the decision in favour of $\mathscr{H}_0$ and $\mathcal{D}_1$ is the decision in favour of $\mathscr{H}_1$. According to (11.4), the optimal detection threshold $\tau^\star(t)$ is the value of $P_w(t)$ when the equality holds, i.e., $\tau^\star(t) \in \left[Q_w(t) + \sigma^2 - \rho, \sigma^2 + \rho\right)$, and minimal detection error probability $\xi^\star(t)$ is given by:

$$\xi^\star(t) = P_r(\mathcal{D}_1|\mathscr{H}_0) + P_r(\mathcal{D}_0|\mathscr{H}_1) = 1 - \frac{Q_w(t)}{2\rho}. \tag{11.5}$$

11.4.2 The Average of Minimal Detection Probability

In the last subsection, the minimal detection error probability $\xi^\star(t)$ with a perfectly known position is derived. Considering the practical issue that its position is with uncertainty, the average of $\xi^\star(t)$ is selected and further studied as the covert communication metric. We first introduce the normalized square of horizontal distance between UAV and warden with position uncertainty as $M(t) \triangleq \frac{1}{\epsilon_w^2}\|\mathbf{q}(t) - \hat{\mathbf{w}} - \Delta\mathbf{w}\|^2$, which follows non-central chi-square distribution with the non-central parameter $\psi(t) = \frac{1}{\epsilon_w^2}(\|\boldsymbol{q}(t) - \hat{\boldsymbol{w}}\|)^2$ and 2 freedom degrees. The average of minimal detection error probability is the expectation of square of horizontal distance between UAV and warden with position $\xi^\star(t)$ with respect to $\psi(t)$, i.e.,

$$\overline{\xi}^*(t) = \int_0^{+\infty} \xi^\star((t))f(M(t))dM(t) = 1 - E_{M(t)}\left[\frac{\beta_0 P(t)}{2\rho\,\epsilon_d^2\left(M(t) + \frac{H^2}{\epsilon_w^2}\right)}\right] \tag{11.6}$$

where $f(M(t))$ is the probability density function of $M(t)$ and $\frac{\beta_0 P(t)}{2\rho\epsilon_d^2\left(M(t)+\frac{H^2}{\epsilon_w^2}\right)}$ is convex function with respect to $M(t)$. As $M(t)$ follows non-central chi-square distribution, the close form expression of $\overline{\xi}^*(t)$ is hard to derive. However, We can obtain a lower bound of $\overline{\xi}^*(t)$ which can replace $\overline{\xi}^*(t)$ to get a stricter covert communication constraint.

Firstly, according to [25], for any random variable $z \in [z_1, z_2]$ with expectation μ_z, the following inequality holds for convex function $g(z)$:

$$E_z[g(z)] \leq \frac{E(|z-\mu_z|)}{2(\mu_x-a)}g(z_1) + \left(1 - \frac{E(|z-\mu_z|)}{2(\mu_z-z_1)} - \frac{E(|z-\mu_z|)}{2(\mu_z-z_2)}\right)g(\mu_z) + \frac{E(|z-\mu_z|)}{2(\mu_z-z_2)}g(z_2) \tag{11.7}$$

For $M(t) \in [0, \infty)$ which follows non-central chi-square distribution with the non-central parameter $\psi(t)$ and 2 freedom degrees, $\mu_{W(t)} = \psi(t) + 2$. According to [14], a non-central chi-square distribution with a large non-central parameter ψ and λ freedom degrees can be precisely approximated to normal distribution $N(\psi + 2, 4\psi + 2\lambda)$ and accordingly $M(t)$ can be approximated to $N(\psi(t)+2, 4\psi(t)+4)$ where $E_{W(t)}\left(\left|W(t) - \mu_{W(t)}\right|\right) = \sqrt{\frac{8}{\pi}(\psi(t)+1)}$. Above all, the upper bound of $E_{M(t)}\left[\frac{\beta_0 P(t)}{2\rho\epsilon_d^2\left(M(t)+\frac{H^2}{\epsilon_w^2}\right)}\right]$ is:

$$E_{M(t)}\left[\frac{\beta_0 P(t)}{2\rho\,\epsilon_d^2\left(M(t)+\frac{H^2}{\epsilon_w^2}\right)}\right] \leq \frac{\beta_1 P(t)}{\psi(t)+2+\frac{H^2}{\epsilon_d^2}} + \sqrt{\frac{2}{\pi}}\frac{1}{\sqrt{\psi(t)+2}}\left(\frac{\epsilon_d^2\,\beta_1 P(t)}{H^2} - \frac{\beta_1 P(t)}{\psi(t)+2+\frac{H^2}{\epsilon_d^2}}\right) \triangleq \chi_1(\psi(t), P(t)) \tag{11.8}$$

where $\beta_1 = \frac{\beta_0}{\rho_w}$. According to (11.8), we have $\overline{\xi}^*(t) \geq 1 - \chi_1(\psi(t), P(t))$ and the stricter covert communication constraint is $\chi_1(\psi(t), P(t)) \leq \rho_w$.

11.4.3 The Maximum Transmit Power and Transformed Problem

To maximize the throughput, the UAV transmit power should be as large as possible under the transmit power constraint (11.2c) and covert communication (11.2e). As $\chi_1(\psi(t), P(t))$ is monotonically increasing with respect to $P(t)$, the maximum transmit power is:

$$P^\star(t) = \min\left\{\frac{\beta_2}{\frac{1}{\upsilon(\boldsymbol{q}(t))+2_w^2+H^2} + \sqrt{\frac{2}{\pi}}\frac{w}{\sqrt{\upsilon(\boldsymbol{q}(t))+2_w^2}}\left(\frac{1}{H^2} - \frac{1}{\upsilon(\boldsymbol{q}(t))+2_w^2+H^2}\right)}, P_{max}\right\} \triangleq \min\{\chi_2(\upsilon(\boldsymbol{q}(t))), P_{max}\} \triangleq \chi_3(\upsilon(\boldsymbol{q}(t))) \tag{11.9}$$

where $\beta_2 = \frac{\rho_w}{\epsilon_w^2 \beta_1}$ and $\upsilon(\mathbf{q}(t)) = \left\|\mathbf{q}(t) - \widehat{\mathbf{w}}\right\|^2$ is the square of horizontal distance between UAV and estimated position of warden. According to (11.8), we can equivalently transform problem (OP) to:

$$\text{(P1):} \quad \max_{\{\boldsymbol{q}(t)\},\{\boldsymbol{s}(t)\}} \min_{k} \int_0^T log_2\left(1 + \frac{\beta_0 s_k(t)\chi_3(\upsilon(\boldsymbol{q}(t)))}{\sigma^2 d_w(\boldsymbol{q}(t))^2}\right)dt \tag{11.10}$$

$$\boldsymbol{q}(0) = \boldsymbol{q}_0, \boldsymbol{q}(T) = \boldsymbol{q}_F \tag{11.10a}$$

$$\dot{\boldsymbol{q}}(t) \leq V, t \in \mathcal{T} \tag{11.10b}$$

$$0 \leq s_k(t) \leq 1, \sum_{k=1}^{K} s_k(t) = 1, k \in \mathcal{K}, t \in \mathcal{T} \tag{11.10c}$$

To solve the non-convex problem (P1) with infinite variables, in the next section, the optimal SHF structure will be adopted to reformulate it into one with finite variables without loss of optimality. And then, an efficiently iterative algorithm based on convex approximation is provided.

11.5 Low Complexity Design

11.5.1 Problem Reformulation

Applying similar mathematical analysis process as Chaps. 1–4, the optimal UAV trajectory solution of problem (P1) still follows the so-called SHF structure, i.e., UAV havers at K hovering points which lasts for the corresponding hovering time, and it flies among those points along certain path with maximum speed V. Consequently, we are motivated to adopt the optimal SHF structure to reformulate the trajectory and power allocation variables to reduce variables without loss of optimality.

In concrete term, we define K hovering points whose position is represented by $\boldsymbol{q}_m, m \in \mathcal{K}$. The corresponding hovering time is τ_m where the set of all hovering times is $\boldsymbol{t} = \{\tau_m\}, m \in \mathcal{K}$. In 2D trajectory design scenario, the flying path with maximum speed V between hovering points is not simply straight line but certain arc. Therefore, we introduce N turning points between two neighbour hovering points which are denoted by $\boldsymbol{q}_{m,n}, n \in \mathcal{N} \triangleq \{1, ..., N\}$ to constrcut $N+1$ line segments for path fitting as shown in Fig. 11.2. Noting that a significant quantity of turning points enhances the accuracy of fitting, albeit at the expense of heightening design complexity. This necessitates a delicate balancing act when determining the value of N, requiring a comprehensive consideration of both performance enhancement

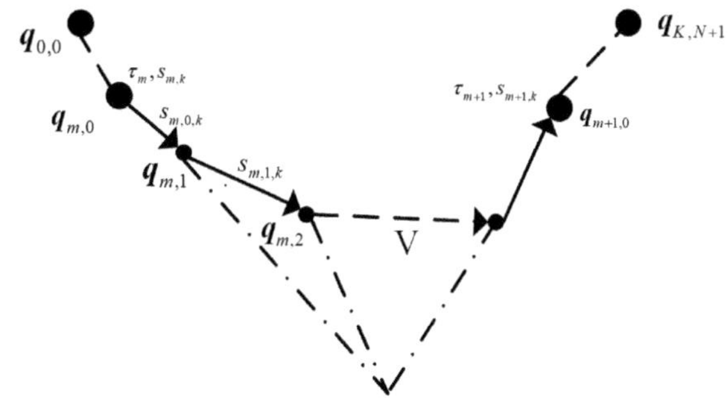

Fig. 11.2 Reformulated variables based on SHF structure with turning points

and complexity reduction. For denotation unification, the hovering point is redefined as $\boldsymbol{q}_{m,0}(\boldsymbol{q}_{m-1,N+1}) = \boldsymbol{q}_m$ such that the reformulated UAV trajectory can be deterministically represented by the set of hovering times $\boldsymbol{t}$ and trajectory points $\boldsymbol{q} = [\boldsymbol{q}_{0,0}, \boldsymbol{q}_{0,1}, \ldots, \boldsymbol{q}_{1,0}(\boldsymbol{q}_{0,N+1}), \boldsymbol{q}_{1,1}, \ldots, \boldsymbol{q}_{K+1,0}(\boldsymbol{q}_{K,N+1})]$ (including the hovering points and turning points).

In the reformulation trajectory, the UAV will sequentially navigate through each trajectory point in $\boldsymbol{q}$, flying with maximum speed V along the line segments connecting these turning points and hovering at specified hovering points for their corresponding hovering time. Defining the distance between two consecutive trajectory points $\boldsymbol{q}_{m,n}$ and $\boldsymbol{q}_{m,n+1}$ as $l_{m,n}(\boldsymbol{q}) = \|\boldsymbol{q}_{m,n+1} - \boldsymbol{q}_{m,n}\|$, and assuming the moment of departure from $\boldsymbol{q}_{m,n}$ as the starting point, the path traced by the UAV between these points can be formulated as $\boldsymbol{q}_{m,n}(\tau) = \boldsymbol{q}_{m,n} + \frac{(\boldsymbol{q}_{m,n+1} - \boldsymbol{q}_{m,n})V\tau}{\|\boldsymbol{q}_{m,n+1} - \boldsymbol{q}_{m,n}\|}$, $\tau \in \left[0, \frac{l_{m,n}(\boldsymbol{q})}{V}\right]$. Based on the reformulation trajectory, the whole task time can be denoted by $T(\boldsymbol{q}, \boldsymbol{t}) = \sum_{m=1}^{K} \tau_i + \sum_{m=1}^{K} \sum_{n=0}^{N} \frac{l_{m,n}(\boldsymbol{q})}{V}$ which is convex in the reformulation variables $(\boldsymbol{q}, \boldsymbol{t})$.

Moreover, motivated by the optimal SHF structure, we are compelled to redefine the power allocation variables, recognizing that the power allocation variable to a single ground user remains constant at each hovering point $\boldsymbol{q}_m$ and while traversing each line segment $l_{m,n}(\boldsymbol{q})$. Denoting the power allocation variable in hovering point $\boldsymbol{q}_m$ and flying line segment $l_{m,n}(\boldsymbol{q})$ for user k by $s_{m,k}$ and $s_{m,n,k}$ respectively and using $\boldsymbol{s}$ to denote the set of all allocation variables, we have the throughput for each user based on the reformulated trajectory and scheduling variables:

$$U_k(\mathbf{q}, \mathbf{t}, \mathbf{s}) = \sum_{m=1}^{K} \log_2\left(1 + \frac{\beta_0 s_{m,k} \chi_3(\upsilon(\mathbf{q}_m))}{\sigma^2 d_k(\mathbf{q}_m)^2}\right)\tau_m$$
$$+ \sum_{m=0}^{K}\sum_{n=0}^{N} \int_0^{\frac{l_{i,j}(\boldsymbol{q})}{V}} \log_2\left(1 + \frac{\beta_0 s_{m,n,k} \chi_3\left(\upsilon\left(\mathbf{q}_{m,n}(\tau)\right)\right)}{\sigma^2 d_k\left(\mathbf{q}_{m,n}(\tau)\right)^2}\right) d\tau \quad (11.11)$$

Under the reformulation variables, the transformed problem (P1) can be reformulated as:

$$\text{(P2)}: \quad \underset{q,t,s}{max}\ \underset{k}{min}\ U_k(\mathbf{q}, \mathbf{t}, \mathbf{s}) \tag{11.12}$$

$$T(\boldsymbol{q}, \boldsymbol{t}) \leq T \tag{11.12a}$$

$$\boldsymbol{t} \geq 0 \tag{11.12b}$$

$$0 \leq \mathbf{s} \leq 1, \sum_{k=1}^{K} s_{m,k} = 1, \sum_{k=1}^{K} s_{m,n,k} = 1 \tag{11.12c}$$

Noting that problem (P2) is still non-convex due to the non-concave $U_k(\mathbf{q}, \mathbf{t}, \mathbf{s})$ and all of the variables are coupled with each other. To make (P2) more addressable, we are motivated to adopt the SCA technique, building up a convex approximation for (P2) by constructing a sequence of lower bound concave function for $U_k(\mathbf{q}, \mathbf{t}, \mathbf{s})$ at the local point $\left(\boldsymbol{q}^{(i)}, \boldsymbol{t}^{(i)}, \boldsymbol{s}^{(i)}\right)$.

11.5.2 Convex Approximation

Problem (P2) is non-convex as the throughput function $U_k(\mathbf{q}, \mathbf{t}, \mathbf{s})$ is non-concave, to tackle this difficulty, we will give a lower-bound concave function $U_k(\mathbf{q}, \mathbf{t}, \mathbf{s})$ at the local point $\left(\boldsymbol{q}^{(i)}, \boldsymbol{t}^{(i)}, \boldsymbol{s}^{(i)}\right)$ to construct a convex problem for (P2). More specifically, $U_k(\mathbf{q}, \mathbf{t}, \mathbf{s})$ is constructed through proposing approximation for throughput at each hovering point, represented by $\theta_{m,k}^{(i)}(\mathbf{q}, \mathbf{t}, \mathbf{s})$ and in flying period in each line segment with maximum speed represented by $\gamma_{m,n,k}^{(i)}(\mathbf{q}, \mathbf{t}, \mathbf{s})$.

Approximation at Hovering Point

According to the convexity of $\log_2(1 + \frac{1}{x})$, we have $\log_2\left(1 + \frac{\beta_0 s_{m,k} \chi_3(\upsilon(q_m))}{\sigma^2 d_k(q_m)^2}\right)$ is monotonically decreasing and convex in $\frac{d_k(q_m)^2}{s_{m,k}\chi_3(\upsilon(q_m))}$ such that at the local point $\left(\boldsymbol{q}^{(i)}, \boldsymbol{t}^{(i)}, \boldsymbol{s}^{(i)}\right)$, the following inequality holds:

$$\log_2\left(1+\frac{\beta_0 s_{m,k}\chi_3(\upsilon(\mathbf{q}_m))}{\sigma^2 d_k(\mathbf{q}_m)^2}\right)\tau_m \geq -A_{1,m,k}^{(i)}\frac{d_k(\mathbf{q}_m)^2}{s_{m,k}\chi_3(\upsilon(\mathbf{q}_m))}\tau_m + A_{2,m,k}^{(i)}\tau_m \quad (11.13)$$

where positive constants $A_{1,m,k}^{(i)} = \frac{\beta_0\,(s_{m,k}^{(i)})^2\,\chi_3\left(\upsilon\left(\mathbf{q}_m^{(i)}\right)\right)^2}{(\sigma^2\,d_k(\mathbf{q}_m^{(i)})^4+\beta_0\,d_k(\mathbf{q}_m^{(i)})^2\,s_{m,k}^{(r)}\,\chi_3\left(\upsilon\left(\mathbf{q}_m^{(i)}\right)\right)\ln 2}$ and $A_{2,m,k}^{(i)} = \log_2\left(1+\frac{\beta_0 s_{m,k}^{(i)}\chi_3\left(\upsilon\left(\mathbf{q}_m^{(i)}\right)\right)}{\sigma^2 d_k(\mathbf{q}_m^{(i)})^2}\right) + A_{1,m,k}^{(i)}\frac{d_k(\mathbf{q}_m^{(i)})^2}{\chi_3\left(\upsilon\left(\mathbf{q}_m^{(i)}\right)\right)}$. Based on the mean inequality, $\forall b_1, ..., b_R \in R^+, b_1 b_2 ... b_R \geq \frac{1}{R}\left(b_1^R +, ..., +b_R^R\right)$ where the equality holds at $b_1 = , \ldots = b_\mathrm{R}$. We have:

$$\begin{aligned}
&-A_{1,m,k}^{(i)}\frac{d_k(\mathbf{q}_m)^2}{s_{m,k}\chi_3(\upsilon(\mathbf{q}_m))}\tau_m + A_{2,m,k}^{(i)}\tau_m\\
&= -\frac{A_{i,k}^{(r)}}{F_{1,m,k}^{(i)}F_{2,m,k}^{(i)}F_{3,m,k}^{(i)}}\frac{F_{1,m,k}^{(i)}F_{2,m,k}^{(i)}F_{3,m,k}^{(i)}d_k(\mathbf{q}_m)^2}{s_{m,k}\chi_3(\upsilon(\mathbf{q}_m))}\tau_m + A_{2,m,k}^{(i)}\tau_m\\
&\geq -\frac{A_{1,m,k}^{(i)}(F_{1,m,k}^{(i)})^3}{4F_{2,m,k}^{(i)}F_{3,m,k}^{(i)}}d_k(\mathbf{q}_m)^8 - \frac{A_{1,m,k}^{(i)}(F_{2,m,k}^{(i)})^3}{4F_{1,m,k}^{(i)}F_{3,m,k}^{(i)}}\frac{1}{s_{m,k}^4}\\
&-\frac{A_{1,m,k}^{(i)}(F_{3,m,k}^{(i)})^3}{4F_{1,m,k}^{(i)}F_{2,m,k}^{(i)}}\frac{1}{\chi_3(\upsilon(\mathbf{q}_m))^4} - \frac{A_{1,m,k}^{(i)}}{4F_{1,m,k}^{(i)}F_{2,m,k}^{(i)}F_{3,m,k}^{(i)}}\tau_m^4 + A_{2,m,k}^{(i)}\tau_m \qquad (11.14)
\end{aligned}$$

where the non-negative constants $F_{1,m,k}^{(i)} = \frac{\tau_i^{(r)}}{d_k(\boldsymbol{q}_m^{(i)})^2}$, $F_{2,m,k}^{(i)} = \tau_m^{(i)} s_{m,k}^{(i)}$, $F_{3,m,k}^{(i)} = \tau_m^{(i)}\chi_3\left(\nu\left(\boldsymbol{q}_m^{(i)}\right)\right)$ are given based on the equal condition of mean equality. Then, to approximate $\frac{1}{\chi_3(\upsilon(\mathbf{q}_m)}$ into upper bound convex form, the mean equality is also applied which gives:

$$\begin{aligned}
\frac{1}{\chi_3(\boldsymbol{q}_m)} \leq \max\Bigg\{&\frac{\beta_3}{\nu(\boldsymbol{q}_m)+2\in_w^2+H^2}\\
&+\frac{\sqrt{2}\in_w\beta_3}{2\sqrt{\pi}H^2}\left(\frac{\nu(\boldsymbol{q}_m)+2\in_w^2}{F_{4,m,k}^{(i)}}+\frac{F_{4,m,k}^{(i)}}{(\nu(\boldsymbol{q}_m)+2\in_w^2+H^2)^2}\right), \frac{1}{P_{max}}\Bigg\}\\
&\triangleq \chi_{4,m}^{(i)}(\boldsymbol{q}_m) \qquad (11.15)
\end{aligned}$$

where $\beta_3 = \frac{1}{\beta_2}$ and $F_{4,m,k}^{(i)} = \left(\nu(\mathbf{q}_m^{(i)})+2\in_w^2+H^2\right)\sqrt{\nu\left(\mathbf{q}_m^{(i)}\right)+2\in_w^2}$. Combining the above approximation steps, we build up the lower bound concave approximation of throughput at hovering point as:

$$\log_2\left(1+\frac{\beta_0 s_{m,k}\chi_3(\upsilon(\mathbf{q}_m))}{\sigma^2 d_k(\mathbf{q}_m)^2}\right)\tau_m$$
$$\geq -\frac{A^{(i)}_{1,m,k}(F^{(i)}_{1,m,k})^3}{4F^{(i)}_{2,m,k}F^{(i)}_{3,m,k}}d_k(\mathbf{q}_m)^8 - \frac{A^{(i)}_{1,m,k}(F^{(i)}_{2,m,k})^3}{4F^{(i)}_{1,m,k}F^{(i)}_{3,m,k}}\frac{1}{s^4_{m,k}}$$
$$-\frac{A^{(i)}_{1,m,k}(F^{(i)}_{3,m,k})^3}{4F^{(i)}_{1,m,k}F^{(i)}_{2,m,k}}\frac{1}{\chi^{(i)}_{4,m}(\mathbf{q}_m)^4} - \frac{A^{(i)}_{1,m,k}}{4F^{(i)}_{1,m,k}F^{(i)}_{2,m,k}F^{(i)}_{3,m,k}}\tau^4_m + A^{(i)}_{2,m,k}\tau_m \triangleq \theta^{(i)}_{m,k}(\mathbf{q},\mathbf{t},\mathbf{s}). \tag{11.16}$$

Approximation in Flying Period

In each flying period among line segment $l_{m,n}(\boldsymbol{q})$, to normalize the upper bound of throughput integral, we introduce $\mathrm{z} = \frac{\mathrm{V}\tau}{\mathrm{l_{m,n}(q)}}$ which gives:

$$\int_0^{\frac{l_{m,n}(\boldsymbol{q})}{V}} log_2\left(1+\frac{\beta_0 s_{m,n,k}\chi_3\left(\upsilon\left(\boldsymbol{q}_{m,n}(\tau)\right)\right)}{\sigma^2 d_k\left(\boldsymbol{q}_{m,n}(\tau)\right)^2}\right)d\tau$$
$$=\frac{l_{m,n}(\boldsymbol{q})}{V}\int_0^1 log_2\left(1+\frac{\beta_0 s_{m,n,k}\chi_3\left(\upsilon\left(\boldsymbol{q}_{m,n}(z)\right)\right)}{\sigma^2 d_k\left(\boldsymbol{q}_{m,n}(z)\right)^2}\right)dz$$
$$\geq \frac{l_{m,n}(\boldsymbol{q})}{V}\int_0^1 A^{(i)}_{1,m,n,k}(z)\frac{d_k\left(\boldsymbol{q}_{m,n}(z)\right)^2}{s_{m,n,k}\chi_3\left(\nu\left(\boldsymbol{q}_{m,n}(z)\right)\right)} + A^{(i)}_{2,m,n,k}(z)dz$$
$$= -\frac{l_{m,n}(\boldsymbol{q})}{Vs_{m,n,k}}\int_0^1 A^{(i)}_{1,m,n,k}(z)\frac{d_k\left(\boldsymbol{q}_{m,n}(z)\right)^2}{\chi_3\left(\nu\left(\boldsymbol{q}_{m,n}(z)\right)\right)}dz + \frac{A^{(i)}_{3,m,n,k}l_{m,n}(\boldsymbol{q})}{V} \tag{11.17}$$

where $\mathbf{q}_{m,n}(z) = \mathbf{q_{m,n}} + z\left(\mathbf{q_{m,n+1}} - \mathbf{q_{m,n}}\right)$, and following constants $A^{(i)}_{1,m,n,k}(z) = A^{(i)}_{1,m,k}|_{\boldsymbol{q}^{(i)}_m=(\boldsymbol{q}_{m,n}(z))^{(i)},s^{(i)}_{m,k}=s^{(i)}_{m,n,k}}$, $A^{(i)}_{2,m,n,k}(z) = A^{(i)}_{2,m,k}|_{\boldsymbol{q}^{(i)}_m=(\boldsymbol{q}_{m,n}(z))^{(i)},s^{(i)}_{m,k}=s^{(i)}_{m,n,k}}$, $A^{(i)}_{3,m,n,k} = \int_0^1 A^{(i)}_{2,m,n,k}(z)dz$ are all positive. Then, the mean inequality is also applied to approximate (11.17) as:

$$
\begin{aligned}
& -\frac{l_{m,n}(\boldsymbol{q})}{Vs_{m,n,k}}\int_0^1 A_{1,m,n,k}^{(i)}(z)\frac{d_k(\boldsymbol{q}_{m,n}(z))^2}{\chi_3(\nu(\boldsymbol{q}_{m,n}(z)))}dz+\frac{A_{3,m,n,k}^{(i)}l_{m,n}(\boldsymbol{q})}{V} \\
&= -\frac{F_{1,m,n,k}^{(i)}F_{2,m,n,k}^{(i)}l_{m,n}(\boldsymbol{q})}{VF_{1,m,n,k}^{(i)}F_{2,m,n,k}^{(i)}s_{m,n,k}}\int_0^1 \frac{A_{1,m,n,k}^{(i)}(z)F_{3,m,n,k}^{(i)}(z)}{F_{3,m,n,k}^{(i)}(z)}\frac{d_k(\boldsymbol{q}_{m,n}(z))^2}{\chi_3(\nu(\boldsymbol{q}_{m,n}(z)))}dz+\frac{A_{3,m,n,k}^{(i)}l_{m,n}(\boldsymbol{q})}{V} \\
&\geq -\frac{(F_{1,m,n,k}^{(i)})^2}{3VF_{2,m,n,k}^{(i)}}\left(\int_0^1 \frac{A_{1,m,n,k}^{(i)}(z)F_{3,m,n,k}^{(i)}(z)}{2\chi_3(\nu(\boldsymbol{q}_{m,n}(z)))^2}+\frac{A_{1,m,n,k}^{(i)}(z)d_k(\boldsymbol{q}_{m,n}(z))^4}{2F_{3,m,n,k}^{(i)}(z)}dz\right)^3 \\
&\quad -\frac{(F_{2,m,n,k}^{(i)})^2}{3VF_{1,m,n,k}^{(i)}}\frac{1}{s_{m,n,k}^3}-\frac{l_{m,n}(\boldsymbol{q})^3}{3VF_{1,m,n,k}^{(i)}F_{2,m,n,k}^{(i)}}+\frac{A_{3,m,n,k}^{(i)}l_{m,n}(\boldsymbol{q})}{V} \\
&\geq -\frac{(F_{1,m,n,k}^{(i)})^2}{3VF_{2,m,n,k}^{(i)}}\left(\int_0^1 \frac{A_{1,m,n,k}^{(i)}(z)F_{3,m,n,k}^{(i)}(z)}{2\chi_{4,m,n}(\nu(\boldsymbol{q}_{m,n}(z)))^2}+\frac{A_{1,m,n,k}^{(i)}(z)d_k(\boldsymbol{q}_{m,n}(z))^4}{2F_{3,m,n,k}^{(i)}(z)}dz\right)^3 \\
&\quad -\frac{(F_{2,m,n,k}^{(i)})^2}{3VF_{1,m,n,k}^{(i)}}\frac{1}{s_{m,n,k}^3}-\frac{l_{m,n}(\boldsymbol{q})^3}{3VF_{1,m,n,k}^{(i)}F_{2,m,n,k}^{(i)}}+\frac{A_{3,m,n,k}^{(i)}l_{m,n}(\boldsymbol{q})}{V}
\end{aligned}
\tag{11.18}
$$

where approximation constants $F_{1,m,n,k}^{(i)}=\frac{l_{m,n}(\boldsymbol{q}^{(i)})}{\int_0^1 A_{1,m,n,k}^{(i)}(z)\frac{d_k(\boldsymbol{q}_{m,n}(z))^2}{\chi_3(\nu(\boldsymbol{q}_{m,n}(z)))}dz}$, $F_{2,m,n,k}^{(i)}=l_{m,n}(\boldsymbol{q}^{(i)})s_{m,n,k}^{(i)}$, $F_{3,m,n,k}^{(i)}(z)=\chi_3\left(\nu\left(\boldsymbol{q}_{m,n}^{(i)}(z)\right)\right)d_k(\boldsymbol{q}_{m,n}^{(i)}(z))^2$ are positive, and $F_{4,m,n,k}^{(i)}=\left(\nu\left(\mathbf{q}_{m,n}^{(i)}(z)\right)+2\in_w^2+H^2\right)\sqrt{\nu\left(\mathbf{q}_{m,n}^{(i)}(z)\right)+2\in_w^2\mathbf{q}}$, $\chi_{4,m,n}\left(\upsilon\left(\boldsymbol{q}_{m,n}(z)\right)\right)=\max\{\frac{\beta_3}{\upsilon\left(\boldsymbol{q}_{m,n}(z)\right)+2\in_w^2+H^2}+\frac{\sqrt{2}\in_w\beta_3}{2\sqrt{\pi}H^2}(\frac{\upsilon\left(\boldsymbol{q}_{m,n}(z)\right)+2\in_w^2}{F_{4,m,n,k}^{(i)}(z)}+$ $\frac{F_{4,m,n,k}^{(i)}(z)}{(\upsilon\left(\boldsymbol{q}_{m,n}(z)\right)+2\in_w^2+H^2)^2},\frac{1}{P_{max}}\}$.

As $l_{m,n}(\mathbf{q})$ is convex in $\mathbf{q}$, the low bound convex approximation of $l_{m,n}(\mathbf{q})$ is $l_{m,n}(\boldsymbol{q})\geq\frac{(\boldsymbol{q}_{m,n+1}^{(i)}-\boldsymbol{q}_{m,n}^{(i)}(\boldsymbol{q}_{m,n+1}-\boldsymbol{q}_{m,n})}{l_{m,n}(\boldsymbol{q}^{(i)})}\triangleq l_{m,n}^{(i)}(\boldsymbol{q})$ which conduct the approximation in flying period:

$$
\begin{aligned}
& \int_0^{\frac{l_{m,n}(\boldsymbol{q})}{V}} log_2\left(1+\frac{\beta_0 s_{m,n,k}\chi_3\left(\upsilon\left(\boldsymbol{q}_{m,n}(\tau)\right)\right)}{\sigma^2 d_k\left(\boldsymbol{q}_{m,n}(\tau)\right)^2}\right)d\tau \\
&\geq -\frac{(F_{1,m,n,k}^{(i)})^2}{3VF_{2,m,n,k}^{(i)}}\left(\int_0^1 \frac{A_{1,m,n,k}^{(i)}(z)F_{3,m,n,k}^{(i)}(z)}{2\chi_{4,m,n}(\upsilon\left(\boldsymbol{q}_{m,n}(z)\right))^2}+\frac{A_{1,m,n,k}^{(i)}(z)d_k(\boldsymbol{q}_{m,n}(z))^4}{2F_{3,m,n,k}^{(i)}(z)}dz\right)^3 \\
&\quad -\frac{(F_{2,m,n,k}^{(i)})^2}{3VF_{1,m,n,k}^{(i)}}\frac{1}{s_{m,n,k}^3}-\frac{l_{m,n}(\boldsymbol{q})^3}{3VF_{1,m,n,k}^{(i)}F_{2,m,n,k}^{(i)}}+\frac{A_{3,m,n,k}^{(i)}l_{m,n}^{(i)}(\boldsymbol{q})}{V} \\
&\triangleq \gamma_{m,n,k}^{(i)}(\mathbf{q},\mathbf{t},\mathbf{s})
\end{aligned}
\tag{11.19}
$$

Iterative Algorithm

Problem (P2) is non-convex as the throughput function $U_k(\mathbf{q}, \mathbf{t}, \mathbf{s})$ is non-concave, to tackle this difficulty, we will give a lower-bound concave function $U_k(\mathbf{q}, \mathbf{t}, \mathbf{s})$ at the local point $\left(\boldsymbol{q}^{(i)}, \boldsymbol{t}^{(i)}, \boldsymbol{s}^{(i)}\right)$ to construct a convex problem for (P2). More specifically, $U_k(\mathbf{q}, \mathbf{t}, \mathbf{s})$ is constructed through proposing approximation for throughput at each hovering point, represented by $\theta_{m,k}^{(i)}(\mathbf{q}, \mathbf{t}, \mathbf{s})$ and in flying period in each line segment with maximum speed represented by $\gamma_{m,n,k}^{(i)}(\mathbf{q}, \mathbf{t}, \mathbf{s})$.

Based on the above approximation, at the local point $\left(\boldsymbol{q}^{(i)}, \boldsymbol{t}^{(i)}, \boldsymbol{s}^{(i)}\right)$, we can obtain the following convex approximation problem of (P2):

$$\text{(P3):} \quad \max_{\mathbf{q},\mathbf{t},\mathbf{s}} \min_{k} \sum_{m=1}^{K} \theta_{m,k}^{(i)}(\mathbf{q}, \mathbf{t}, \mathbf{s}) + \sum_{m=0}^{K} \sum_{n=0}^{N} \gamma_{m,n,k}^{(i)}(\mathbf{q}, \mathbf{t}, \mathbf{s}) \tag{11.20}$$

$$T(\boldsymbol{q}, \boldsymbol{t}) \leq T \tag{11.20a}$$

$$\boldsymbol{t} \geq 0 \tag{11.20b}$$

$$0 \leq \mathbf{s} \leq 1, \sum_{k=1}^{K} s_{m,k} = 1, \sum_{k=1}^{K} s_{m,n,k} = 1 \tag{11.20c}$$

In conclusion, for a given local point $\left(\boldsymbol{q}^{(i)}, \boldsymbol{t}^{(i)}, \boldsymbol{s}^{(i)}\right)$, we can always reconstruct problem (P2) into the correspondingly convex problem (P3) through the above steps, and then, by efficiently solving this problem via convex optimization method, we can obtain a better point which promotes us to propose an efficient iterative algorithm in the next subsection to solve problem (P2).

Before introducing the iterative algorithm, it is worthwhile to note that acquiring a satisfying feasibly initial point $\left(\boldsymbol{q}^{(0)}, \boldsymbol{t}^{(0)}, \boldsymbol{s}^{(0)}\right)$ can conspicuously improve the algorithm convergence speed. We first select locations of the K users as the horizontal coordinates of K initial hovering points. Then, the shortest path passing through all the hovering points once is found. The N + 1 bisection points along the path between two neighbour initial N hovering points is select as the turning points where all of the initial hovering points and turning points construct the initial trajectory points. Next, the remaining task time is equally allocated to the initial hovering durations $\mathrm{t}^{(0)}$. Finally, the power allocation variable is $\boldsymbol{s}^{(0)}$ initialized as $\mathrm{s} = \frac{1}{\mathrm{K}}$ for all hovering durations and flying period with maximum speed.

After initializing the trajectory and user scheduling variables, an efficient iterative algorithm is provided to solve problem (P3). Specifically, in ith iteration, we first construct the lower-bound concave approximation of $U_k(\mathbf{q}, \mathbf{t}, \mathbf{s})$ following in the above convex approximation steps in the last subsection, and subsequently build up the convex problem (P3). Then the convex optimization methods like the ellipsoid method is applied to solve (P3) efficiently, where the current optimal trajectory and user scheduling variables $(\boldsymbol{q}^*, \boldsymbol{t}^*, \boldsymbol{s}^*)$ is obtained which is then set as the local point $\left(\boldsymbol{q}^{(i+1)}, \boldsymbol{t}^{(i+1)}, \boldsymbol{s}^{(i+1)}\right)$ for the next iteration and repeat the above steps until the result converges to the threshold η.

Whereafter, applying the ellipsoid method to address the convex problem (P3) in each iteration, we follow the analysis in [26, 27] to give the overall complexity of our algorithm. First, note that in our reformulated problem, there contain K hovering points, K hovering durations, $(K + 1)N$ turning points and $K(K + (K + 1)(N + 1))$ power allocation variables where each trajectory points contains two coordinates, leading the total variable number as $M_{var} = K^2N + 2K^2 + 3KN + 4K + 2N$. when applying the ellipsoid method in each iteration, it $\mathcal{O}\left(\frac{1}{\eta}M_{var}^2\right)$ costs for ellipsoid generation. And in each call, calculating the objective function and constraints will brings about $\mathcal{O}\left(M_{var}^2\right)$ and $\mathcal{O}(M_{var}M_{con})$ computing cost where $M_{con} = (2K + 1)(K + (K + 1)(N + 1)) + K + 1$ denotes the constraints number. Denoting the iteration number by Φ, the total complexity of proposed algorithm is $\mathcal{O}\left(\left(\frac{\Phi}{\eta}M_{var}^2\left(M_{var}^2 + M_{var}M_{con}\right)\right)\right)$.

11.6 Simulation

We mainly conducted experiments on comparison of the UAV trajectory design algorithm based on conventional SCP-line [27] where the continuous trajectory is discretized into a sequence of flying segments, each characterized by a distinct constant speed. This discretization process results in the trajectory being represented by a finite set of N_s endpoints, which mark the boundaries of these segments.

We first start with Fig. 11.3 to evaluate the convergence of proposed algorithm where the current throughput in each iteration with different covert communication requirement ρ_w and warden distribution variance ε_w^2 is plotted. It is shown that as the iteration progresses, the values of throughput gradually increase and converge around 20 iterations, which validates the convergence of proposed scheme. Besides, it is shown that the throughput achieved under a loose covert communication requirement ρ_w and perfect knowledge of warden is better than with stricter covert communication requirement and more dispersed distribution of warden. This is because stricter covert communication requirement and more dispersed distribution of warden leads to a smaller transmit power and further a smaller channel capacity.

Next, we investigate the differences in optimization results and algorithm performance between the proposed algorithm and the benchmark. In Fig. 11.4, we present a comparative analysis of the optimized trajectories attained by proposed SHF-based scheme and the benchmark under varying number of turning points N and endpoints N_s. It is obvious that the paths achieved by benchmark exhibit a notable sensitivity to the number of endpoints. Conversely, the proposed SHF-based scheme demonstrates remarkable stability, with little trajectory variation upon increasing the turning points number. By focusing on a detailed enlargement at user 3, we observe that an increase in turning points number results in closer hovering points to the serving user, which is

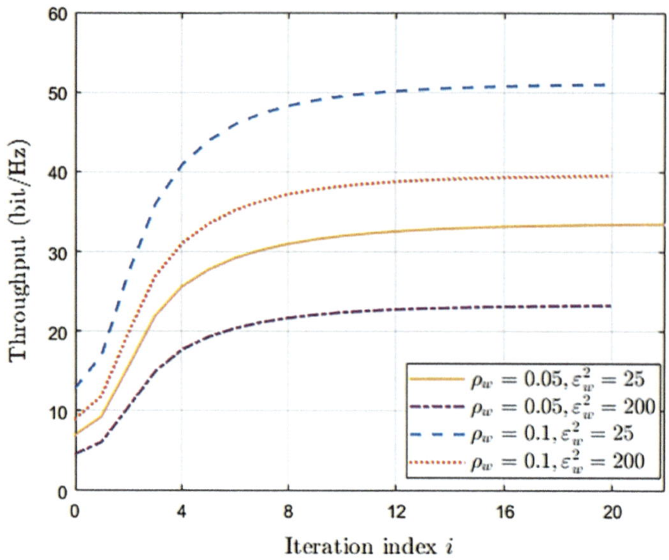

Fig. 11.3 Convergence evaluation of proposed algorithm

due to more accurate trajectory reformulation. To further emphasize the advantages of proposed SHF-based scheme, Fig. 11.5 showcases a comparative study between the throughput and the computing time achieved by proposed scheme when N = 1, versus the benchmark's performance when N_s ranges from 10 to 40. As N_s increases, the throughput of benchmark gradually converges towards that of proposed scheme, yet its computing time significantly surpasses that of proposed scheme, underlining the superior algorithmic efficiency and performance of proposed scheme.

Then, we delve into the impact of the turning points number N in proposed SHF-based scheme on both throughput and computing time performance, aiming to provide insights for network design optimization as illustrated in Fig. 11.6. It reveals that increasing the number of turning points generally enhances throughput but concurrently leads to larger the computation time. Notably, a significant throughput boost is observed upon introducing 1 turning point between two hovering points, which validates the effectiveness of introducing the turning point in the optimal SHF structure. However, when further increase the turning points number, it yield diminishing the growth rate of throughput while computing time escalates exponentially. This throughput variation can be rationally explained by the influence of turning points on power allocation variable adjustments, as depicted in Fig. 11.7. Specifically, power allocation is optimized more efficiently with N = 1 compared to N = 0, as when N = 0, the power allocation is not a binary distributed during some flying segments as under a worse precision, the throughputs for the nearest two users must be considered at the same time. The inaccurate user scheduling reformulation leads to the performance loss on throughput. And when N = 1, the power allocation relatively completely performs as binary variable. This observation indicates that setting

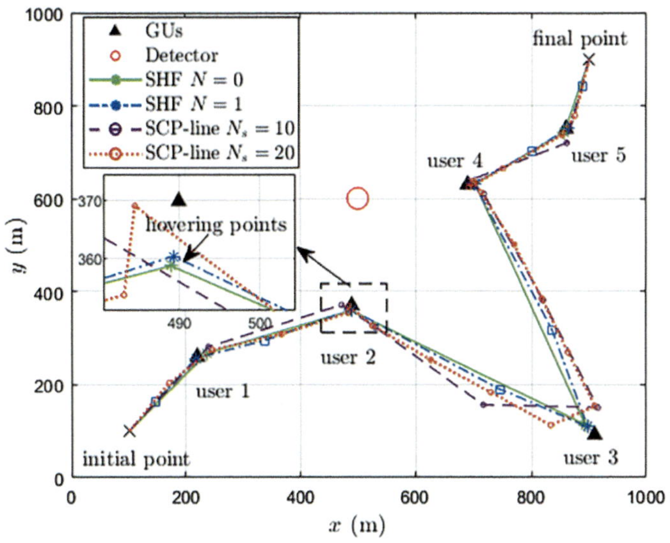

Fig. 11.4 Optimized trajectory with different number of turning points and endpoints

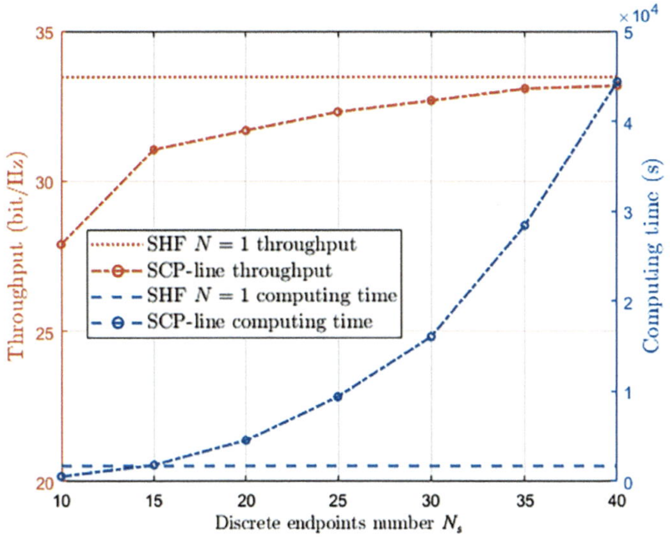

Fig. 11.5 Throughput and computing time comparison with different number of endpoints

N = 1 can already achieve a satisfying reformulation precision on power allocation variable and balance the throughput and computing time performance in practical system design.

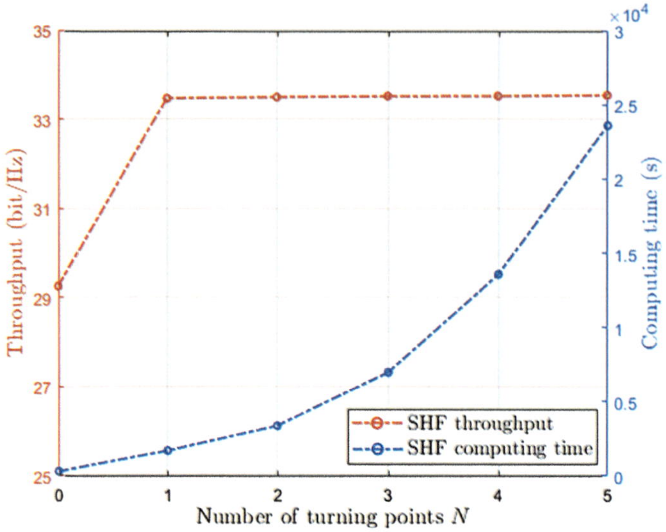

Fig. 11.6 Throughput and computing time comparison with different number of turning points

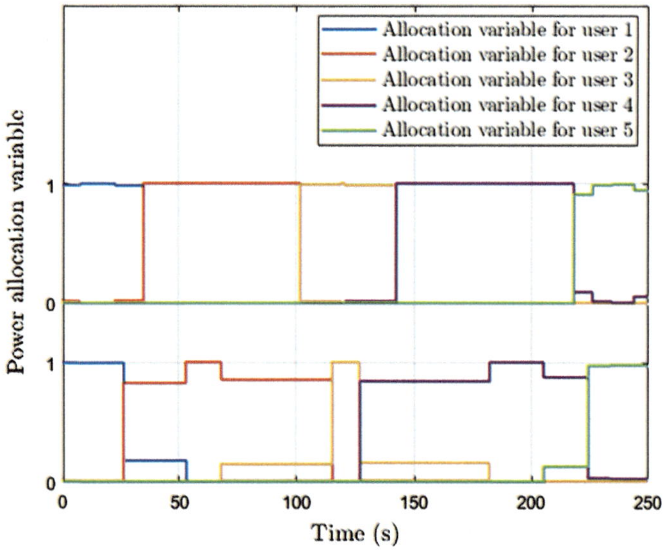

Fig. 11.7 Optimized power allocation variables with different number of turning points

Finally, we turn on to focus on elucidating the impact of the estimated position of warden. Specifically, we investigate how the proximity of warden affects the UAV trajectory and transmit power when it is surveying users. As shown in Fig. 11.8, when a near warden is present, the UAV does not circumnavigate all users comparing with

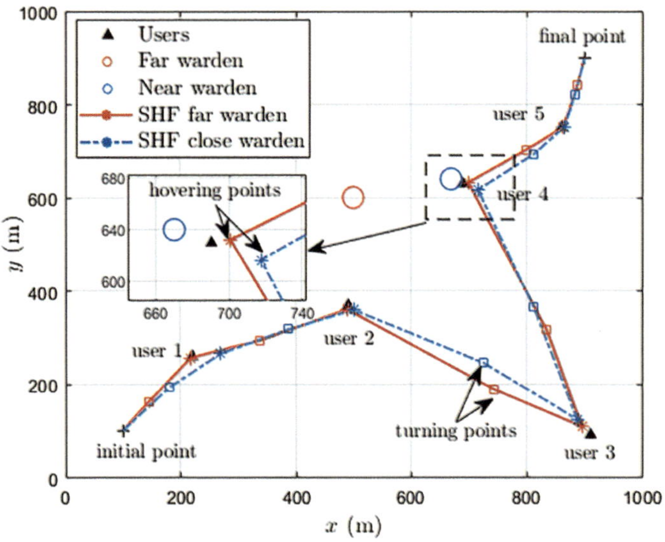

Fig. 11.8 Optimized UAV trajectory under different warden's location

a far warden. Instead, it is more likely to save the flying time and allocate it to the hovering duration around the user close to the warden. Furthermore, an intriguing tendency emerges where the UAV prefers to maintain a greater distance from the warden, even when it is in close proximity to a user, suggesting a potential avoidance mechanism or optimization strategy based on the estimated position of the warden.

This observation can be revealed in Fig. 11.9 which offers valuable insights into the influence of the warden's estimated position on transmit power dynamics over time. It becomes evident that in contrast to scenario with a far warden, where the hovering times are relatively uniformly allocated, the UAV's trajectory with a nearby warden primarily hovering around the user closest to the warden, significantly altering its movement pattern, where prolonging hovering time around the user adjacent to the warden is driven by the significantly reduced transmit power requirements to follow the cover communication constraint. Specifically, when compared to scenario where the warden is positioned far from users, the transmit power in the immediate area surrounding the user near the detector approaches zero. This necessitates a longer hovering time by the UAV to ensure equitable performance among all users, addressing fairness concerns within the system.

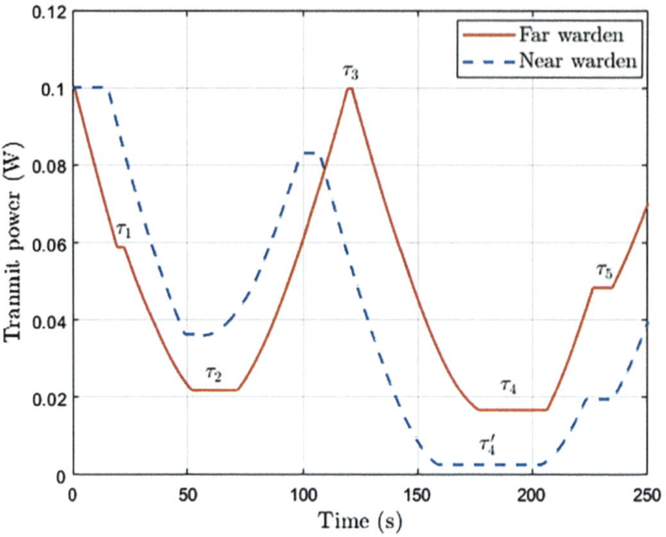

Fig. 11.9 UAV transmit power over time under different warden's location

11.7 Conclusion

This chapter zeroed in on a covert communication framework leveraging UAVs. Amidst constraints pertaining to stealth, UAV speed limitations, and mission timelines, we delved into a joint optimization problem involving UAV trajectory, transmit power, and power allocation strategies, with the objective of maximizing the minimum throughput across users. To tackle the complexity stemming from the vast array of variables in this joint design, we firstly derived the close form expression of optimal transmit power, grounded in the covertness constraint, thereby transforming the original problem into one which focused joint trajectory and power allocation task. Notably, the optimal SHF structure subsequently facilitated the reformulation of the original problem with numerous variables into a more manageable form with a limited set of variables. Faced with the non-convex nature of the reformulated problem, we employed a convex approximation technique and devised an iterative algorithm, ensuring efficient resolution. Through simulations, we demonstrated that our proposed solution achieves commendable performance with significantly reduced computational overhead, outperforming benchmark approaches. These findings underscore the significance and necessity of research.

References

1. Ye, R., Peng, Y., Al-Hazemi, F., & Boutaba, R. (2024). A Robust cooperative jamming scheme for secure UAV communication via intelligent reflecting surface. *IEEE Transactions on Communications, 72*, 1005–1019.
2. Yang, H., Liu, S., Xiao, L., Zhang, Y., Xiong, Z., & Zhuang, W. (2024). Learning-based reliable and secure transmission for UAV-RIS-assisted communication systems. *IEEE Transactions on Wireless Communications, 23*, 6954–6967.
3. Wu, P., Yuan, X., Hu, Y., & Schmeink, A. (2024). Trajectory and user assignment design for UAV communication network with no-fly zone. *IEEE Transactions on Vehicular Technology, 73*, 15820–15825.
4. Gupta, L., Jain, R., & Vaszkun, G. (2016). Survey of important issues in UAV communication networks. *IEEE Communications Surveys & Tutorials, 18*, 1123–1152.
5. Shahzad, K., Zhou, X., Yan, S., Hu, J., Shu, F., & Li, J. (2018). Achieving covert wireless communications using a full-duplex receiver. *IEEE Transactions on Wireless Communications, 17*, 8517–8530.
6. Jiang, X., Yang, Z., Zhao, N., Chen, Y., Ding, Z., & Wang, X. (2021). Resource allocation and trajectory optimization for UAV-enabled multi-user covert communications. *IEEE Transactions on Vehicular Technology, 70*, 1989–1994.
7. Wang, N., Wang, P., Alipour-Fanid, A., Jiao, L., & Zeng, K. (2019). Physical layer security of 5G wireless networks for IoT: Challenges and opportunities. *IEEE Internet of Things Journal, 6*, 8169–8181.
8. Wu, Q., Mei, W., & Zhang, R. (2019). Safeguarding wireless network with UAVs: A physical layer security perspective. *IEEE Transactions on Wireless Communications, 26*, 12–18.
9. Guo, H., Liu, J., Ren, J., & Zhang, Y. (2020). Intelligent task offloading in vehicular edge computing networks. *IEEE Transactions on Wireless Communications, 27*, 126–132.
10. Kang, H., Chang, X., Mišić, J., Mišić, V. B., Fan, J., & Bai, J. (2022). Improving Dual-UAV aided ground-UAV bi-directional communication security: Joint UAV trajectory and transmit power optimization. *IEEE Transactions on Vehicular Technology, 71*, 10570–10583.
11. Sun, G., Li, J., Wang, A., Wu, Q., Sun, Z., & Liu, Y. (2022). Secure and energy-efficient UAV relay communications exploiting collaborative beamforming. *IEEE Transactions on Communications, 70*, 5401–5416.
12. Na, Z., Ji, C., Lin, B., & Zhang, N. (2022). Joint optimization of trajectory and resource allocation in secure UAV relaying communications for Internet of Things. *IEEE Internet of Things Journal, 9*, 16284–16296.
13. Wu, J., Yuan, W., & Hanzo, L. (2023). When UAVs meet ISAC: Real-time trajectory design for secure communications. *IEEE Transactions on Vehicular Technology, 72*, 16766–16771.
14. Zhou, X., Yan, S., Hu, J., et al. (2019). Joint optimization of a UAV's trajectory and transmit power for covert communications. *IEEE Transactions on Signal Processing, 67*, 4276–4290.
15. Goeckel, D., Bash, B., Guha, S., & Towsley, D. (2016). Covert communications when the warden does not know the background noise power. *IEEE Communications Letters, 20*, 236–239.
16. Zhang, J., Chen, X., Li, M., & Zhao, M. (2021). Optimized throughput in covert millimeter-wave UAV communications with beam sweeping. *IEEE Wireless Commun. Lett., 10*, 720–724.
17. Du, H., Niyato, D., Xie, Y.-A., Cheng, Y., Kang, J., & Kim, D. I. (2022). Performance analysis and optimization for jammer-aided multiantenna UAV covert communication. *IEEE Journal on Selected Areas in Communications, 40*, 2962–2979.
18. Zhou, X., Yan, S., Shu, F., Chen, R., & Li, J. (2021). UAV-enabled covert wireless data collection. *IEEE Journal on Selected Areas in Communications, 39*, 3348–3362.
19. Rao, H., Xiao, S., Yan, S., Wang, J., & Tang, W. (2022). Optimal geometric solutions to UAV-enabled covert communications in line-of-sight scenarios. *IEEE Transactions on Wireless Communications, 21*, 10633–10647.
20. Wang, H., Bai, H., Li, F., Chen, J., & Wang, L. (2024). Throughput maximization for covert UAV relaying system. *IEEE Transactions on Vehicular Technology, 73*, 4429–4434.

21. Wang, C., et al. (2023). Covert communication assisted by UAV-IRS. *IEEE Transactions on Communications, 71*, 357–369.
22. Qian, Y., Yang, C., Mei, Z., Zhou, X., Shi, L., & Li, J. (2023). On joint optimization of trajectory and phase shift for IRS-UAV assisted covert communication systems. *IEEE Transactions on Vehicular Technology, 72*, 12873–12883.
23. Tatar Mamaghani, M., & Hong, Y. (2022). Aerial intelligent reflecting surface-enabled terahertz covert communications in beyond-5G Internet of Things. *IEEE Internet Things Journal, 9*, 19012–19033.
24. Hamamreh, J. M., Furqan, H. M., & Arslan, H. (2019). Classifications and applications of physical layer security techniques for confidentiality: A comprehensive survey. *IEEE Communications Surveys & Tutorials, 21*, 1773–1828.
25. Ben-Tal, A., & Hochman, E. (1972). More bounds on the expectation of a convex function of a random variable. *The Journal of Applied Probability, 9*, 803–812.
26. Hu, Y., Yuan, X., Xu, J., & Schmeink, A. (2019). Optimal 1D trajectory design for UAV-enabled multiuser wireless power transfer. *IEEE Transactions on Communications, 67*, 5674–5688.
27. Wu, P., Yuan, X., Hu, Y., & Schmeink, A. (2024). Joint power allocation and trajectory design for UAV-enabled covert communication. *IEEE Transactions on Wireless Communications, 23*, 683–698.

Chapter 12
Collaborative AI-Driven Intrusion Detection and Response for UAV Networks

Hassan Jalil Hadi, Yue Cao, and Waleed Omar Paracha

Abstract Unmanned Aerial Vehicles (UAVs) are utilized extensively across multiple domains. UAVs offer a range of services to users, such as monitoring, logistics and sensing due to their adaptable deployment as well as dynamic reconfigurability. Nonetheless, UAV networks have become increasingly vulnerable to malicious threats due to their multi-connectivity and openness. Significant effort has been expended to create an efficient Intrusion Detection System (IDS) utilizing machine-learning methodologies for UAVs. Unfortunately, limitations of standalone IDS make them inadequate for protecting UAV networks from potential security threats. The lack of accurate identification for compromised UAV nodes in UAV networks is a significant security vulnerability, jeopardizing the integrity of the entire network due to the compromise of a single node. In this chapter, an autonomous collaborative UAV-CIDS utilizing a Convolutional Neural Network (CNN) is presented. This system effectively finds zero-day vulnerabilities with high precision. The suggested approach considers encoded Wi-Fi traffic logs from three prevalent UAV types: DBPower UDI, Parrot Bebop and DJI Spark. The evaluation findings demonstrate that proposed FFCNN model has achieved exceptional performance on the UAVIDS dataset, with an accuracy of 97.21% in comparison to previous models. Following the identification of attacks, their mitigation is equally crucial. Furthermore, real-time incident response protocols are created and executed for cyber-attacks on UAV networks. The incident response management will aid in mitigating the impact of a security breach, addressing vulnerabilities and thoroughly securing the entire UAV networks.

Keywords CIDS · UAV networks · Incident response · CNN

H. J. Hadi · Y. Cao (✉)
School of Cyber Science and Engineering, Wuhan University, Wuhan, China
e-mail: yue.cao@whu.edu.cn

W. O. Paracha
Arab Open University, Al-Ardiya, Kuwait

Y. Hu et al. (eds.), *Reliable and Efficient Trajectory Design in UAV-Aided Wireless Networks*, https://doi.org/10.1007/978-981-96-5988-3_12

Abbreviations

CNN	Convolutional Neural Network
CPS	Cyber-Physical Systems
DBN	Deep Belief Networks
DCNN	Deep Convolutional Neural Networks
DL	Deep Learning
DRL	Deep Reinforcement Learning
GCS	Ground Control Station
GPS	Global Positioning System
GRUs	Gated Recursive Units
IDS	Intrusion Detection System
IG	Information Gain
IoT	Internet of Things
KNNs	K-Nearest Neighbors
LSTM	Long Short-Term Memory Networks
ML	Machine Learning
MLP	Multi-layer Perceptron
NGN	Next-Generation Networks
UAV	Unmanned Aerial Vehicle
UAV-CIDS	Unmanned Aerial Vehicle Collaborative IDS

12.1 Introduction

12.1.1 Background

UAVs have become prevalent aerial devices, capable of identifying and communicating with other devices by exchanging information through smart gadgets like tablets and cell phones. UAVs have surged in popularity and are employed for diverse applications [1]. The primary applications encompass aerial mapping, package delivery, environmental management, aerial photography, irrigation, monitoring and rescue operations. Wireless communication and the reliability of UAVs are of paramount importance in such critical applications. In UAV networks, essential security elements are guaranteed via security protocols and intrusion detection systems [2].

Further, UAV networks typically consist of a Ground Control Station (GCS), which enables human oversight of UAVs, and the UAVs themselves [1]. Every element of UAV networks may be susceptible to cyber-attacks, including the infiltration of an unauthorized UAV, as seen in Fig. 12.1. Regular attacks on UAV system components include optical navigation, spoofing, jamming, traffic analysis, hijacking [3]. Furthermore, the heightened utilization of UAVs is compromised by the threat

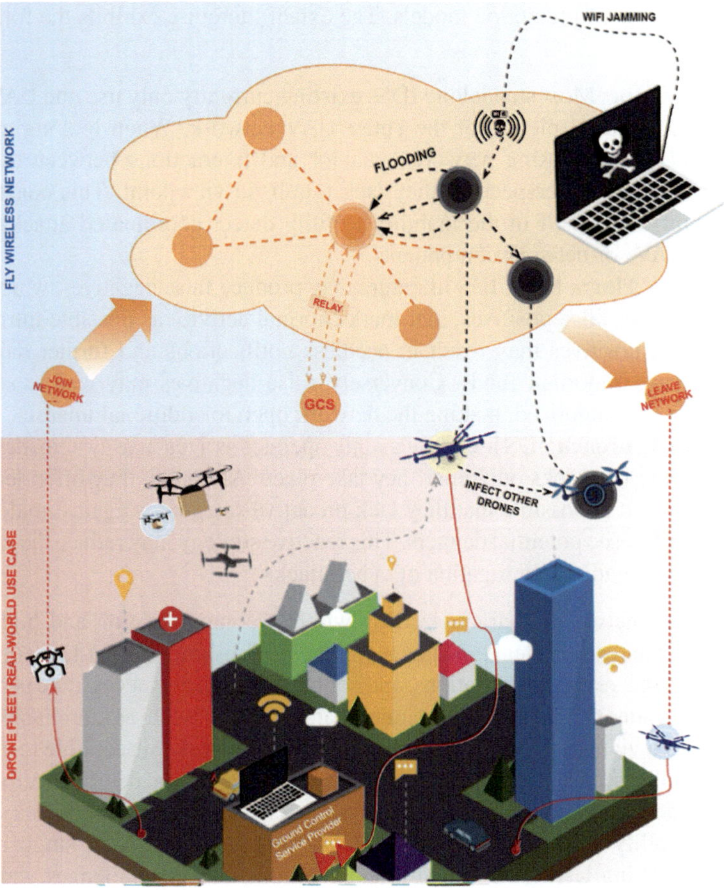

Fig. 12.1 Components of networked UAVs and potential cyber-attacks [6]

of attacks [4]. Hence, the IDS created for UAV networks is important to boost attack identification information and decrease network security threats. The IDS is crucial for securing UAV networks by identifying unforeseen threats. However, attacks are becoming increasingly difficult to detect and more intricate with time. IDS reliant on Deep Learning (DL) for anomaly detection are currently getting attention from the research community [5].

Moreover, the primary objective of these systems is to improve the accuracy of attack detection. The objective of employing deep learning is to train modules that enable UAVs to autonomously detect and counteract hostile activities. Tao et al. [7] proposed a Deep Reinforcement Learning (DRL) method for detecting hostile intrusions in UAVs. Nevertheless, literature continues to encounter numerous obstacles, especially in real-time situations. These issues arise from aspects including the selection of appropriate features, feature extraction, computational duration, and the attributes of prediction models. The extant literature exhibits the following limitations:

- **Narrow Scope**: Most standalone IDSs existing currently only use one UAV, they just offer a partial picture of the entire UAV network. When looking at UAV networks, encompassing network behavior and interactions between different UAVs in a global perspective, they lack a holistic viewpoint. This constrained viewpoint may result in the inability to fully detect coordinated attacks using several UAVs or network components.
- **High False Alarm Rate**: IDS literature may produce false negatives, which miss real attacks or false positives, that marks normal activity as possible intrusions. Here, false positives may result in needless notifications and further inquiries, squandering important funds. Conversely, false negatives may lead to security lapses that go unnoticed, making the network open to additional misuse.
- **Reactive Approach**: IDS literature mainly operates as a reactive system, detecting and reacting to attacks only after they take place. Although, they offer detection capabilities, it is possible that they lack proactive safeguards against malevolent attacks or foresee potential dangers. This reactive strategy may reduce the efficacy of defenses against or mitigation of cyberattacks.

Also, UAV networks characterized by wireless communication and high node mobility encounter significant challenges in the establishment of ad-hoc networks. Restricted range, resource limitations, and continuous mobility necessitate persistent collaboration among nodes. The unique characteristics, such as unreliable links, contribute to a vulnerable topology, rendering UAV networks susceptible to attacks.

Additionally, security solutions primarily emphasize external detection rather than internal response and comprehensive network mechanisms. This results in a significant vulnerability in the overall network security, particularly when a single node is breached. Existing UAV IDS in real-time environments face challenges, including suboptimal feature selection and decreased accuracy. The inability to identify zero-day attacks and to recognize distinctive characteristics in UAV traffic presents security vulnerabilities. The generation of false positives and inefficient incident response handling continue to be significant challenges.

12.1.2 Motivation

Therefore, this chapter introduces a computationally efficient and precise Collaborative Intrusion Detection System (UAV-CIDS) for UAV networks utilizing deep learning. The CIDS is designed to address the limitations of standalone IDS in effectively protecting UAV networks from potential security threats. The constraints encompass restricted visibility, absence of context, occurrence of false positives, and constrained response capabilities. A comprehensive solution is necessary to ensure total safety for UAV networks [8]. The CIDS has the capability to correlate security incidents from various sources, analyse network traffic across multiple segments and implement preventative measures to counteract threats. In the development of our CIDS, an established dataset known as UAVIDS [9] is utilized. This dataset includes actual network traffic data from three well-known UAV models: DJI Spark, DBPower UDI and Parrot Bebop.

12.1.3 Main Contributions

The main contributions of this work can be summarized as follows:

(1) To address the challenge of low detection rates for zero-day attacks, an IDS utilizing CNN architecture is develop and evaluated. The latest significant dataset (UAVIDS) is utilized to classify encoded Wi-Fi traffic logs for networked UAVs.
(2) A CIDS is designed and implemented, utilizing CNN architecture to provide defence against various attacks on UAV networks. An efficient and accurate detection system is constructed using a distributed CIDS framework. This approach addresses the challenge of incomplete detection of coordinated attacks by safeguarding the entire UAV networks. Furthermore, if a single node is compromised, proposed distributed CIDS effectively contains the breach, preventing it from propagating throughout the entire network or impacting other nodes. An event correlation engine is crafted to minimize the effects of false detections, resulting in enhanced accuracy and reliability in intrusion detection.
(3) This work proposes and implements an incident response handling system for addressing cyberattacks on UAV networks. This approach will address the challenges posed by the reactive characteristics of current IDS methodologies, effectively preventing or alleviating cyber-attacks on complete UAV networks.

12.1.4 Structural Organization

This chapter is organized as follows: Sect. 12.2 discusses the preliminaries. Section 12.3 reviews the literature on Machine Learning (ML) methods for UAV-CIDS. Section 12.4 details the proposed CIDS for UAV networks, along with an evaluation of the conducted experiments and an in-depth discussion of the results. Section 12.5 introduces the designed automated cyber incident response. Finally, the study's conclusions are presented in Sect. 12.6.

12.2 Preliminaries

12.2.1 Intrusion Threat in UAV Networks

UAVs have a wide range of uses with significant benefits for improving the Quality of Services (QoS). When mobile devices and their users proliferate in UAV networks, intruders decrease the endurance of UAVs by depleting their power through different attacks. The malicious attacks, such as impersonation attacks, jamming attacks and intrusion attacks are described below and shown in Fig. 12.2.

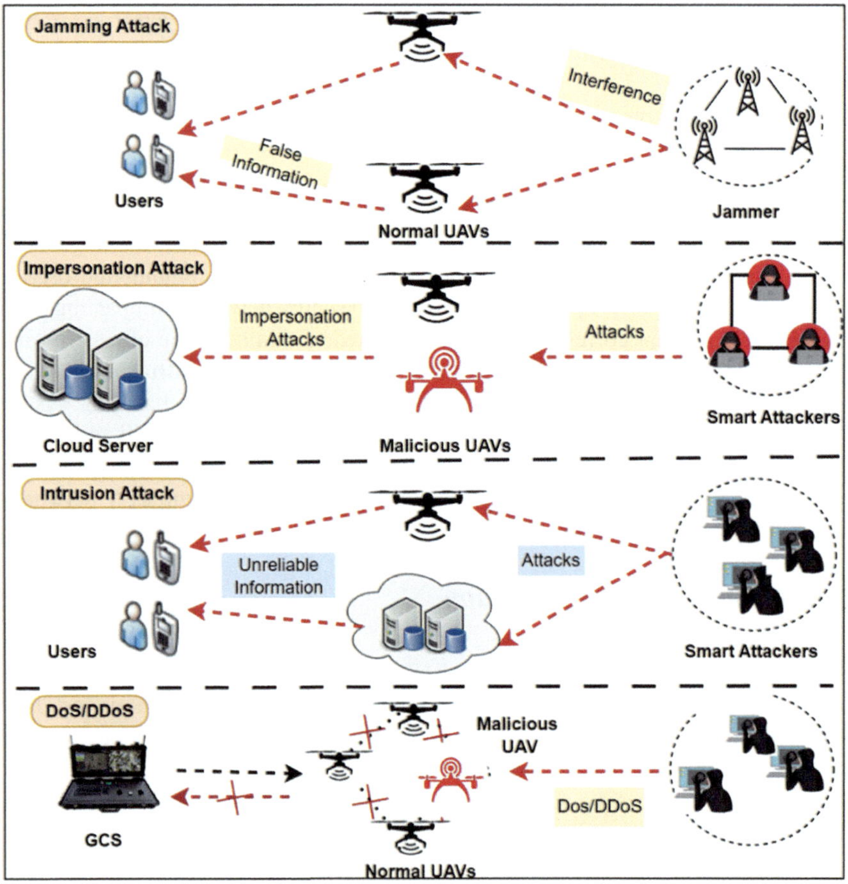

Fig. 12.2 Intrusion threats in UAV networks

- **Jamming Attacks**: Malicious users launch jamming UAVs to interfere with communications in the UAV networks. The modified UAVs transmit jamming power for preventing UAVs from accessing services or data. Malicious users launch jamming UAVs to interfere with communications in the UAV networks.
- **Impersonation Attacks**: Impersonation attacks can destroy UAVs network security by influencing UAVs [7]. Hackers can control UAVs and act as normal UAVs for delivering services, like sensing and relaying. Malicious UAVs can tamper or steal data during transmission and relay for illegal purposes.
- **Intrusion Attacks**: Attackers with abnormal intentions might break into UAVs. They can secretly program UAVs with malicious software to collect personally identifiable information. Consequently, users may receive inaccurate information, leading to a low QoS [7].
- **Denial of Service (DoS)**: attacks in UAV networks can compromise the communication infrastructure by inundating it with a flood of unauthorized traffic [10]. The overwhelming influx of data clogs channels, making them unable to facilitate authentic commands and data transfers. As UAVs rely on instantaneous data flow, such attacks can lead to communication breakdowns, hindering mission-critical operations. The possible ramifications of DoS attacks on UAV networks are complex. From a safety standpoint, disrupted communications can result in a loss of control, potential mid-air collisions, or even unauthorized takeovers. Furthermore, such attacks have the potential to undermine the integrity of data streams, resulting in erroneous decision-making and possibly serious repercussions.

12.2.2 Convolutional Neural Network (CNN)

In Image Processing (IP) and Natural Language Processing (NLP) domains, CNN is used to automatically extract high level information from images. The CNN can automatically extract high-level features, eliminating manual extraction of features from data. The time required to train the CNN model from scratch is reduced because the learned model can recognize new patterns. A CNN also has a few hidden layers other than input and output layers. The CNN applies several mathematical techniques across these hidden layers to extract important attributes from raw images. In CNN, lower-level layers oversee the extraction of fundamental features, shown in Fig. 12.3. Likewise, the upper-level layer extracts advanced features based on fundamental features [11].

To extract high-level attributes, the CNN layer sequentially executes two operations. The first operation is the data convolution operation. In convolutional operation, a particular feature can be extracted from data using a sliding function. For this type of extraction, various convolutional filters have been defined. The second operation employed by CNN is the activation function. There are numerous activation functions available, where the frequently used activation function in CNN is the Rectified Linear Activation Function (ReLU). For negative input values, this function returns a value of zero, and for other inputs, it returns the input itself. Contrary to other

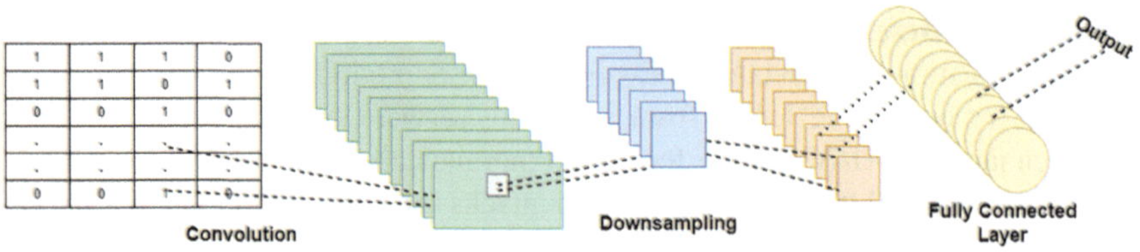

Fig. 12.3 Convolutional Neural Network

functions like the sigmoid function, this function aids in achieving superior learning performance.

Next, to provide a short output and cut-down connections to the next layer, down sampling is performed using a polling procedure. The features are changed to be location-insensitive after this operation. As a result, polling operations shorten the time needed for the learning process, permitting CNN to learn features regardless of where they are. In literature, average and max polling are two frequently used polling techniques. The average polling approach depends on the feature average value, whereas in max polling highly active features are focused.

12.3 Related Work

12.3.1 IDS for UAV

Analyzing real-time network traffic is essential for identifying intrusions that may target UAVs while they are in flight. Implementing an IDS for UAVs facilitates the identification of various intrusion classes. Common categories of intrusion encompass flight signal tampering, routing attacks, malware [12] and message forging attacks [13]. Generally, IDS for UAV networks was designed using ML approaches. There are three primary categories of machine learning algorithms including unsupervised learning, supervised learning and semi-supervised learning. The implementation of these categories is aimed to enhance the security of UAV networks. For instance, the authors in [14] employed various deep learning techniques to recognize unconventional attack types and to detect network intrusion traffic. The researchers implemented ten-fold cross-validation by dividing the training and testing datasets into 80–20 or 70–30 splits.

Additionally, the authors in [15] merged the CIC-IDS2018 and CICIDS2017 datasets to investigate the generalizability of an IDS dataset. A total of twelve learning algorithms from different families of supervised learning were employed to assess the performance. Nonetheless, the presumption that categorical features possess an identical quantity of novel values in datasets represents a limitation in this research.

Furthermore, the IDS for UAVs is implemented to identify a range of anomalies and threat types, such as GPS spoofing, UAV capturing/spoofing and routing attacks [16]. There has been a notable increase in the development of cybersecurity solutions utilizing machine learning approaches to protect the Internet of Things (IoT) and ensure the security of their communications, along with Cyber-Physical Systems (CPS) [17]. Besides, Vanitha and Ganapathi [18] performed a comprehensive evaluation of ML algorithms to address challenges associated with UAVs, including channel modeling, resource management, and security. This study offered a comprehensive overview of the taxonomy of deep learning-based intrusion detection techniques along with pertinent research publications. The study involved testing four deep learning models: a feedforward neural network, an autoencoder, a deep belief network and a long short-term memory network. In order to evaluate all four models, the researchers utilized the KDD Cup 1999, NSL-KDD, CICIDS2017 and CICIDS2018 datasets [19]. The primary constraint of this study lies in the utilization of KDD Cup 1999 and NSL-KDD, as both datasets are outdated and possess recognized deficiencies.

Also, in the literature [20], the authors developed a CNN utilizing the DoS dataset from CIC-IDS-2018 and KDD Cup 1999, employing Python and TensorFlow for implementation. The ratio for training and testing was set at 70–30 for the two datasets involved. The authors in [21] utilized historical datasets, including KDD Cup 1999, NSL-KDD, and ISCX2012, to evaluate the IDS system through machine learning techniques. In [22] and [23], several attacks along with countermeasures were examined across different network layers, starting from the physical layer and concluding at the application layer. The study provided an overview of methods based on machine learning and deep learning, including K-Nearest Neighbors (KNNs) [24], Support Vector Machines (SVMs) [21], Random Forests (RFs) [24], Q-Learning Algorithms (QLs) [17], Genetic Algorithms (GAs) [17] and Artificial Neural Networks (ANNs) [18]. These methods are applied to tackle issues such as GPS spoofing, jamming mitigation, eavesdropping detection, interference, trespassing detection and the interception of malicious UAVs, as detailed in Table 12.1.

Additionally, a DL-based intrusion detection model called DeepFed was introduced in [32]. In this model, the authors utilized CNN and Gated Recursive Units (GRUs) to detect cyber threats targeting industrial CPSs. The method utilizing federated learning allows multiple industrial cyber-physical systems to collaboratively create a comprehensive intrusion detection model while ensuring privacy is maintained. A distinctive multistage optimized machine learning-based network intrusion detection system framework has been introduced in the literature [33]. This framework reduced computational complexity while maintaining detection performance. A system was developed by Lu et al. [34] that utilizes distributed DL CNNs and analytic URLs for the detection of network attacks. The previously mentioned study utilized machine learning algorithms to identify intrusion data; however, it overlooked real-time security issues during the design of the intrusion detection system.

Table 12.1 Comparing the existing machine learning schemes for UAVs IDS

Ref-Year	Domain	Attacks on UAVs	Dataset	Detection models	CIDS (Detection)	Incident response handling
[25]	6G Networks	N/A	NSL-KDD, CIC-IDS2017	SVM, RF	X	X
[26]	UAV Networks	N/A	CIC-IDS2017	Incremental Learning	X	X
[27]	UAV Networks	N/A	KDDCup99	DDPM	X	X
[28]	ToN-IoT	N/A	ToN-IoT	SVMs	X	X
[29]	UAV Networks	N/A	KDD-CUP99	CNN-LSTM	X	X
[30]	UAV Networks	N/A	Private Dataset	SVM, MLP, GNB, Random Forest	X	✓
[4]	6G Networks	N/A	UAV-5G, UAVIDS-2020	Fusion-DNN	X	X
[31]	UAV Networks	Different attacks	CIC-IDS2017	CGAN	X	X
[31]	UAV Networks	Jamming	CSE-CIC IDS-2018	KNN, LR, DT, K-mean	X	X
[6]	UAV Networks	Cyber-physical attacks	UAVIDS-2020	FFCNN	X	X
This work	UAV Networks	All	UAVIDS	CNN	✓	✓

12.3.2 IDS for UAV Network

Analyzing real-time network traffic is essential for identifying intrusions that may target UAVs while they are in flight. Implementing an IDS for UAVs will facilitate the identification of various intrusion types. Common intrusion categories encompass flight signal tampering, routing attacks, malware, and message forging attacks [35]. For instance, the Denoising Diffusion Probabilistic Models (DDPMs) are introduced in [25] for the purpose of detecting anomalies in UAV network traffic. This IDS demonstrated impressive accuracy and minimal false alarm rates, even when faced with noise and incomplete data. Nonetheless, the IDS showed limited emphasis on the classification of attacks and the management of imbalanced datasets. One-class SVM is trained exclusively on a single class of data, specifically normal network traffic. Similarly, a self-adaptive IDS for UAV networks inspired by the human immune system is presented in [26]. This IDS can identify a range of malicious attacks within simulated UAV networks; however, additional assessment in real-world scenarios is necessary.

Furthermore, in [36], the authors introduced the application of a Long Short-Term Memory (LSTM) neural network for detecting anomalies in UAVs. A prediction model was initially developed using normal data, enabling the forecasting of future data points and the estimation of prediction uncertainty. This approach successfully identified point anomalies within actual UAV sensor data. It is important to recognize that a singular neural network approach may lack the robustness found in a hybrid architecture. The integration of diverse methods and techniques has the potential to improve the performance of UAV anomaly detection. This entails the application of a hybrid framework aimed at enhancing the identification and mitigation of unconventional threats. Moreover, the authors in [37] presented a sophisticated intrusion detection system tailored for UAV swarms, aimed primarily at detecting in-flight anomalies and preventing network attacks. This advanced IDS was created to address the security issues linked to UAV technology, providing a wider range of attack detection, safeguarding privacy, and ensuring hardware independence. It is crucial to highlight a possible gap in the investigation concerning the evaluation of the practical efficiency and real-world effects of the system on hardware components.

The authors in the literature [36] introduced a framework for intrusion detection in UAV networks, utilizing mobile edge computing technology. An optimized random forest model was employed on specialized UAV network servers to effectively identify a range of attacks [4]. The dependence on a singular DNN algorithm indicates that integrating various secure techniques might enhance the precision of attack detection even more. Additionally, in the literature [38], the authors introduced an optimized intrusion detection system for military operations within Internet of Drones (IoD) networks, employing an effective random forest classifier. The model demonstrated superior performance in simulations, however, to achieve greater accuracy, it is advisable to investigate a combination of secure methodologies rather than depending exclusively on one classifier.

Next, studies have explored distributed ML techniques [6], emphasizing their ability to simplify the complexity of high-dimensional data. The advantages of employing machine learning algorithms for pre-processing in intrusion detection systems to manage large volumes of data efficiently have been emphasized. Nonetheless, deep learning algorithms advance the analysis by revealing concealed features during the anomaly classification stage across various targets. They consistently demonstrate exceptional performance regarding false alarms, false negatives, and detection rates, exceeding the capabilities of leading technologies while ensuring efficiency, especially when utilized across various datasets.

12.4 Proposed Intrusion Detection and Response System for UAV Networks

12.4.1 Proposed Intrusion Detection

Evaluation of Dataset (UAVIDS-2020)

The UAVIDS-2020 dataset serves as a training resource for the IDS model, enabling the identification of cyber-attacks within UAV networks. The dataset includes unique and current encoded Wi-Fi traffic logs. Additionally, binary output tags are provided for all logs, indicating 1 for anomalous and 0 for normal UAVs. In addition to the testing and training logs, the dataset includes supplementary metadata. The metadata comprises logs detailing the computational dependencies among different features, as well as logs documenting the computational generation of each feature. The UAVIDS-2020 dataset is available to the public in the form of MAT files, containing a suitable quantity of distinct, non-repeating samples. This distinction enhances the reliability of the detection procedure, as the binary classifier gains increased protection through frequent logs.

Furthermore, the UAVIDS-2020 dataset serves to uncover intricate UAV traffic patterns and potential cyber threats. It constitutes effective features and computational statistics for the binary classification of UAV traffic. The dataset can be effectively tailored, recreated, expanded and analyzed to categorize UAV traffic patterns. The following outlines three primary justifications for utilizing this dataset:

- The dataset is current as it was introduced in 2020.
- The dataset demonstrates self-expression through the real-world network traffic generated by three widely used UAVs. The UAVs in question comprise the DJI Spark, DBPower UDI and Parrot Bebop. It also includes a diverse array of anomalies in the attack vector.

In UAVIDS-2020, legitimate (normal) and anomalous (illegal) networked UAV traffic are analyzed through multiple interpretations. Legitimate data packets are the standard data samples processed by networked UAVs. This dataset is characterized by its creation from encrypted Wi-Fi traffic logs. Three distinct UAVs are evaluated for this purpose: DBPower UDI, DJI Spark and Parrot Bebop. The actual dataset encompasses two distinct modes of UAV communication flow. One mode is bi-directional flow, which encompasses downlink flow, uplink flow and overall traffic flow. The alternative is the uni-directional flow, which encompasses only the overall traffic flow. The main data sources for all flow types consist of packet inter-arrival times and packet sizes. The total number of samples is 68,736. The dataset provides traffic distribution statistics, which are presented in Table 12.2. The dataset for uni-directional flow comprises 18 attributes, in addition to 1 label. In contrast, the dataset for bi-directional flow consists of 54 attributes, also accompanied by 1 label.

Table 12.2 Traffic distribution of UAVIDS-2020

Flow mode	DBPower	DJI spark	Parrot	Total
Bidirectional	17,256	5500	19,380	42,136
Unidirectional	14,864	73	11,663	26,600
Total	32,120	5573	31,043	68,736

DL-Based Empowered CIDS

This chapter also presents an effective cybersecurity solution that blocks most of the threats targeting the traffic of UAVs. The workflow diagram of the UAV-CIDS system development process is presented in Fig. 12.4. This diagram illustrates the relationship and process among five subsystems.

- The first component is the subsystem responsible for data collection. The collection comprises eight uni/bidirectional datasets for UAV-CIDS, which include the DBPower dataset, Parrot dataset and DJI Spark datasets.
- The second component is the data pre-processing subsystem, where the collected datasets undergo a series of preprocessing steps. The outlined steps encompass data distribution, importation, generation, padding, encoding, transposon, reshaping and data normalization.
- The third component is the FNN subsystem, which consists of four distinct phases. The processes involved include training, five-fold cross-validation, optimization and testing. All the phases have been applied to pre-processed datasets using our FNN.
- The fourth component is the model evaluation subsystem, which is tasked with assessing the performance of the system for each experiment. To achieve this, it utilizes a range of performance metrics, including detection rate, detection accuracy, detection sensitivity and the confusion matrix.
- The fifth component is the implementation subsystem for intrusion detection. After selecting the optimal system configuration for the communication mode

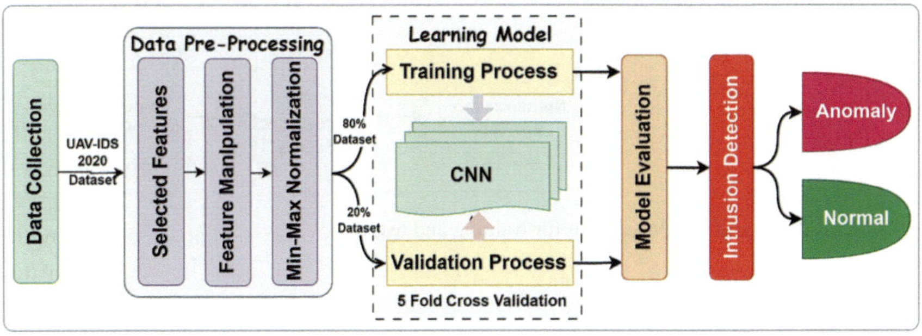

Fig. 12.4 Process flow of the proposed UAV-CIDS architecture

of each UAV, the training and testing model can commence. The deployment is executed to provide predictions of UAV-CIDS categorized as either "anomalous" or "normal" based on the input features.

Data Pre-processing Method

This section outlines the procedures for pre-processing the dataset and the recommended deep learning models. The procedure for pre-processing to create a dataset for the IDS is illustrated in Fig. 12.5. Several features that are initially considered to be extraneous in the decision-making process are eliminated. Two methodologies, encompassing domain expertise and the significance of features, are employed to assess the necessity of features. Understanding the specific field aids in selecting features that are more probable to enhance the effectiveness and accuracy of the predictive model. Furthermore, the significance of features guarantees that the chosen attributes impact the model's predictions, enhancing both accuracy and predictive capability. The size of the dataset is condensed by using random selection. The removal of duplicated records, along with all zero or null values, is executed to improve overall performance. Then two sub-datasets are derived from the original dataset by the application of min–max and logarithmic normalization techniques.

Table 12.2 presents the various traffic types found in the new sub-datasets. Following the pre-processing of raw data, the total instances are presented in Table 12.3. The following outlines the sequence of pre-processing actions performed on the dataset.

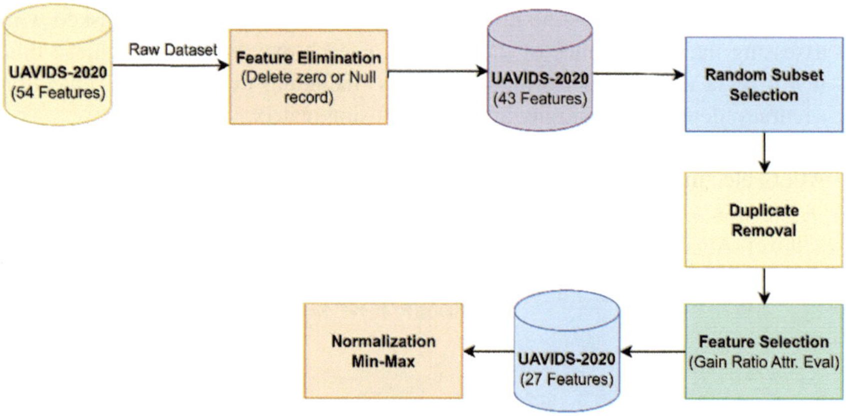

Fig. 12.5 Raw dataset pre-processing for training and testing

Table 12.3 Total instance count after applying feature selection

Class	Number of instances	Duplicates removed
Normal	26,966	26,484
Malicious	41,770	41,335
Total	68,736	67,819

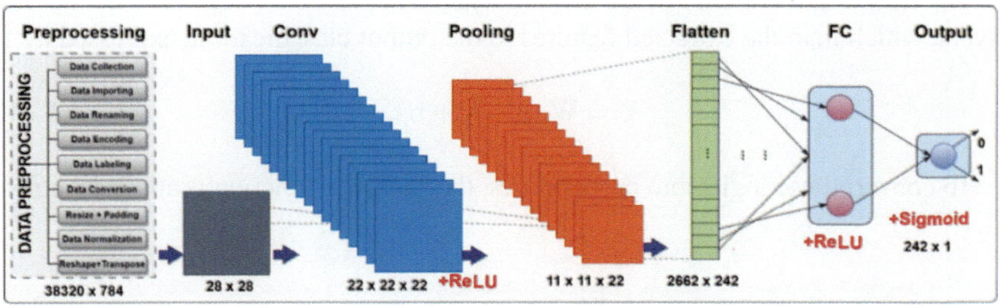

Fig. 12.6 Architecture diagram of the ConvNet model utilized in UAV-CIDS

Model Creation

The suggested system employs the Deep Convolutional Neural Networks (DCNN) approach to improve detection accuracy and minimize false positives through the identification of malicious UAVs. This chapter specifically shows how the DCNN is used to build suggested IDS for UAV communication networks. The UAV-CIDS initiates with a data pre-processing layer and concludes with a detection layer. The detection layer features a singular output node responsible for classifying data samples into normal or abnormal categories. Figure 12.6 presents the model encompassing each layer for the UAV detection system utilizing CNN.

To develop a Convolutional Neural Network (CNN)-based Intrusion Detection System (CIDS) for UAV networks, several mathematical formulations are employed. These formulations ensure robust feature extraction, effective classification and efficient learning for real-time detection in UAV networks. The convolution operation is the first step, where features are extracted from input traffic data. This is achieved by sliding a filter (kernel) across the input feature matrix and computing the dot product between the filter weights and the input values. The result, along with a bias term, forms the output feature map, as shown in Eq. (12.1):

$$Z_{i,j,k} = \sum_{m=0}^{M-1} \sum_{n=0}^{N-1} X_{i+m,j+n} \cdot W_{m,n,k} + b_k \quad (12.1)$$

The output feature map is then passed through a Rectified Linear Unit (ReLU) activation function to introduce non-linearity, defined in Eq. (12.2):

$$\mathrm{A}_{\{i,j,k\}} = \max\left(0, \mathrm{Z}_{\{i,j,k\}}\right) \quad (12.2)$$

Following this, a pooling operation, typically max pooling, is applied to reduce the spatial dimensions of the feature map while retaining the most significant features. For max pooling, the output is given in Eq. (12.3):

$$P_{\{i,j,k\}} = max\left(Z_{\{p:q,r:s,k\}}\right) \tag{12.3}$$

The resulting feature maps are then flattened and passed through fully connected layers, which map the extracted features to the output classification, expressed as:

$$\mathrm{Y} = \mathrm{W}_{\{\mathrm{FC}\}} \cdot \mathrm{A} + \mathrm{b}_{\{\mathrm{FC}\}} \tag{12.4}$$

To convert these logits into probabilities, the SoftMax function is applied, shown as:

$$P(y_i) = \frac{e^{z_i}}{\sum_{j=1}^{C} e^{z_j}} \tag{12.5}$$

The model is trained using the cross-entropy loss function, which measures the difference between the predicted probabilities and the ground truth labels, as defined in:

$$L = -\frac{1}{N}\sum_{i=1}^{N}\sum_{c=1}^{C} y_{i,c}\log\left(P\left(y_{i,c}\right)\right) \tag{12.6}$$

Finally, the CNN is optimized using backpropagation and gradient descent. The weights are updated iteratively based on the gradients of the loss function with respect to the weights, as shown in Eq. (12.7):

$$W_{t+1} = W_t - \eta \cdot \nabla L(W_t) \tag{12.7}$$

The ConvNet scheme given in Fig. 12.6 shows that data logs of UAVs must pass from seven separate computational and processing layers before the output layer. This layer executes a singular node of the Sigmoid activation function. Furthermore, this can yield detection outcomes categorized as either normal or anomalous. The prediction model aims to differentiate between benign and anomalous UAV communication traffic within networks. Additionally, the proposed prediction model has been developed utilizing MATLAB 2020b. Besides, a five-fold cross-validation method is employed, partitioning the dataset into training and testing sets (80%:20%) to ensure a robust validation process. Five tests are conducted following the randomization of data samples for each test. The distribution of different datasets for training and testing is carried out based on the previously stated percentages.

12.4.2 Proposed Incident Response System for UAV Networks

The UAV Cyber Incident Response Method is a systematic approach designed to monitor, detect, validate, and respond to cyber incidents within UAV networks. Each UAV client node is equipped with an IDS that continuously monitors network traffic, classifying events as either benign or alerts. Significant alert events are logged and flagged for further analysis. These events are then transmitted to a UAV-CIDS Node of Consensus, where their validity is checked. Invalid events lead to a reduction in the trust score of the originating node, ensuring accountability within the network. Valid events are forwarded to the centralized CIDS controller for further investigation and response. Based on the severity of the validated events, the controller initiates appropriate measures: low-severity events trigger remediation actions, medium-severity events prompt additional measures, and high-severity events necessitate emergency responses as shown in Fig. 12.7. This collaborative and hierarchical approach ensures that threats are addressed efficiently, while logs are maintained for network-wide transparency and incident tracking.

Once the CIDS is deployed in UAV networks, it continuously monitors network traffic using distributed sensors. When a potential intrusion is detected, denoted as Spe_e the CIDS generates alerts and provides information about the threat to the

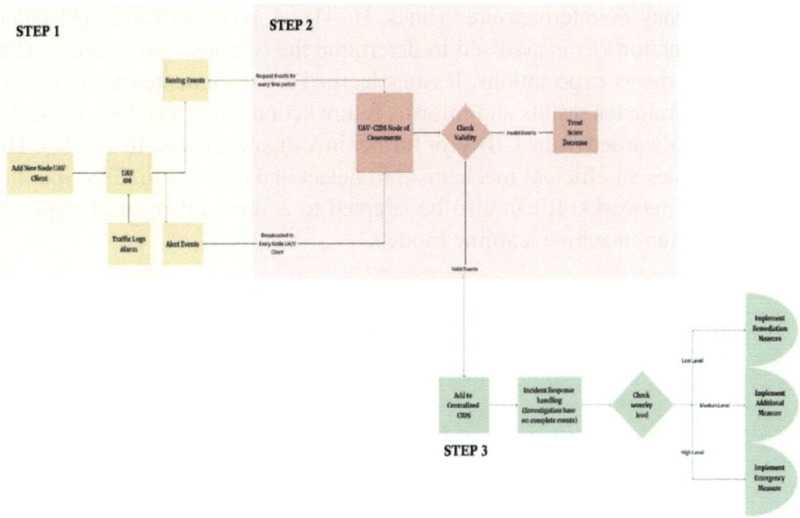

Fig. 12.7 UAV cyber incident response method

centralized server. The server takes appropriate actions to mitigate the threat, such as isolating the malicious UAV, updating security policies, or notifying the human operator. Algorithm is specifically designed to detect and respond to GPS Spoofing attacks, as shown in Fig. 12.7. When nodes are classified as attackers (triggered by E_{Triger}) the feature vector is pre-processed (line 4). This vector is then passed into an FNN model to predict the probability of a GPS Spoofing attack (line 5). If the predicted probability exceeds a predefined threshold, Algorithm initiates a countermeasure, such as activating a backup GPS or altering the UAV's flight path.

Algorithm processes nodes in the network contextually. For each node N^{j}_{uc}, the algorithm calculates the total bits that are set B_{set} and unset B_{unset} in malicious node results. Using these values, it determines the status of nodes as either normal or compromised (lines 6–12). The severity level of the threat is defined using a function that considers the attack's impact, type, and duration.

For high-severity attacks, the response mechanism is triggered to ensure immediate action, as responding to all alerts may not be feasible (lines 14–17). Based on the severity level ('low', 'medium', or 'high'), the algorithm either proceeds with the mission, modifies the mission, or aborts it and recalls all UAVs. It also generates an attack alert for the UAV operator or relevant personnel (lines 18–30).

Incident response is managed by devising a mitigation plan according to the severity level. Low-severity attacks are addressed with basic remediation measures, medium-severity attacks require additional responses, and high-severity attacks demand emergency countermeasures (lines 31–41). A progress value, (f) calculated using a function (f) is assessed to determine the response effectiveness (line 42). If progress meets expectations, lessons learned are documented using a function (f) which evaluates results and informs future actions (line 44). Otherwise, the actual result is forwarded to the CIDS for further investigation (lines 46 and 47). This algorithm provides an efficient mechanism to detect and respond to GPS Spoofing attacks in UAV networks. It can also be adapted to address other attack types by integrating different machine learning models.

Algorithm 1 Response to Potential Security Incident

Input: Get N_u^e result & feature vector from UAV networks
Result: Identify the Malicious Activity on UAV Clients Node Activity

Record N_u^e according to environment conditions;
for all Environment Condition **do**
  for all N_u^e **do**
    // Pre-process feature vector for attack classification $E_{trigger}$;
    Input pre-processed feature vector to FFCNN Model;
    $B_{\text{set}} \leftarrow$ Number of bits set in N_u^e result (identify potential security incident from Algorithm(4));
    $B_{\text{unset}} \leftarrow$ Number of bits not set in N_u^e result
    if $B_{\text{set}} = B_{\text{unset}}$ **then**
      $Status \leftarrow$ "Normal";
    else
      $Status \leftarrow$ "Attack";
    end if
  end for
  if $Status =$ "Attack" **then**
    $Severity\text{-}level \leftarrow f(\text{Impact, type, duration})$;
    Response based upon *Severity-level*;
  end if
  if *Severity-level* = *low* **then**
    proceed with a mission;
  else
    if *Severity-level* = *medium* **then**
      Modify mission plans;
    else
      if *Severity-level* = *high* **then**
        Abort mission & recall all UAVs;
      else
        Notify relevant personnel of the attack;
      end if
    end if
  end if
  if *Severity-level* = *low* **then**
    Implement remediation measures;
  else
    if *Severity-level* = *medium* **then**
      Implement additional measures;
    else
      if *Severity-level* = *high* **then**
        Implement emergency measures;
      end if
    end if
  end if
  $Progress \leftarrow f(\text{time, resources})$;
  if *Progress* is not satisfactory **then**
    $lesson\text{-}learned \leftarrow \{\text{outcomes, action-taken}\}$;
  else
    Send actual result to CIDS for further investigation;
  end if
end for

12.5 Experimental Results and Performance Analysis

A two-class Confusion Matrix (CM) analysis method, on the testing dataset, to evaluate the effectiveness of proposed system is used. The CM is employed to evaluate the effectiveness of the classification model, considering metrics such as True Positives, False Positives, True Negatives and False Negatives. The general structure of components in a two-class classification model utilized to evaluate the effectiveness of the proposed solution is illustrated in Fig. 12.5. Additionally, further performance evaluation metrics, such as Detection Accuracy (DA), Detection Error Rate (DE), F1-score and Detection Precision (DP), are computed utilizing the parameters from the confusion matrix. Given that error rate and detection accuracy are critical factors for assessing the effectiveness of our model, we monitor all variables across 100 epochs of testing as well as training datasets.

12.5.1 UAVs Traffic Flow Bi-directional Mode

The detection accuracy trajectory of our binary classifier is documented across 50 epochs of testing and validation for bidirectional mode. The detection accuracy rate is initially quite low, as illustrated in Fig. 12.7. The recorded accuracy values for the DBPower UDI UAV as well as Parrot Bebop UAV are 59.23 and 37.43%, respectively. In a similar vein, the detection accuracy rate for combined UAVs stands at 58.42%, while for DJI Spark UAVs, it is 57.34%. The detection accuracy plots consistently show a steady upward trend as the number of iterations increases.

Moreover, the overall prediction accuracy following 100 iterations of the testing and training process stands at 96.13, 98.35, and 94.36%, respectively. Figure 12.8 illustrates the overall error rate performance of the binary classifier through a graphical representation. This performance has been documented in a bidirectional mode following 100 iterations. All misprediction rates ultimately decreased when the binary prediction is process advanced. The trajectory of the Parrot Bebop UAV demonstrates the lowest misprediction rate while utilizing the fewest epochs, specifically 18 epochs.

After approximately 100 testing and training epochs, the misprediction rate for each dataset stabilizes, showing <3.0–5.0% of incorrect predictions. When 100 iterations of training and testing are completed, the model exhibited testing error rates of 0.03, 4.38, 2.66 and 2.21% for the individual UAVs, in addition to the overall combined Bi-directional communication mode. Figure 12.6 presents the findings of the CM analysis conducted on the IDS dataset, detailing each type of UAV separately, along with the aggregated results for the combined UAVs in bi-directional flow mode. Likewise, Table 12.4 presents a summary of the performance evaluation metrics derived from the parameters of CM, offering additional context for the

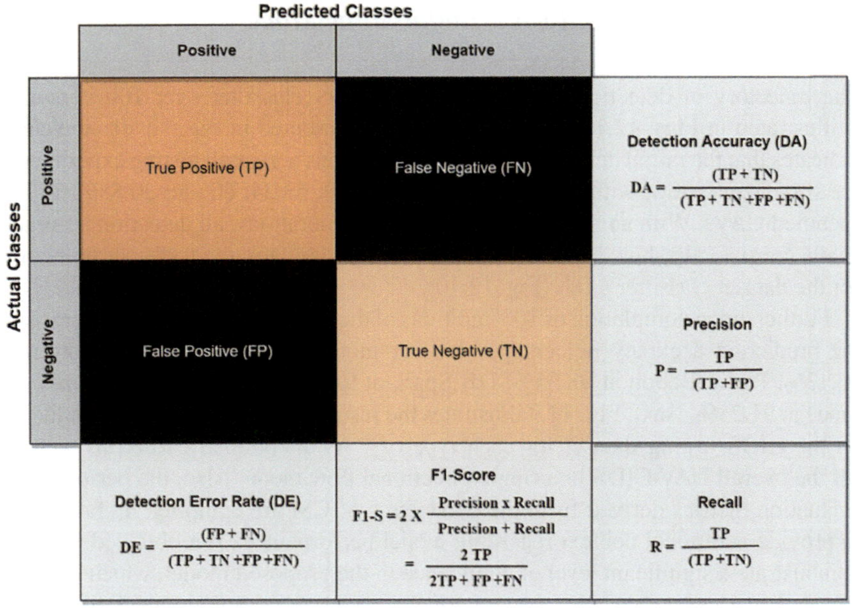

Fig. 12.8 Two-class confusion matrix [6]

Table 12.4 A summary of the performance analysis metrics is compile for a model using bi-directional

Metric	Parrot Be-bop	DBPower UDI	DJI Spark	Collective UAVs
Recall (%)	100	99.97	99.89	99.95
Precision (%)	100	99.29	97.81	99.74
False alarm	0	0.007	0.2	0.002
F1-Score	1	0.99	0.98	0.99
Accuracy (%)	99.7	95.62	97.34	97.79
Error rate (%)	0.03	4.38	2.66	2.21

model's performance. Also, the findings indicate that the proposed model demonstrates a significant degree of reliability. The model demonstrated impressive sensitivity, precision and detection accuracy, accompanied by a minimal occurrence of false alarms as well as detection errors.

12.5.2 Uni-directional UAV Traffic Flow Mode

The trajectory of detection accuracy for our binary classifier over 100 iterations is illustrated in Figs. 12.10 and 12.11. The data presented in Fig. 12.10a–d clearly indicates that the initial detection accuracy proportions across all testing experiments are significantly low, with recorded values of 68.49, 63.34, 60 and 30.88% for the specified UAVs. With an increase in the number of iterations, all detection accuracy charts consistently show a gradual upward trend, culminating in a steeper trajectory for the dataset classifier's DA (Fig. 12.10).

Further, upon completion of 100 iterations of the data training and testing process, the prediction accuracy percentages for our model are as follows: DBPower at 96.13%, Parrot Bebop at 98.35%, DJI Spark at 94.36% and the overall combined mode at 91.23%. Next, Fig. 12.9 illustrates the results of the CM analysis conducted on the CIDS testing dataset for each type of UAV mentioned earlier, as well as for the overall UAV-CIDS in a single directional flow mode. Also, the performance evaluation metrics derived from the parameters of CM are compiled in Table 12.5 to provide additional context regarding model performance. The obtained findings demonstrate a significant level of confidence in the proposed model, which exhibits elevated metrics for detection accuracy, precision and sensitivity, alongside a minimal detection error and a low False Alarm Rate (FAR).

Moreover, utilizing various architectures of FFCNN, we employ a range of techniques in our experiments, adjusting the sizes of convolutional filters and the necessary number of neurons. Additionally, this chapter implements various polling techniques, activation functions, dropout layers and dropout proportions. As a result, it is found that training a comparable FFCNN architecture for 200 epochs, with a dropout rate of 75%, enhances the accuracy trajectory for both bidirectional and unidirectional datasets of networked UAVs. The enhanced performance of the detection system achieves a general detection accuracy of 98.23% following 105 training iterations (Fig. 12.12).

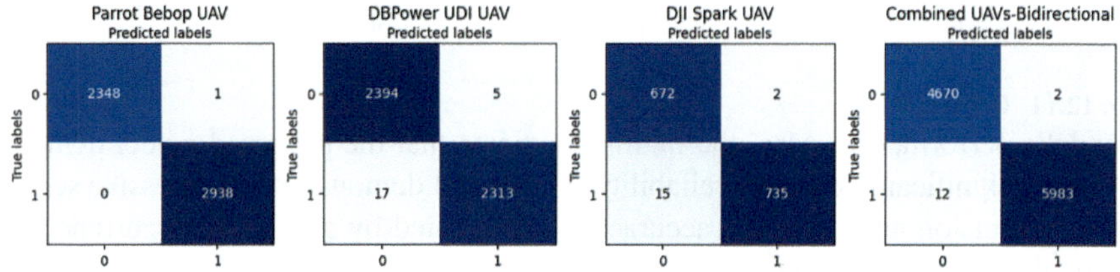

Fig. 12.9 CM analysis for a bi-directional communication-based

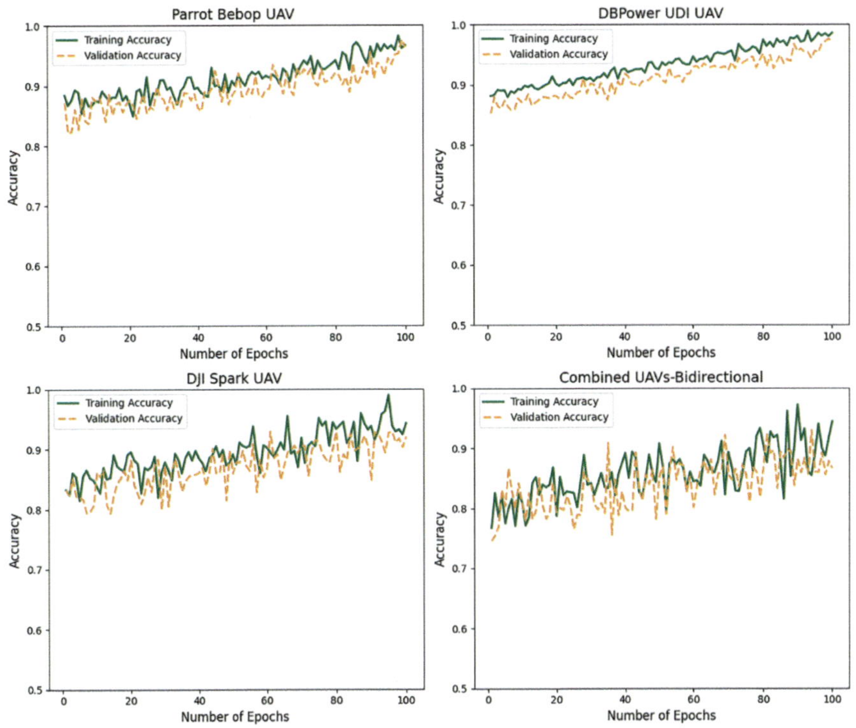

Fig. 12.10 Progression of prediction accuracy and misprediction proportions over epochs using bi-directional communication mode

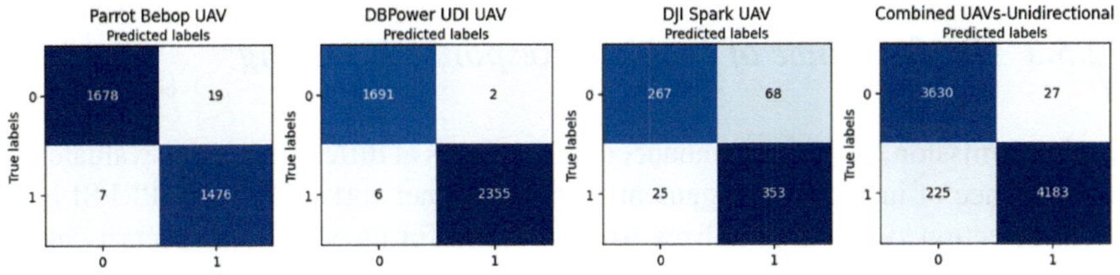

Fig. 12.11 CM analysis for uni-directional communication in CIDS

Table 12.5 A summary of the performance analysis metrics is compile for a model using unii-directional

Metric	UAV Parrot Bepob	UAV DBPower UDI	UAV spark	Collective UAVs
Recall (%)	98.88	99.87	79.7	99.26
Precision (%)	99.95	99.64	91.4	94.16
F1-Score	0.99	0.99	0.84	0.96
False alarm	0.004	0.002	0.066	0.051
Accuracy (%)	96.13	98.35	94.36	91.23
Error rate (%)	3.87	1.65	5.64	8.77

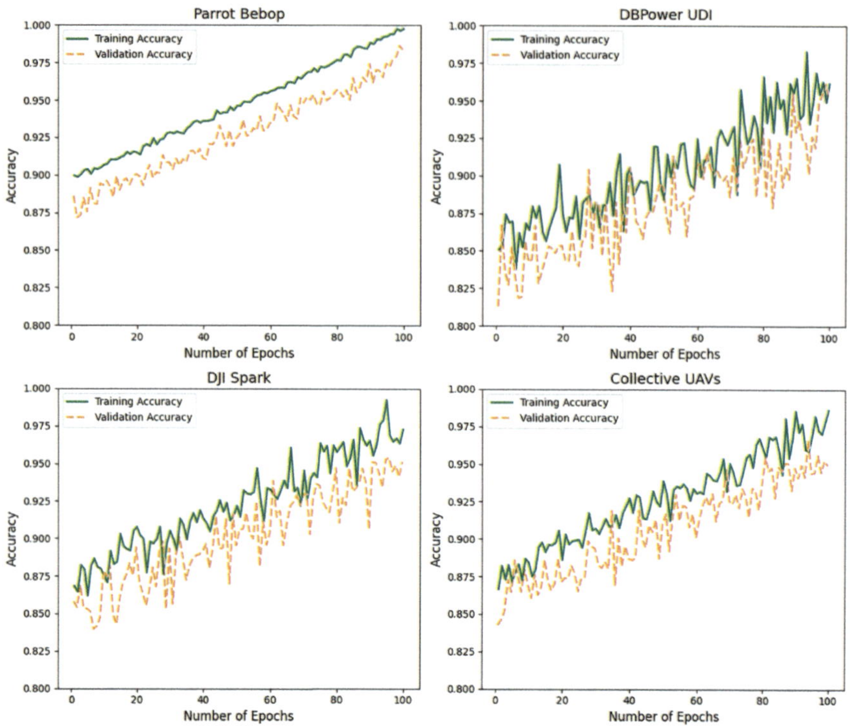

Fig. 12.12 Progression of prediction accuracy and misprediction proportions over epochs using uni-directional communication mode

12.5.3 Performance of Incident Response Handling

The transmission of events is enhanced by detectors at different rates to evaluate the performance of incident management within UAV networks. The UAV-PRPSI is an event detection system that utilizes Apache Storm for its operations. Security alerts are generated by incident detectors following the processes of filtering, handling, and correlating events. The alerts are subsequently sent to the response module. The correlation engine collects the alerts, subsequently generating a file and transmitting the alerts to the security orchestrator.

This chapter also provides a series of 100 alarm events at different frequencies such as 5, 10 and 20 Hz to conduct experiments. When these events are forwarded to assess the efficacy of the components involved in monitoring, response and detection, a security alert is triggered. The incident detector produces 100 alerts at the specified rate. This would enable the filtering of numerous events and help avoid system overload. We conduct a simulation of a particularly challenging workload scenario for the incident detector, orchestrator and incident response, with the test results illustrated in Fig. 12.13. The processing time for each event by the incident detector is illustrated on the y-axis. The response time of the event detector remains consistent,

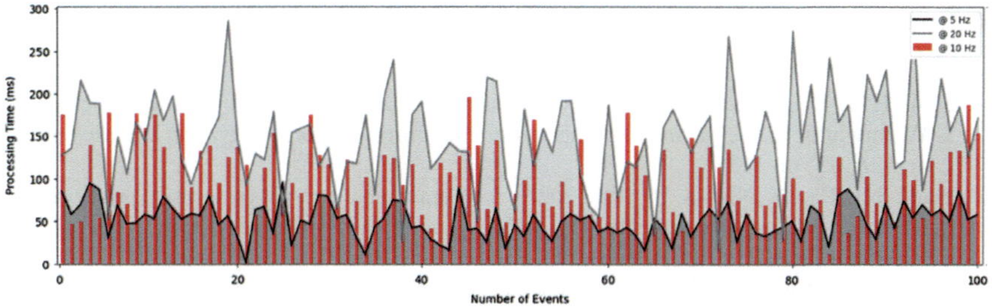

Fig. 12.13 Incident response performance under bursts of 100 events at varying frequencies [6]

even though there are noticeable peaks in each test. A peak response of 260 ms is observed at 10 Hz in event #78. The incident detector promptly resumes its normal function despite these irregularities. It is important to observe that the duration for stability to return is extended at 20 Hz, as seen in the transition from event #36 to #46.

12.6 Conclusion

This book chapter presents an innovative, autonomous UAVCIDS that uses a FFCNN. It is designed to accurately and rapidly detect both zero-day as well as known attacks in real time. The proposed methodology includes a distributed CIDS framework, an event correlation engine and a real-time incident response handling mechanism to enhance the security of UAV networks in a thorough manner. The evaluation results highlight the success of our methodology, with the FFCNN model attaining an impressive accuracy of 98.23% using the UAVIDS dataset. This signifies a notable improvement over current models, which generally achieve accuracy levels between 80 and 95%. Moreover, our UAV-CIDS demonstrates an impressive 99% attack detection rate, highlighting its exceptional capability in recognizing and addressing security threats. Besides, there is an opportunity to enhance the UAVCIDS technique in the future, to create a multi-path route planning system, a capability that has not been incorporated in this study.

References

1. Hadi, H. J., Du, X., & Cao, Y. (2022). Critical analysis of security and privacy challenges for UAV networks. In *Secure and digitalized future mobility: shaping the ground and air vehicles cooperation* (Vol. 115).
2. Hadi, H. J., Cao, Y., Nisa, K. U., et al. (2023). A comprehensive survey on security, privacy issues and emerging defense technologies for UAVs. *Journal of Network and Computer Applications, 213*, Article 103607.
3. Khan, M. A., Kumar, N., Mohsan, S. A. H., et al. (2022). Swarm of UAVs for network management in 6G: A technical review. *IEEE Transactions on Network and Service Management, 20*(1), 741–761.
4. Hadi, H. J., Cao, Y., Li, S., et al. (2024). Real-time fusion multi-tier DNN-based collaborative IDPS with complementary features for secure UAV-enabled 6G networks. *Expert Systems with Applications, 252*, Article 124215.
5. Khan, M. A., Kumar, N. (2024). Security and privacy issues and solutions for UAVs in B5G networks: A review. *IEEE Transactions on Network and Service Management.*
6. Hadi, H. J., Cao, Y., Li, S., et al. (2024). Real-time collaborative intrusion detection system in UAV networks using deep learning. *IEEE Internet Things Journal.*
7. Tao, J., Han, T., & Li, R. (2021). Deep-reinforcement-learning-based intrusion detection in aerial computing networks. *IEEE Network, 35*(4), 66–72.
8. Li, W., Meng, W., & Kwok, L. F. (2021). Surveying trust-based collaborative intrusion detection: State-of-the-art, challenges, and future directions. *IEEE Communications Surveys & Tutorials, 24*(1), 280–305.
9. Alipour-Fanid, A., Dabaghchian, M., Wang, N., et al. (2019). Machine learning-based delay-aware UAV detection and operation mode identification over encrypted Wi-Fi traffic. *IEEE Transactions on Information Forensics and Security, 15*, 2346–2360.
10. Hadi, H. J., Hayat, U., Musthaq, N., et al. (2022). Developing realistic distributed denial of service (DDoS) dataset for machine learning-based intrusion detection system. In *2022 The international conference on Inte Things: Systems, Management, and Security (IOTSMS)* (pp. 1–6).
11. Gurung, G., Bendiab, G., Shiaele, M., & Shiaeles, S. (2022). CIDS: Collaborative intrusion detection system using blockchain technology. In *2022 IEEE International Conference on Cyber Security and Resilience (CSR)* (pp. 125–130).
12. Hadi, H. J., Cao Y., Li, S., et al. (2024). FCG-MFD: Benchmark function call graph-based dataset for malware family detection. *The Journal of Network and Computer Applications*, 104050.
13. Li, G., & Jung, J. J. (2023). Deep learning for anomaly detection in multivariate time series: Approaches, applications, and challenges. *Inf. Fusion, 91*, 93–102.
14. Elsaeidy, A. A., Jagannath, N., Sanchis, A. G., et al. (2020). Replay attack detection in smart cities using deep learning. *IEEE Access, 8*, 137825–137837.
15. Choudhary, S., & Kesswani, N. (2020). Analysis of KDD-Cup'99, NSL-KDD and UNSW-NB15 datasets using deep learning in IoT. *Procedia Computer Science, 167*, 1561–1573.
16. Kim, J., Kim, J., Kim, H., Shim, M., & Choi, E. (2020). CNN-based network intrusion detection against denial-of-service attacks. *Electronics, 9*(6), 916.
17. Gamage, S., & Samarabandu, J. (2020). Deep learning methods in network intrusion detection: A survey and an objective comparison. *Journal of Network and Computer Applications, 169*, 102767.
18. Vanitha, N., & Ganapathi, P. (2020) Traffic analysis of UAV networks using enhanced deep feed forward neural networks (EDFFNN). In *Handbook of research on machine and deep learning applications for cyber security* (pp. 219–244). IGI Global.
19. Panice, G., Luongo, S., Gigante, G., Pascarella, D., Di Benedetto, C., Vozella, A., & Pescape, A. (2017). A SVM-based detection approach for GPS spoofing attacks to UAV. In *2017 23rd International Conference on Automation and Computing (ICAC)* (pp. 1–11). IEEE.

20. Lin, Z., Lu, X., Dai, C., Sheng, G., & Xiao, L. (2019). Reinforcement learning based UAV trajectory and power control against jamming. In *International conference on machine learning for cyber security* (pp. 336–347). Springer.
21. Xiao, L., Xie, C., Min, M., & Zhuang, W. (2017). User-centric view of unmanned aerial vehicle transmission against smart attacks. *IEEE Transactions on Vehicular Technology, 67*(4), 3420–3430.
22. Wang, A., Wang, W., Zhou, H., & Zhang, J. (2021). Network intrusion detection algorithm combined with group convolution network and snapshot ensemble. *Symmetry, 13*(10), 1814.
23. Liu, G., Zhang, R., Wang, C., & Liu, L. (2019). Synchronization-free gps spoofing detection with crowdsourced air traffic control data. In *2019 20th IEEE international conference on Mobile Data Management* (MDM) (pp. 260–268). IEEE.
24. Arthur, M. P. (2019). Detecting signal spoofing and jamming attacks in UAV networks using a lightweight IDS. In *2019 international conference on Computer, Information and Telecommunication Systems (CITS)* (pp. 1–5). IEEE.
25. Wang, Y., Ding, J., He, X., et al. (2023). Intrusion detection method based on denoising diffusion probabilistic models for UAV networks. *Mobile Networks and Applications*, 1–10.
26. Fotohi, R., Abdan, M., & Ghasemi, S. (2022). A self-adaptive intrusion detection system for securing UAV-to-UAV communications based on the human immune system in UAV networks. *J. Grid Comput., 20*(3), 22.
27. Fu, R., Ren, X., Li, Y., et al. (2023). Machine-learning-based UAV-assisted agricultural information security architecture and intrusion detection. *IEEE Internet of Things Journal, 10*(21), 18589–18598.
28. Michelena, Á., Ramírez, C., Vega, J., et al. (2023). Hybrid classification model based on supervised techniques for denial of service attacks detection over CoAP protocol. In *International Symposium on Distributed Computing and Artificial Intelligence* (pp. 1–10).
29. Alotaibi, A., Barnawi, A. (2023). IDSoft: A federated and softwarized intrusion detection framework for massive Internet of Things in 6G networks. *Journal of King Saud University—Computer and Information Sciences, 35*(6), 101575.
30. Michelena, Á., Aveleira Mata, J., Jove, E., Alaiz Moretón, H., Quintián, H., & Calvo-Rolle, J. L. (2023). Development of an intelligent classifier model for denial of service attack detection.
31. He, X., Chen, Q., Tang, L., Wang, W., & Liu, T. (2022). CGAN-based collaborative intrusion detection for UAV networks: A blockchain-empowered distributed federated learning approach. *IEEE Internet of Things Journal, 10*(1), 120–132. Shrestha, R., Omidkar, A., Roudi, S. A., Abbas, R., & Kim, S. (2021). Machine-learning-enabled intrusion detection system for cellular connected UAV networks. *Electronics, 10*(13), 1549.
32. Liu, J., Yang, W., Xu, S., Liu, J., & Zhang, Q. (2019). Q-learning based UAV secure communication in presence of multiple UAV active eavesdroppers. In *2019 11th international conference on Wireless Communications and Signal Processing (WCSP)* (pp. 1–6). IEEE.
33. Sedjelmaci, H, Senouci, S. M., & Ansari, N. (2017). A hierarchical detection and response system to enhance security against lethal cyber-attacks in UAV networks. *IEEE Transactions on Systems, Man, and Cybernetics: Systems, 48*(9), 1594–1606.
34. Lu, X., Xu, D., Xiao, L., Wang, L., & Zhuang, W. (2017). Anti-jamming communication game for UAV-aided VANETs. In *GLOBECOM 2017–2017 IEEE global communications conference* (pp. 1–6). IEEE.
35. Manesh, M. R., Kenney, J., Hu, W. C., Devabhaktuni, V. K., & Kaabouch, N. (2019). Detection of GPS spoofing attacks on unmanned aerial systems. In *2019 16th IEEE annual Consumer Communications & Networking Conference (CCNC)* (pp. 1–6). IEEE.
36. Mitchell, R., & Chen, R. (2013). Adaptive intrusion detection of malicious unmanned air vehicles using behavior rule specifications. *IEEE Transactions on Systems, Man, and Cybernetics: Systems, 44*(5), 593–604.
37. Li, Z., Lu, Y., Shi, Y., Wang, Z., Qiao, W., & Liu, Y. (2019). A Dyna-Q-based solution for UAV networks against smart jamming attacks. *Symmetry, 11*(5), 617.
38. Yenugunti, C., & Yau, S. S. (2020). A blockchain approach to identifying compromised nodes in collaborative intrusion detection systems. In *2020 IEEE Intl Conf on Dependable, Autonomic*

and Secure Computing, Intl Conf on Pervasive Intelligence and Computing, Intl Conf on Cloud and Big Data Computing, Intl Conf on Cyber Science and Technology Congress (DASC/PiCom/CBDCom/CyberSciTech) (pp. 87–93). IEEE.

Chapter 13
Multi-tier MIHIDS: Integrating CTI for UAV Security Enhancement

Abdul Majid Jamil and Yue Cao

Abstract Unmanned Aerial Vehicles (UAVs) face distinct security challenges due to their distributed nature and vulnerability to evolving threats. Traditional IDS often fall short in addressing real-time and zero-day attacks in these environments, highlighting the need for specialized solutions. This chapter introduces a Unified Intrusion Detection System (UIDS) designed to enhance UAV security using a Lightweight Multi-tier IDS architecture. The proposed framework deploys multiple detection units across UAV networks, improving detection accuracy while conserving resources. By optimizing the Aho-Corasick algorithm and integrating Honeypot Intelligence, the system not only detects known and emerging threats but also supports IDS signature creation through collaborative knowledge sharing and honeypot integration. Experimental results show that our approach significantly improves detection rates and reduces false positives, achieving a remarkable 99.71% overall detection rate against UAV intrusions. Additionally, the Lightweight Multi-Tier IDS demonstrates enhanced response times and proactive threat detection capabilities.

Keywords Cyber security · Cyber threat intelligence · Lightweight IDS · UAV security

Abbreviations

GCS	Ground Control Station
IDS	Intrusion Detection Systems
IoV	Internet of Vehicles
MISP	Malware Information Sharing Platform
SGE	Signature Generation Engine

A. M. Jamil · Y. Cao (✉)
School of Cyber Science and Engineering, Wuhan University, Wuhan, China
e-mail: yue.cao@whu.edu.cn

A. M. Jamil
e-mail: majidjamil@whu.edu.cn

Y. Hu et al. (eds.), *Reliable and Efficient Trajectory Design in UAV-Aided Wireless Networks*, https://doi.org/10.1007/978-981-96-5988-3_13

STIX	Structured Threat Information Expression
TAXII	Trusted Automated Exchange of Indicator Information
UAV	Unmanned Aerial Vehicles

13.1 Introduction

Unmanned Aerial Vehicles (UAVs) have become ubiquitous airborne platforms, enabling seamless communication with other devices through smart interfaces like tablets and mobile phones [1]. Their adoption spans a wide range of applications, including aerial mapping, package delivery, environmental monitoring, photography, irrigation management, and search and rescue operations. Given the critical roles played by UAVs, ensuring robust wireless connectivity and operational reliability is essential. Security within UAV networks is maintained through the implementation of security protocols and intrusion detection mechanisms. The typical network architecture includes a Ground Control Station (GCS) and multiple UAVs, each of which is vulnerable to potential cyber-attacks, such as the infiltration of rogue UAVs [2]. Common attack vectors against UAV systems include visual navigation compromise, spoofing, jamming, traffic analysis, and hijacking, among others [3].

As UAV utilization continues to rise, so does the risk of malicious intrusions, underscoring the crucial role of Intrusion Detection Systems (IDS) designed for UAV environments [4]. These systems are vital in strengthening UAV networks by identifying novel and complex attacks. However, the evolving nature of cyber threats presents ongoing challenges, requiring continuous advancements in detection capabilities [5]. Communication infrastructure is one of the most impacted areas [6], due to its dynamic topology and frequent disruptions. In this context, UAVs serve as intermediaries between ground users, network devices, intelligent vehicles, and smart homes, offering a wide range of network services to ensure seamless operations. However, the increased usage and connectivity also raise the risk of unauthorized access to UAV systems, making UAVs more susceptible to cyberattacks.

To counter these threats, the deployment of IDS is imperative [6]. Initially conceptualized by Jim Anderson with a basic architecture (see Fig. 13.1), IDSs have evolved into various models tailored to meet the specific security needs of UAVs. These systems are designed to detect and respond to unauthorized access attempts, providing operators with timely alerts to mitigate potential threats. IDSs generally fall into three main categories: Host-based Intrusion Detection Systems (HIDS), Network-based Intrusion Detection Systems (NIDS), and Hybrid IDS. HIDS monitor both network and UAV states and events, while NIDS focus exclusively on network states and events. Hybrid IDS combine features of both HIDS and NIDS.

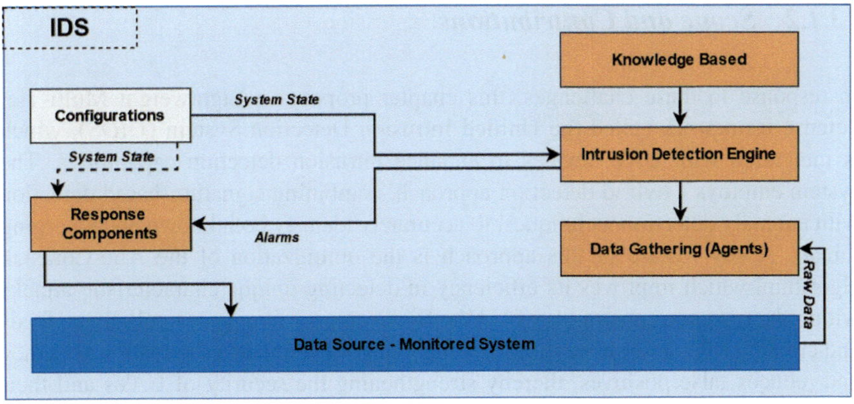

Fig. 13.1 Basic architecture of IDS

13.1.1 Problematic and Limitations

The rapid growth of network activity has posed significant challenges for existing Intrusion Detection Systems (IDS) in effectively detecting malicious intrusions. Additionally, the exploration and application of multiple IDSs are still incomplete and insufficient. Traditional IDSs require security systems to closely monitor network traffic or system logs to identify suspicious activity.

The following limitations are evident in the existing literature:

(1) **Effective Detection of Zero-Day Attacks**: Traditional signature-based detection methods struggle to identify zero-day attacks, which can unpredictably alter their behavior before patches are developed. This challenge underscores the need for innovative detection techniques that can identify novel threats in real-time and proactively.
(2) **Resource Efficiency in Intrusion Detection**: As the number of datasets and rules grows, especially on hardware-constrained UAV devices, the efficiency and performance of IDSs decline. Addressing this issue requires the development of lightweight algorithms and data structures that can process network traffic effectively while minimizing resource usage.
(3) **Optimizing Detection Accuracy within Resource Constraints**: Signature-based IDSs face the challenge of accurately identifying malicious UAV attacks while minimizing false positives and negatives, all within the resource limitations of UAV devices. Overcoming this challenge demands inventive strategies that enhance detection precision while respecting resource constraints.

13.1.2 Scope and Contributions

In response to these challenges, this chapter proposes a Lightweight Multi-Tier defense framework called the Unified Intrusion Detection System (UIDS), which is integrated with UAV devices to enhance intrusion detection capabilities. The system employs a hybrid detection approach, combining signature-based detection with anomaly detection techniques, to accurately identify both known and emerging threats. A key feature of this approach is the optimization of the Aho-Corasick algorithm, which improves its efficiency in detecting unique characteristic attacks within the resource constraints of UAVs. By leveraging honeypot intelligence feeds and creating rules from honeypot logs, this integration enhances detection accuracy and reduces false positives, thereby strengthening the security of UAVs and their networks.

This presents three significant contributions to the field of UAV network security:

(1) A Lightweight IDS was introduced that combines anomaly detection with signature-based detection, offering an effective approach to detecting both attacks and anomalies in UAV networks.
(2) The proposed Multi-tier Unified IDS includes the optimization of the Aho-Corasick algorithm, aimed at reducing computational overhead. This optimization enhances UAV network security by efficiently identifying and categorizing unique characteristic attacks, providing a robust defense mechanism.
(3) To address challenges faced by IDS, such as uncertain attack classification, false positives, and resource constraints, the UIDS integrates rules from honeypot threat intelligence feeds. This integration improves accuracy and reduces false positives by creating rules from network traffic logs, forwarding them to Honeypot Intelligence, and generating unique patterns that are subsequently converted into a robust signature engine.

13.2 Related Work

This chapter presents a survey to identify relevant studies for the proposed work and performs a comprehensive comparison (see Table 13.1), evaluating various parameters to position the proposed approach within the current research landscape. One of the honeypots available for UAV, known as HoneyDrone [7, 8]. HoneyDrone is described by its creators as a medium-interaction, portable solution designed to operate on portable devices. It mimics various aspects of a drone's network interface, including services such as Telnet, SSH, FTP, and MAVLink. Additionally, it simulates a range of sensors like GPS, motion sensors, battery status, and the internal file system. The primary focus of this chapter is on the intricacies of system architecture and protocol design for honeypots within IoT and UAV networks. However, there is a noticeable gap in exploring multi-tiered honeypot-based defense mechanisms specifically tailored for UAVs, an area that warrants further investigation.

Table 13.1 Literature review

References	Honeypot type	Domain	ML/signature/ Sat-based IDS	CTI rule creation	Signature verification	Simulated environment test	Attacks classification
2018 [7]	Medium	UAV	Statistical	✗	✗	✓	✗
2018 [8]	Medium	UAV	✗	✗	✗	✗	✗
2022 [12]	Low	UAV	ML	✗	✗	✗	✗
2022 [13]	High	Cyber-physical	✗	✗	✗	✗	✗
2022 [10]	Low	IoT	ML	✗	✗	✗	✗
2022 [11]	Low	IoT	ML	✗	✗	✗	✗
2023 [9]	High	ICS	Signature	✗	✗	✗	✗
2023 [14]	High	IoV	Signature	✗	✗	✓	✗
2023 [15]	Low	Wi-Fi 802.11	✗	✗	✗	✓	✗
This work	**Medium**	**UAV**	**Hybrid**	✓	✓	✓	✓

The author of ICSvertase proposed a framework to enhance ICS high-interaction honeypots [9]. However, existing approaches often overlook adversary behavior and lack suitable classification schemes. When compared to UAV honeypots with IDS, limitations may include challenges in accurately replicating complex UAV protocols, detecting intrusions, and integrating seamlessly with existing infrastructures. Another work introduces IoTZeroJar, a honeypot system designed for zero-day attack detection in IoT environments [10, 11]. IoTZeroJar uses honeypots to collect malicious traffic and employs Elasticsearch for log data processing. However, its focus on IoT environments may limit its applicability to UAV network IDS.

Another methodology was proposed for deploying a honeypot framework within Internet of Vehicles (IoV) environments to analyze attack propagation patterns [14]. The results demonstrated the successful simulation of vehicle capabilities and the efficacy of the proposed pattern analysis in identifying common attack patterns on vehicle systems, geolocating attackers, and evaluating honeypot node usage from an attacker's perspective. Nevertheless, a challenge may arise in accurately modeling the diverse and dynamic behaviors within IoV environments, despite the proposed medium-interaction honeypot network approach. Additionally, the authors introduced WiFi-NID, a tool for detecting intrusions from WiFi networks, particularly focusing on 802.11 networks [15]. WiFi-NID automates the detection of illegal network scanning attacks and is validated through realistic testbed scenarios. However, its limitations include the lack of emphasis on developing updated signature generation and the tool's unsuitability for UAV-specific IDS to detect attacks.

13.3 Background

13.3.1 R-Tree Rule-Based Filtering

A R-tree is a spatial data structure used for storing and retrieving spatial data, such as rectangles or points. It consists of root nodes, internal nodes, and leaf nodes, each containing a fixed-size collection of Minimum Bounding Rectangles (MBRs) that enclose child node MBRs, as shown in Fig. 13.2a. MBRs correspond to nodes, with root and internal nodes marked by dotted lines and leaf nodes by solid lines, as shown in Fig. 13.2a. The root node covers the largest region, internal nodes point to child nodes, and leaf nodes store actual data. Searching in R-trees starts at the root node and involves checking each rectangle in a node for overlap with the search rectangle. The search continues recursively until all overlapping nodes are examined. This structure allows the search algorithm to focus on relevant regions, optimizing the search process. Insertions involve recursively traversing the tree to find a suitable node for the new spatial object. If a leaf node is full, it splits into two, and this split propagates upward, potentially causing parent nodes to split. The challenge is to find an optimal splitting procedure that minimizes overlap and maintains tree balance.

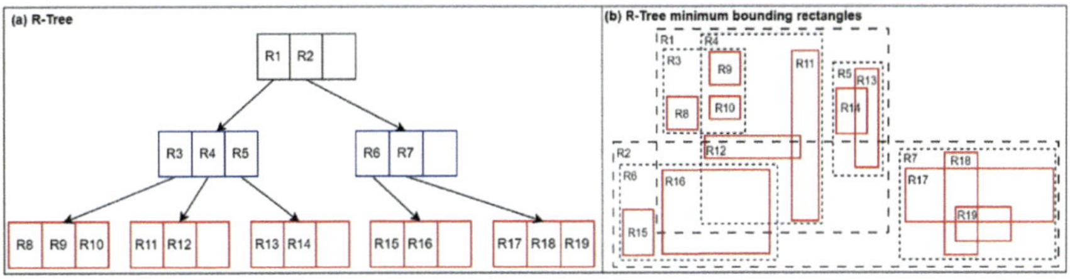

Fig. 13.2 **a** R-Tree and **b** R-Tree minimum bounding rectangles

These aspects are particularly important for a lightweight IDS for UAV devices, where efficient spatial data processing is essential.

13.3.2 Aho-Corasick Algorithm

The Aho-Corasick algorithm [6] is a string-searching technique used to find every instance of a set of keywords in a text string. It utilizes a finite state machine built from the predetermined set of keywords, K. The algorithm uses three functions to control state transitions: the goto-function *g(i,c),* the fail-function *f(i),* and the success-function *s(i)*. The goto-function *g(i,c)* determines the next state based on the current state i and input character ccc. The fail-function *f(i)* handles cases where the goto-function does not define a transition for the current character. The success-function *s(i)* identifies states that correspond to pattern matches (Fig. 13.3).

The goto-, fail-, and success-functions of an Aho-Corasick state machine for a set of keywords $K =$ *{/bin/sh, /bin/dash, /bin/bash}* are illustrated. To ensure constant time complexity, the state machine is typically represented as a prefix tree. This tree structure incorporates extensions for suffix and dictionary transitions, introducing cycles into the state graph's goto-function when input symbols either fail to match or result in a successful pattern match. Constructing the state machine has a complexity of $O(|K|L)$, where L is the average size of the keywords. Processing an input string of size n has a complexity of $O(n+\{max\}(K))$, as each transition in the state machine

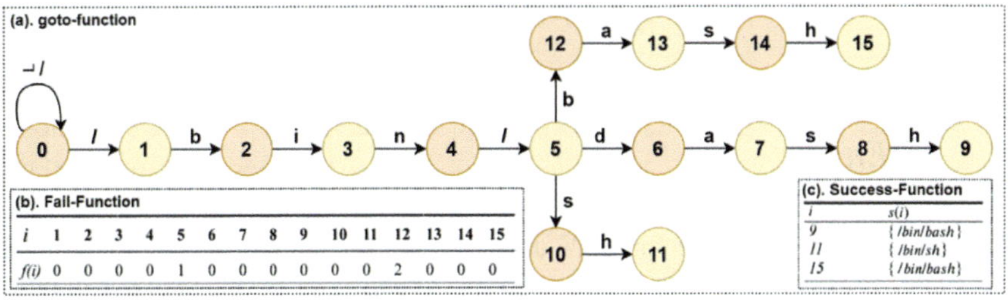

Fig. 13.3 goto-function

can be performed in constant time. Due to its efficiency, the Aho-Corasick algorithm is well-suited for deployment in a lightweight IDS, where quick and accurate pattern matching is critical for real-time threat detection in resource-limited environments.

13.4 UIDS Proposed Scheme

13.4.1 UIDS Architecture

This section outlines the methodology of the proposed CTI-UIDS for UAVs and its components. The key contribution is the development of a unified intrusion detection system (UIDS) leveraging CTI. Unlike traditional systems, it integrates CTI within UAV systems for precise detection and security. Each IDS segment monitors subsystems, networks, or protocols, collaborating to detect intrusions. Honeypots log malicious activities and feed data to the IDS for signature-based detection. Traffic not matching known signatures undergoes further analysis using MISP CTI, generating new signatures for future detection.

UAV traffic also undergoes ongoing analysis through honeypots, which generate logs shared with MISP to update threat intelligence. The Signature Generation Engine updates the Signature Pool, enhancing detection capabilities. By combining honeypots, IDS, and CTI platforms like MISP, the system proactively adapts to emerging threats and strengthens UAV security.

Unlike existing machine learning-based IDS approaches that focus on specific components, this method leverages honeypot-generated flight logs, providing broader attack detection without separate IDS instances. UIDS agents deployed across sensors, systems, and communication protocols ensure continuous security analysis. This distributed setup enables coordinated threat detection and efficient UAV protection.

Figure 13.4 depicts the interconnection between the UIDS modules. In this illustration, each management module is linked to a Honeypot-based IDS, enabling the UAV to gather data. The UAV is connected to the IDS through a deployed agent. Subsequently, they interface with the communication module, facilitating the transfer and exchange of data with the central UIDS core for analysis and attack detection.

13.4.2 Developed Lightweight Pattern Matching Detection Engine

A lightweight IDS suitable for UAV devices, the efficiency of the IDS is crucial, especially considering the limited processing capability of UAVs. One key component of an IDS is rule-based filtering and pattern-matching, where the system scans captured network packets for predefined patterns. To enhance efficiency in processing

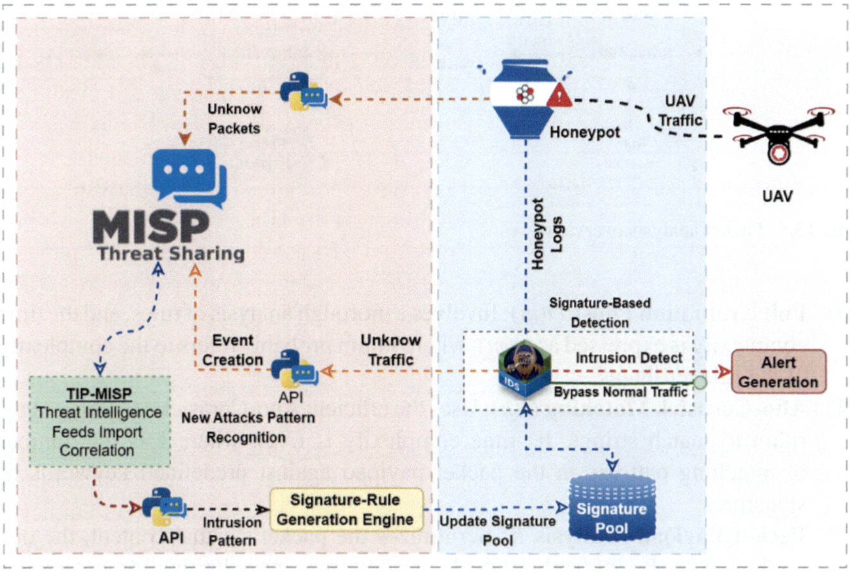

Fig. 13.4 UIDS architecture workflow diagram

observed network traffic, appropriate data structures and algorithms are essential. An effective solution involves incorporating the concept of R-trees, hierarchical data structures commonly used in spatial databases. R-trees organize spatial objects in a tree-like structure, and in the context of an IDS, they can be adapted to efficiently store and retrieve rules or patterns associated with network packet inspection. This incorporation of R-trees aligns with the overall goal of optimizing the lightweight Aho-Corasick algorithm, a multi-pattern matching algorithm used in the IDS for its ability to match multiple patterns simultaneously. The lightweight nature of the Aho-Corasick algorithm is particularly beneficial for UAV devices, ensuring efficient throughput and content-matching time in the intrusion detection process.

13.4.3 Optimized Detection Engine Packet Analysis for UAV

In the comprehensive packet analysis framework of the lightweight detection engine, several key phases contribute to the system's efficacy in real-time threat detection, as shown in Fig. 13.5.

(1) **R-tree Filtering** $O(logn)$: Effectively handles and sorts rules according to the number of rules, using R-trees with a logarithmic time complexity to improve speed and accuracy in the matching of subsequent rules.
(2) **Rule Header Analysis**: Analyses and organizes rule header data, an essential stage in understanding and classifying rule characteristics.

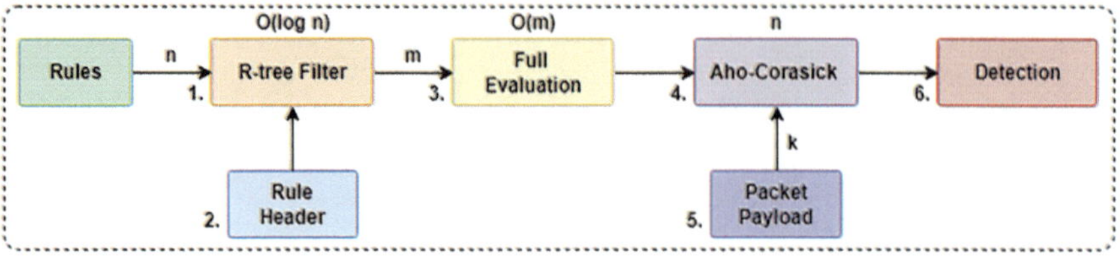

Fig. 13.5 Packet analysis overview

(3) **Full Evaluation Phase** $O(m)$: Involves a thorough analysis of rules, and the time complexity is expressed as $O(m)$, where mmm probably refers to the complexity of the rule attributes.
(4) **Aho-Corasick Matching** $O(k)$: Uses the efficient Aho-Corasick Matching algorithm to match strings. Its time complexity is $O(k)$, where k is the number of matching patterns in the packet payload against predefined keywords or signatures.
(5) **Packet Payload Analysis** K: Scrutinizes the packet's actual content, the one marked with K, in-depth to find any patterns, irregularities, or potentially malicious signatures that point to intrusions.
(6) **Detection**: Through the integration of rule header analysis, Aho-Corasick matching, R-tree filtering, and payload analysis, it is possible to effectively detect potential threats or intrusions within the packet.

The systematic approach employed by the lightweight detection engine, which incorporates these numerical points, indicates its efficacy as a lightweight yet robust tool for real-time intrusion detection in dynamic and resource-limited settings.

13.5 Integrating CTI Indicators for IDS Signature

13.5.1 MISP CTI Integration with UIDS

MISP is an open-source platform designed for aggregating, retaining, and sharing cybersecurity indicators and threat data. It connects various public MISP communities via its API, enabling global exchange of threat intelligence.

The platform involves three main components: Distributed CTI, IDS, and Honeypot (Fig. 13.6). At its core, the MISP CTI Platform securely stores data from multiple sources in an encrypted, distributed manner and shares this data across connected systems. Multiple MISP instances collaborate to exchange real-time threat data across integrated modules.

MISP is favored for its smooth integration with IDS and Honeypot systems, allowing for efficient threat intelligence sharing. Its flexible architecture supports customization, ensuring seamless compatibility with a range of security tools. MISP

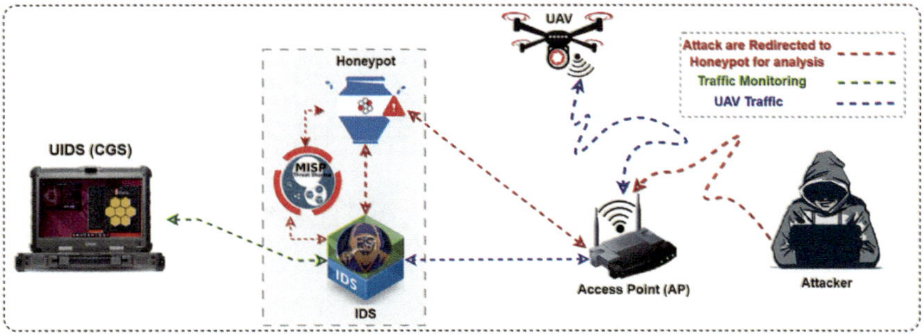

Fig. 13.6 MISP CTI integration with UIDS overview

adheres to industry standards like STIX and TAXII, making it easy to share threat data across various organizations and sectors. Its open-source licensing, extensive documentation, and active community contribute to its widespread adoption.

In the UIDS framework, MISP enables continuous sharing of network traffic and attack logs between IDS and CTI. The system forwards unknown threats to MISP for further analysis, enhancing rule sets and improving the detection of novel attacks.

Honeypots are also strategically deployed to collect intelligence on emerging threats. MISP triggers these Honeypots, ensuring real-time updates to the IDS and CTI systems, thereby enhancing the detection and mitigation of evolving cyber threats.

13.5.2 CTI Event Exchange with UIDS

The UIDS components use a standardized interface via a customized MISP Object in JSON format. This MISP object acts as a structural framework for storing and managing shared threat events. Its adaptability allows for easy expansion to include additional details about specific threat occurrences. We developed a specialized UIDS-MISP template to encapsulate UIDS logs.

The process begins with MISP fetching updates from external MISP feeds [1]. Through an API, UIDS and MISP exchange data, temporarily storing MISP JSON in raw storage [2]. The data is then filtered and stored in a designated database [16]. UIDS retrieves these processed MISP events in real-time via API [5] for analysis and to identify emerging threats [3]. MISP shares new attack patterns with the Signature Generation Engine (SGE) for immediate deployment [6, 17]. This collaborative approach ensures timely detection and response within the UAV security ecosystem, as illustrated in the Fig. 13.7.

MISP generates attack patterns by leveraging intelligence-sharing capabilities and structured workflows as shown in Fig. 13.8. It aggregates data from network traffic and attack logs shared by UIDS, analyzes patterns, and identifies potential threats through advanced techniques such as statistical analysis and machine learning.

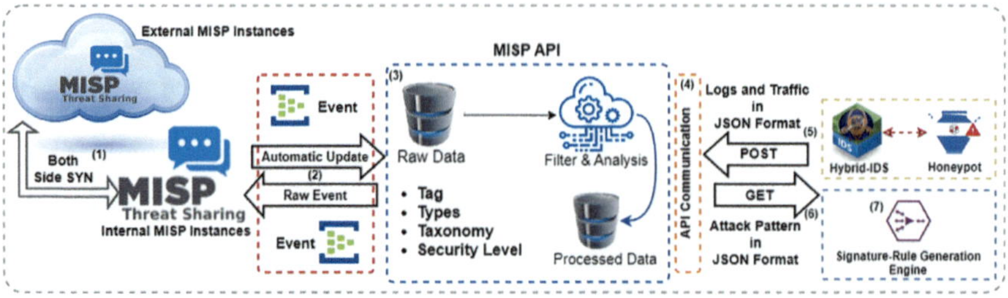

Fig. 13.7 UIDS-MISP threat event sharing overview

Fig. 13.8 MISP attack pattern

MISP collaborates with other threat intelligence platforms to validate its findings, cross-referencing data with external sources for enriched context. Once new attack patterns are identified, they are shared with UIDS's SGE, ensuring integration of threats into the detection mechanisms of the Multi-tier UIDS, enabling proactive defense.

13.5.3 Customized UIDS MISP Object

To simplify the processing, storage, and analysis of security-related data, UIDS logs and network traffic data in JSON (JavaScript Object Notation) format. JSON, a lightweight data format, is widely used for sharing structured data between systems.

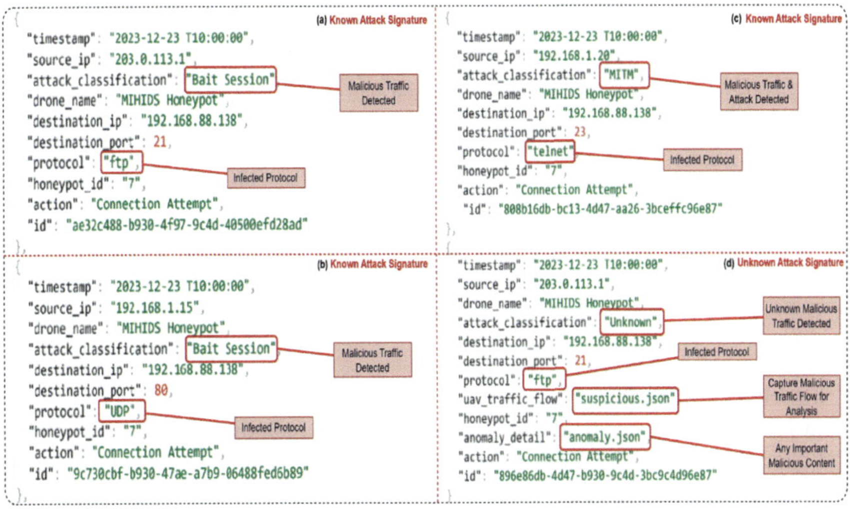

Fig. 13.9 UIDS generated JSON format logs

UIDS ensures that logs and network traffic data are consistently presented in JSON format, making them machine-readable and easy to handle, modify, and understand.

Additionally, UIDS captures detailed security event information, including timestamps, source and destination IP addresses, protocols, event types, and other attributes. This structured format allows for efficient querying, filtering, and analysis of security events, enabling UIDS to identify threats, detect anomalies, and generate real-time alerts.

In the example illustrated in Fig. 13.9, four instances are shown:

- (a) and (b) detect known "Bait Session" attacks on FTP and UDP ports, identified by the *attack_classification* identifier.
- (c) detects a "MITM" attack on an FTP port.
- (d) captures unknown malicious traffic, also identified by *attack_classification*. In this case, UIDS logs the IDS and Honeypot traffic flow as `suspicious.json` and identifies anomalous content in `anomaly.json` from the "uav_traffic". These logs are then shared with MISP CTI for further analysis and attack pattern confirmation.

The key elements of the Custom UIDS MISP object are presented in Table 13.2, highlighting components critical to collaborative UIDS operation.

- **UAV_traffic_flow**
  This element captures communication between the UAV and CGS. It is essential for IDS threat detection and contributes to the platform's data-sharing process. Once threats are confirmed, it enriches the security knowledge base. Due to its sensitivity, this data can be encrypted for secure sharing with trusted peers.

Table 13.2 The key elements of the custom UIDS MISP object

Name	Type	Description
creationdate	Date and time	Threat event date
destincation_ip	IP-Destination	Destination IP
destination_port	Port	Destination Port
source_ip	IP-Source	Source IP
uav_traffic_flow	Attachment	Suspicious traffic flow between two nodes
attack_classification	Text	Attack Classification provided by UIDS in JSON format
anomaly_details	Attachment	Optional Anomaly flow in JSON format

- **attack_classification**
  In JSON format, this field contains threat details, including classification events linked to UAV traffic. For example, if an MITM attack is detected, this entry includes the attack label, confidence score, and detection algorithm's signature.
- **anomaly_details**
  This optional JSON fragment provides aggregate statistics and additional information to help UIDS characterize threats. It may include summaries or signatures from *uav_traffic*.

As shown in Fig. 13.10, the UIDS logs analysis led to the generation of a new attack pattern by MISP. The generated pattern, named "UAV_Server_FTP_Command_Injection," is designed to detect potential command injection attacks on drone communication modules. This pattern focuses on FTP protocol traffic, specifically targeting destination port 21, which is standard for FTP.

The pattern checks payload signatures for the presence of a specific byte sequence (|20 3a 20|), indicating a possible command injection attempt. Additionally, regular expressions are used to search the payload for case-insensitive occurrences of certain command keywords, such as "ping," "nc" (netcat), or "telnet." If any of these conditions are met, the system triggers an alert (trigger_alert), notifying of the detected attack.

13.6 Signature Generation Engine (SGE)

The SGE processes attack patterns from the MISP and includes three components: the Rule Parser, Analysis Engine, and Output Generation module, as shown in Fig. 13.11.

- **Rule Parser**
- The Rule Parser tokenizes and lemmatizes rules from a rule file into three components: *keyname*, *value*, and *modifier*. The *keyname* indicates the rule type, the *value* provides details, and the *modifier* lists relevant keywords. Processed blocks are signature generation.

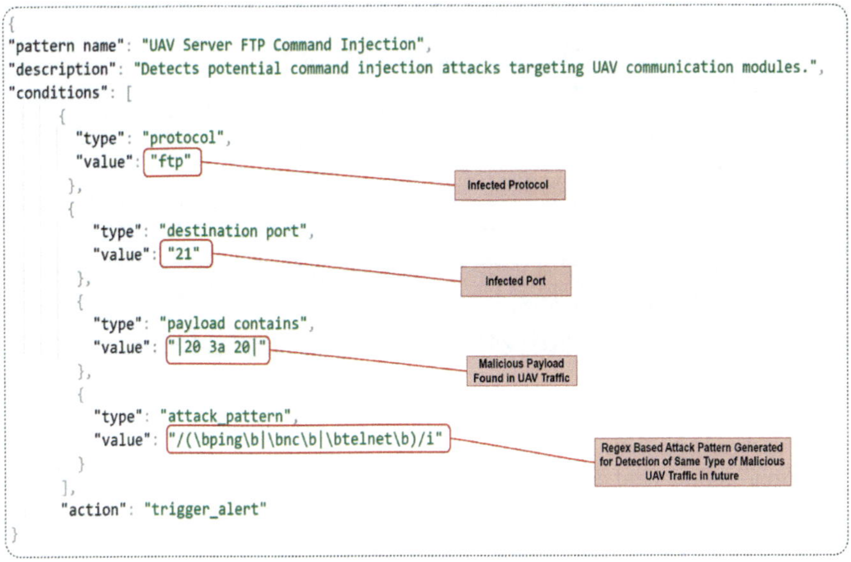

Fig. 13.10 Generated attack pattern by MISP

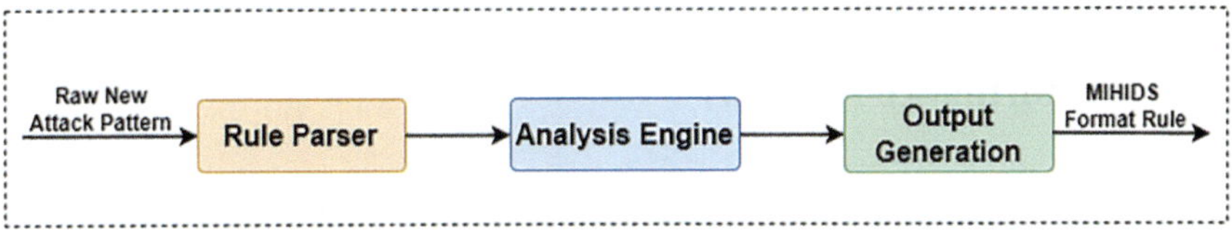

Fig. 13.11 Signature generation engine component

- **Analysis Engine**
  The Analysis Engine (AE) analyzes new attack patterns against existing signatures, controlling redundancy and ensuring quality. It categorizes results as **equal**, **overlap**, or **partial** based on rule similarity. This analysis considers actions, protocols, and IP addresses for thorough comparisons. The AE's findings are then incorporated into the Output Generation module, enhancing the effectiveness of the generated attack signatures for the intrusion detection system.
- **Output Generation**
  The Output Generation module finalizes the analysis by categorizing results and creating UIDS format rules, as shown in Fig. 13.12. It identifies redundant rules and semantic matches while ensuring compatibility with the IDS.

This output acts as a repository for rules, allowing seamless updates to the signature pool and improving the system's ability to detect and respond to new threats. The Multi-Tier UIDS signature generated targets port 21 (FTP) for detecting malicious command execution. Key fields include a message indicating "Malicious Command

```
---MIHIDS-Siganture-Generation Engine--

The rule is Generated for MIHIDS based on MISP Attack Pattern:
Output:
alert tcp any any -> any 21
  (
  msg:"FTP malicious string attempt"; flow:to_server,
  established; content:"%"; pcre:"/\s+.*?%.*?%/smi";
  reference:"MISP-2-MIHIDS",
  classtype:string-detect; sid:10070; rev:1;
  )

MIHIDS Signature Pool successfully updated with new rule have sid:10070
```

Fig. 13.12 New rule generated for UIDS

Execution," a flow field directing connections to the client, and a content field with hexadecimal patterns for commands like "ping" and "telnet." The reference field cites the MISP identifier, while the revision *(rev)* and signature ID (sid) fields provide versioning and uniqueness. Overall, the rule aims to ensure timely detection and response to suspicious activity.

13.7 Performance Analysis

13.7.1 Lightweight Pattern Matching Engine Optimization

The performance of the Aho-Corasick (AC) algorithm in pattern matching is closely tied to memory requirements and access overheads. In UAV devices lacking dedicated GPUs or traditional CPUs, optimizing the AC algorithm's performance becomes essential. As the number of patterns increases, memory usage escalates, particularly with larger state transition tables, reducing cache efficiency and slowing pattern matching for substantial datasets. This work focuses on maximizing performance within the constraints of UAV hardware, emphasizing efficient resource utilization.

To address memory challenges, local memory access patterns within the UAV are optimized, as these devices lack the shared memory architectures typically found in systems with dedicated GPUs. The proposed solution, the MHIDS UAV-AC algorithm, is specifically designed for lightweight UAVs and demonstrates significant performance enhancements over the original AC algorithm.

The comprehensive performance analysis reveals notable reductions in processing time, particularly with larger pattern datasets. This optimization enhances threat detection, streamlines data processing, and improves memory efficiency in UAVs. The results demonstrate approximately a 20% speedup for smaller datasets and a significant 40% improvement for larger, more complex ones, as illustrated in

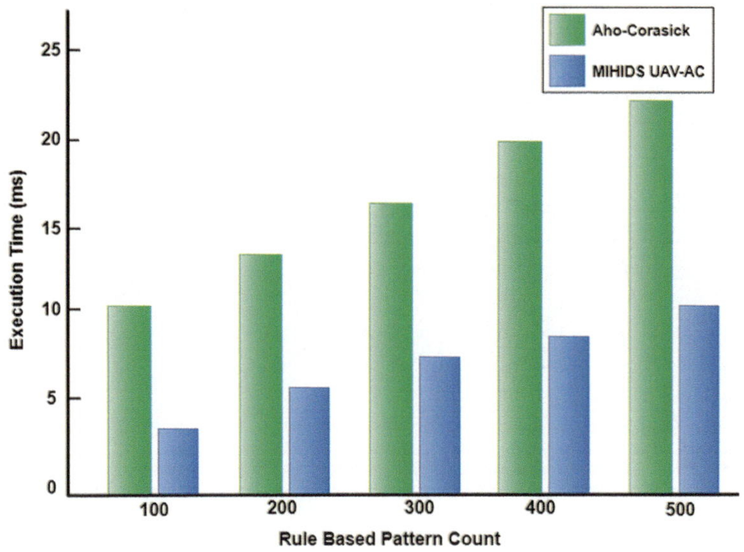

Fig. 13.13 UAV-AC pattern matching engine optimization

Fig. 13.13. The analysis highlights the quantitative gains achieved and the strategic trade-offs between performance and memory usage.

13.7.2 UAV Attack Signature Detection Efficacy

A specialized signature set was developed for detecting UAV device attacks, achieving a detection rate of 99.5% while minimizing false positives. As shown in Fig. 13.14, increasing the number of signatures results in improved attack detection, aligning with expectations. A more comprehensive signature set facilitates the identification of a broader range of attacks.

The graph illustrates the signature set's effectiveness in protecting UAV devices, maintaining high performance while minimizing false alarms.

13.7.3 UIDS Signature Generation Efficiency with MISP CTI

MISP generates signatures based on CTI threats. It processes events to create attack patterns, which are used for signature generation. Figure 13.15 illustrates MISP's efficiency, handling 100 events and generating patterns in under 50 milliseconds, demonstrating MISP's quick and effective event handling performance.

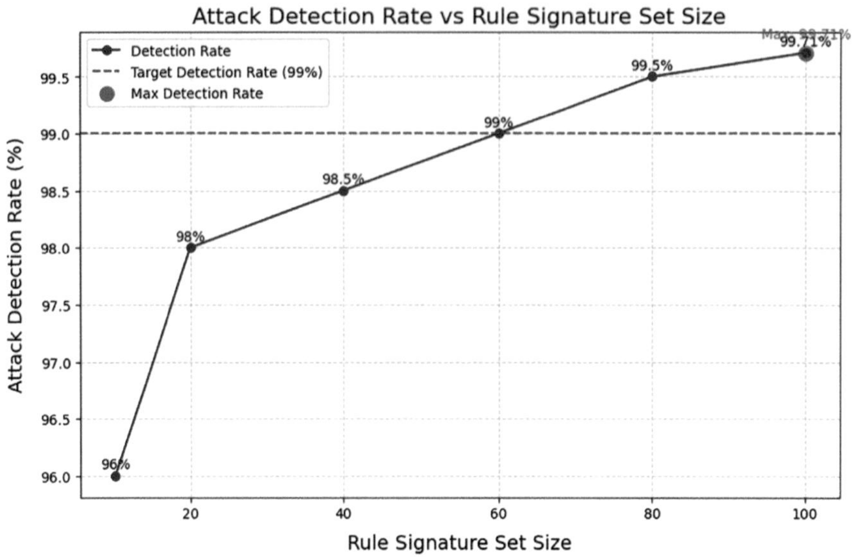

Fig. 13.14 UAV attack signature detection efficacy

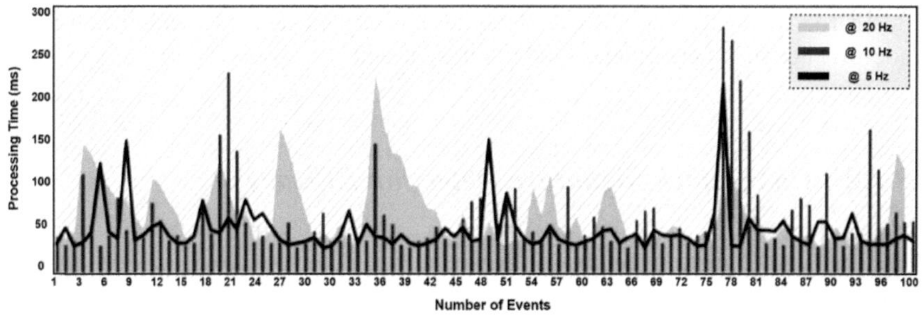

Fig. 13.15 MIHIDS signature generation efficiency with MISP CTI

13.8 Conclusion

In today's fast-evolving technological landscape, the integration of UAVs is widespread, but securing UAV communication networks remains a critical challenge. Existing security measures, such as ML-based or signature-based IDS, struggle to effectively detect novel or zero-day attacks, exposing a significant research gap. This research introduces and evaluates a Multi-tier UIDS specifically designed for UAV environments. UIDS enhances anomaly detection by optimizing the Aho-Corasick algorithm, facilitating efficient detection and classification of UAV-specific attacks. Moreover, the integration of CTI feeds improves detection accuracy and minimizes false positives, offering stronger protection against emerging threats. These advancements lay the foundation for further exploration of UIDS implementation and evaluation, closing the research gap and providing a robust solution to safeguard UAVs from cyber threats.

References

1. Khan, M. A., Kumar, N., Mohsan, S. A. H., et al. (2023). Swarm of UAVs for network management in 6G: A technical review. *IEEE Transactions on Network and Service Management, 20*(1), 741–761.
2. Rodday, N. M., de O Schmidt R., & Pras, A. (2016) Exploring security vulnerabilities of unmanned aerial vehicles. In *Proceedings of IEEE/IFIP Network Operations and Management Symposium (NOMS 2016)* (pp. 993–994).
3. Hadi, H. J., Cao, Y., Nisa, K. U., et al. (2023). A comprehensive survey on security, privacy issues, and emerging defense technologies for UAVs. *The Journal of Network and Computer Applications, 213*, Article 103607. https://doi.org/10.1016/j.jnca.2023.103607
4. Jamil, A. M., Hadi, H. J., Li, S. et al. (2023). Detection of targeted attacks using medium-interaction honeypot for unmanned aerial vehicle. In *International Conference on Digital Forensics and Cyber Crime* (pp. 164–185). Springer Nature Switzerland.
5. Abu, Al-Haija Q., & Al, Badawi A. (2022). High-performance intrusion detection system for networked UAVs via deep learning. *Neural Computing & Applications, 34*(13), 10885–10900.
6. Hadi, H. J., & Cao, Y., Cyber attacks and vulnerabilities assessment for unmanned aerial vehicles communication systems. In *Proceedings of 2022 International Conference on Frontiers of Information Technology (FIT)* (pp. 213–218). https://doi.org/10.1109/FIT57066.2022.00047.
7. Vasilomanolakis, E., Srinivasa, S., Cordero, C. G., et al. (2016). Multi-stage attack detection and signature generation with ICS honeypots. In *Proceedings of IEEE/IFIP NOMS 2016* (pp. 993–994).
8. Daubert, J., Boopalan, D., Mühlhäuser, M., et al. (2018). HoneyDrone: A medium-interaction unmanned aerial vehicle honeypot. In *Proceedings of IEEE/IFIP NOMS 2018* (pp. 1–6).
9. Edwin, G. K., Edwards, S. V., Kathrine, G. J. W., et al. (2022). Honeypot-based intrusion detection system for cyber-physical system. In *2022 International Conference on Augmented Intelligence and Sustainable Systems (ICAISS)* (pp. 958–962).
10. Anastasiadis, M., Moschou, K., Livitckaia, K., et al. (2023). A novel high-interaction honeypot network for internet of vehicles. In *2023 31st Mediterranean Conference on Control and Automation (MED)* (pp. 281–286).
11. Pitre, P., Gandhi, A., Konde, V., et al. (2022) An intrusion detection system for zero-day attacks to reduce false positive rates. In *2022 International Conference for Advancement in Technology (ICONAT)* (pp. 1–6).
12. Daubert, J., Boopalan, D., & Mühlhäuser, M. (2018). Don't steal my drone: Catching attackers with an unmanned aerial vehicle honeypot. In *Proceedings of IEEE/IFIP NOMS 2018* (pp. 1–2).
13. Miao, S., Li, Y., & Pan Q. (2022). Honeypot active defense technology for UAV cyber range. In *2022 China Automation Congress (CAC)* (pp. 30–35).
14. Kempinski, S., Ichaarine, S., Sciancalepore, S., et al. (2023). ICSvertase: A framework for purpose-based design and classification of ICS honeypots. In *Proceedings of 18th International Conference on Availability, Reliability and Security* (pp. 1–10).
15. Ellouh, M., Ghaleb, M., & Felemban, M. (2022). IoTZeroJar: Towards a honeypot architecture for detection of zero-day attacks in IoT. In *2022 14th International Conference on Computational Intelligence and Communication Networks (CICN)* (pp. 765–771).
16. Pleban, J.-S., Band, R., & Creutzburg, R. (2014). Hacking and securing the AR. Drone 2.0 quadcopter: Investigations for improving the security of a toy. In *Mobile devices and multimedia: Enabling technologies, algorithms, and applications* (pp. 191–202).

17. Nawrocki, M., Wählisch, M., Schmidt, T. C., et al. (2016). *A survey on honeypot software and data analysis*. arXiv Preprint arXiv:1608.06249
18. Vasilomanolakis, E. (2013). This network is infected: Hostage—a low-interaction honeypot for mobile devices. In *Proceedings of the Third ACM Workshop on Security and Privacy in Smartphones & Mobile Devices* (pp. 1–10).
19. Koutras, D., Dimitrellos, P., Kotzanikolaou, P., et al. (2023). Automated WiFi incident detection attack tool on 802.11 networks. In *2023 IEEE Symposium on Computers and Communications (ISCC)* (pp. 464–469).

Chapter 14
Trust Management for UAV-Assisted VANET

Mansi Zhang and Yue Cao

Abstract With the development of next-generation wireless communication and Internet of Things, Vehicular Ad Hoc Network (VANET) has emerged as a cornerstone for enhancing road safety and user experience. However, the performance of VANET is challenged by network partition and barriers, leading to connectivity issues. Unmanned Aerial Vehicle (UAV)-assisted VANET address these limitations by leveraging the high mobility and extended coverage of UAVs, thereby enhancing communication reliability and network coverage. Despite the benefits, UAV-assisted VANET faces security threats due to vulnerable terrestrial vehicles and exposed wireless channels. Although traditional cryptographic algorithms (e.g., key agreement, message authentication codes, and digital signatures) can resist external attacks effectively by authentication and encryption, they have no ability to counter internal attacks. To address the limitations of cryptographic solutions, trust management has emerged as a promising alternative to ensure the security. This chapter outlines the network architecture of UAV-assisted VANET, discusses various attacks, classifies trust management schemes, and highlights unresolved issues.

Keywords Network security · Unmanned Aerial Vehicle-assisted Vehicular Ad Hoc network · Trust management

Abbreviations

ACO	Ant Colony Optimization algorithm
AI	Artificial Intelligence
APT	Advanced Persistent Threat
BA	Blackhole Attack
BPT	Bidding Price-based Transaction method
DoS	Denial of Service attack

M. Zhang · Y. Cao (✉)
School of Cyber Science and Engineering, Wuhan University, Wuhan, China
e-mail: yue.cao@whu.edu.cn

Y. Hu et al. (eds.), *Reliable and Efficient Trajectory Design in UAV-Aided Wireless Networks*, https://doi.org/10.1007/978-981-96-5988-3_14

DSRC	Dedicated Short-Range Communication
EA	Eavesdropping Attack
FMIA	False Message Injection Attack
HT	Hybrid Trust model
IoT	Internet of Things
IPFS	InterPlanetary File System
IUAV	Internet of Unmanned Aerial Vehicle
JA	Jamming Attack
LOS	Line-of-Sight communication link
MAC	Media Access Control layer of the network
MCT	Mode-Centric Trust model
NCT	Node-Centric Trust model
NS2	Network Simulator
OBU	On-Board Unit
OOA	On-Off Attack
RA	Recommendation Attack
RN UAV	Relaying Node Unmanned Aerial Vehicle
RRAN UAV	Remote Radio Access Node Unmanned Aerial Vehicle
RSU	Road-Side Unit
SA	Spoofing Attack
SINR	Signal to Interference plus Noise Ratio
SUMO	Simulation of Urban Mobility simulator
TA	Timing Attack
TCP SYN flooding	Transmission Control Protocol SYNchronize segment flooding
U2I	Unmanned aerial Vehicle-to-Infrastructure communication link
U2U	Unmanned aerial Vehicle-to-Unmanned aerial vehicle communication link
U2V	Unmanned aerial Vehicle-to-Vehicle communication link
UAV	Unmanned Aerial Vehicle
V2I	Vehicle-to-Infrastructure communication link
V2V	Vehicle-to-Vehicle communication link
VANET	Vehicular Ad-hoc Network

14.1 Introduction

With the rapid development of next-generation wireless communication and Internet of Things (IoT), Vehicular Ad-hoc Network (VANET) has become as a cornerstone of intelligent transportation systems. VAENT can enable real-time communication among vehicles and infrastructure to reduce accidents as well as enhance user experiences in Intelligent Transportation Systems (ITS). However, the performance of

VANET degrades in extreme conditions due to network partition and barriers on roads, leading to connectivity issues [1].

To address the limitations of traditional VANET, Unmanned Aerial Vehicle (UAV)-assisted VANET has gained significant attention. By leveraging the high mobility, flexibility, as well as extended coverage of UAVs, these systems enhance communication reliability, expand network coverage, and provide real-time message dissemination [2]. Despite these benefits, UAV-assisted VANETs remain susceptible to security threats due to vulnerable terrestrial vehicles and exposed channels [3, 4].

Based on whether the attacker has a legal identity in the network, the attacker can be divided into an external attacker and an internal attacker [5]. The external attacker typically eavesdrops the network and disrupts regular communications. Cryptographic schemes (e.g., encryption, hash code and digital signature) can effectively counter external attackers by preventing unauthorized access and tampering [6]. On the other hand, the internal attacker is a legitimate member of the network with authorized access to resources, which can't be resisted by cryptographic solutions [7]. These inner attacks (e.g., black hole attacks, false message injection, and selective forwarding) can severely disrupt communication, compromise data integrity, and jeopardize the reliability of transportation systems [8].

To mitigate such threats, trust management schemes have been proposed as a vital security mechanism [9–11]. Trust management evaluates and maintains the credibility of network entities by analyzing their behaviors and messages. By incorporating factors such as reputation, historical performance, and contextual information, these systems can identify malicious nodes, ensuring robust network security as well as enhancing the overall reliability of UAV-assisted VANET. These schemes not only protect against inner attacks but also fosters a secure and trustworthy communication environment dynamically [12].

This chapter classifies and introduces the existing trust management schemes in UAV-assisted VANET, and discuss existing problems. The overall structure is organized as follow. Section 14.2 provides the network architecture of UAV-assisted VANET. Section 14.3 discusses attacks in UAV-assisted VANET. Section 14.4 classifies the existing trust management schemes for UAV-assisted VANET. Section 14.5 discusses unresolved open issues.

14.2 The Architecture of UAV-Assisted VANET

This section introduces the architectural design of UAV-assisted VANET. Firstly, it provides the analysis of three core components: vehicles, infrastructures, and UAVs. Afterwards, this section explains how these components communicate with each other. Moreover, the architecture of UAV-assisted VANET is shown in Fig. 14.1. Through the discussion, it highlights how the UAV-assisted VANET architecture overcomes challenges such as coverage limitations and network congestion.

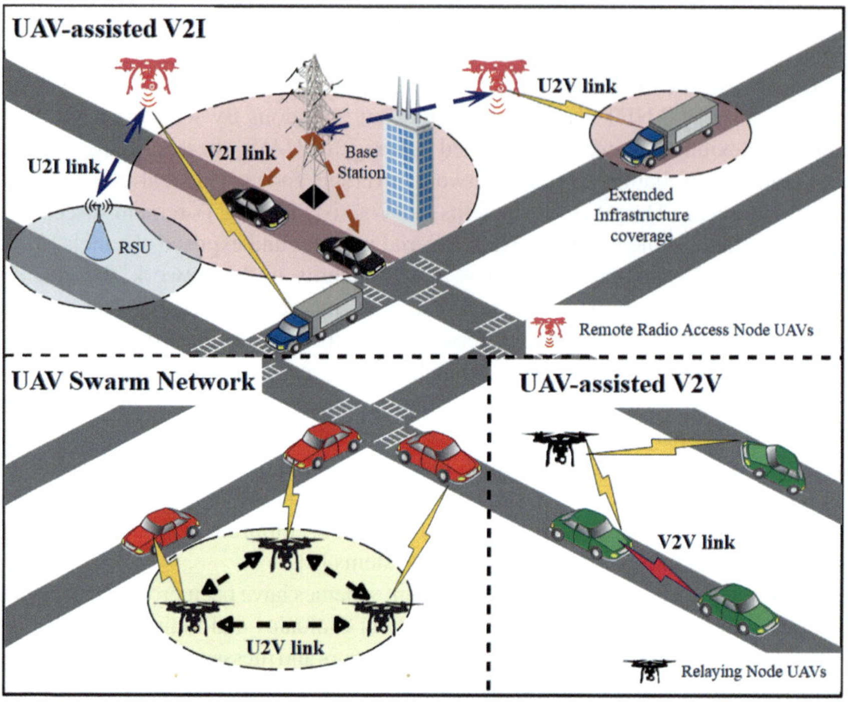

Fig. 14.1 The architecture of UAV-assisted VANET

14.2.1 Network Components

The architecture of UAV-assisted VANET integrates vehicles, UAVs, and infrastructures to enhance network connectivity as well as efficiency. Each component plays a distinct role, which is detailed as follows:

- **Vehicles**: Vehicles in the network are equipped with On-Board Units (OBUs) which facilitate communication with other vehicles through Dedicated Short-Range Communication (DSRC). These OBUs also incorporate message processing modules, allowing seamless message exchange across various heterogeneous networks.
- **Infrastructures**: Road-Side Units (RSUs) and base stations provide connectivity within their coverage zones using DSRC or cellular communication technologies. In coverage gaps or areas with insufficient infrastructure, UAVs serve as relays to maintain Vehicle-to-Infrastructure (V2I) connectivity.
- **UAVs**: UAVs are classified as Relaying Node (RN) UAVs and Remote Radio Access Node (RRAN) UAVs. RN UAVs relay Vehicle-to-Vehicle (V2V) communication, while RRAN UAVs enhance V2I communication by dynamically positioning themselves to address real-time coverage and capacity demands.

Advanced UAVs can switch between RN and RRAN roles as needed to improve adaptability.

14.2.2 The Communication in UAV-Assisted VANET

The communication framework in UAV-assisted VANET ensures reliable message exchange between vehicles, UAVs, and infrastructure, overcoming challenges like limited coverage and network congestion.

- **UAV-assisted V2V Communication**: RN UAVs significantly improve V2V communication by establishing Line-of-Sight (LOS) links, addressing the limitations of traditional VANET that rely on non-LOS connections. RN UAVs dynamically adjust their positions based on traffic topology to maintain robust UAV-to-Vehicle (U2V) links, enabling efficient message relays and message broadcasting.
- **UAV-assisted V2I Communication**: RRAN UAVs extend the coverage and capacity of existing infrastructure by dynamically deploying to high-demand areas or coverage gaps. They establish stable UAV-to-Infrastructure (U2I) links, ensuring high-throughput and low-latency V2I communication. RRAN UAVs adapt to fluctuating network demands, enhancing system reliability in both urban and rural scenarios.
- **UAV Swarm Network**: In areas lacking traditional infrastructure, such as remote or disaster-affected regions, UAVs form a swarm network. The swarm network creates UAV-to-UAV (U2U) links, enabling temporary aerial connectivity for vehicles. The swarm also allocates additional spectrum resources, addressing bandwidth shortages in dense vehicular environments.

14.3 Attacks in UAV-Assisted VANET

This section encompasses a range of attacks that can compromise the integrity, availability, and confidentiality of the network. These threats include passive attacks such as Eavesdropping Attack (EA), active attacks like Jamming Attack (JA), and more sophisticated attacks such as Spoofing Attack (SA) as well as False Message Injection Attack (FMIA). Furthermore, the section discusses the impact of Blackhole Attack (BA), Timing Attack (TA), and Denial of Service (DoS) attack on network performance. Additionally, this section also highlights the importance of addressing Recommendation Attack (RA) and On-Off Attack (OOA) that target at trust management systems.

14.3.1 Eavesdropping Attack (EA)

The EA is known as an Advanced Persistent Threat (APT). The main purpose of the attack is to gather information in preparation for a further devastating attack. EA occurs when a dishonest vehicle or UAV with a valid certificate behaves as a spy, therefore gathering all possible information. Although the impact of such a passive attack on the network is not instantaneous or very evident, it can nevertheless be the cause of many attacks on the privacy and confidentiality of network [13].

14.3.2 Jamming Attack (JA)

The JA is a kind of active attacks. It occupies the communication channel of network nodes by sending high power signals, resulting in reduced Signal to Interference plus Noise Ratio (SINR), the failure of normal communication, and the loss of availability. The JA can be present at ground receivers (downlink jamming), or directed at the UAVs and RSU mixed with an effective carrier (uplink jamming) [14]. Uplink jamming is mainly transmitting strong signals from the ground to air, which may interfere with channels of the UAVs or RSU to interrupt signals. Uplink jamming is easy to implement, but it is difficult to detect and deal with. This is because that the jammer is mobile and it can be located anywhere, and it is infeasible to locate the jammer during broadcast. Downlink jamming is mainly a threat to the information flowing from air to the ground.

14.3.3 Spoofing Attack (SA)

The attacker conducting the SA forges the identities of legitimate vehicles or UAVs to jointly report the occurrence of an event (e.g., traffic jam), aiming to gain a higher degree of trust. Additionally, the SA attacker can also launch a Denial of Service (DoS) attack through multiple identities. The SA can be launched through the cooperation of multiple identities, so as to achieve a higher attack success rate and achieve a variety of malicious purposes [15].

14.3.4 False Message Injection Attack (FMIA)

The attacker conducting the FMIA can create fake messages to report false information to others. In the autonomous driving scenario, the vehicle will change its behaviour after receiving messages related to traffic events. Fake traffic notification messages will cause accidents or congestions under specific circumstances. Moreover, in the disaster rescue scenario, malicious vehicles generate false messages (e.g., fabricated disaster related events) and propagate them throughout the network. By replicating this source message, a large number of infected false messages is generated and forwarded by other vehicles and UAVs, misleading nodes and trust authority to perceive an incorrect disaster situation. FMIA results in wasted resources on life-saving efforts and increased costs in disaster rescue.

14.3.5 Blackhole Attack (BA)

Attackers conducting the BA drop messages during the transmission phase to reduce the delivery rate of network or save their energy consumption selfishly [16]. If an insufficient number of valid messages is collected, the event-related information fused by vehicles (e.g., the location of event) becomes inaccurate, or some events are entirely missing. This ultimately weakens the situation-aware ability of vehicles and UAVs. When is combined with FMIA, the proportion of false messages increases, significantly raising the risk of misinformation.

14.3.6 Timing Attack (TA)

The principle of TA is that the dishonest vehicles store the transmitted messages for a period of time before relaying them again. As a result, the user at the destination will receive the message behind the time. The delay in the message delivery process can be even more dangerous than actually dropping these messages [17]. It may result in mishaps for vehicles that get the message in the later time. For example, when the attacker receives a message showing that there is an accident on the road, it adds some schedule slots to the message before relaying it. This behavior may prevent vehicles from changing their routes immediately upon reaching the accident site, thereby potentially exacerbating the severity of the traffic accident.

14.3.7 Denial of Service (DoS)

The attacker will transmit as many messages as possible (e.g., forging and replaying messages) to consume resources of normal vehicles. If the buffer space or bandwidth resources are occupied, the attacked vehicle cannot perform normal communication functions, thereby suffering a DoS attack. DoS attacks are intended to make network services unavailable by denying legitimate users access to specific network resources. They may lead to system overloads and prevent some or all legitimate requests from being satisfied. Moreover, due to the limited processing capacity of the mobile vehicles and UAVs, the system will be easily overwhelmed.

DoS attacks can occur in all levels of network protocols. In the physical layer, DoS attack includes interference [18]. The Media Access Control (MAC) layer is also susceptible to DoS attacks, which can manifest as conflicts, queries, and message replays [19]. In the network layer, DoS attacks may involve spoofing, replays, and manipulation of routing traffic [20]. In the transport layer, TCP (Transmission Control Protocol) SYN (SYNchronize) flooding is a tactic used to deplete connection buffer resources [21].

14.3.8 Recommendation Attack (RA)

Different from other attacks, the RA is oriented at the trust management system. The RA involves providing inaccurate trust assessments to others, including Bad-Mouthing Attack (BMA) and Ballot-Stuffing Attack (BSA). Attackers conducting BMA spread unfair negative trust assessments towards others, deliberately undermining the trust value of targeted nodes. This collusive behavior in BMA can disrupt trust evaluation results, leading to the obstruction of effective message paths. On the other hand, attackers conducting BSA disseminate unjustified positive recommendations for poorly performing nodes. The goal of these colluding BSA attackers is to disguise themselves as trustworthy, and evade detection by the trust management system.

14.3.9 On-Off Attack (OOA)

OOA is a random attack in which a malicious vehicle or UAV randomly provides good and bad service, so as to evade risk of being labeled as a malicious node with low trust value. By delivering service quality in an unpredictable manner, it sustains a moderate trust level. This approach keeps its chances of being selected as a service provider consistently high. After being selected, it can again collude with other malicious nodes to perform BMA or BSA, or itself provide poor services to the requesting node [22].

14.4 The Classification of Trust Management in UAV-Assisted VANET

Malicious vehicles and UAVs usually conduct the attack described in Sect. 14.3 to destroy the network or achieve their own selfish purposes, leading to a huge waste of network resources. Detecting such nodes and taking certain measures are very important to maintain network security. However, the high mobility of vehicles and UAVs increases the challenge of malicious nodes detection. A possible solution to the problem is to provide the system with trust management mechanism that can detect and avoid nodes with suspicious behaviours. This section divides the trust management in UAV-assisted VANET into 3 types, and introduces them in detail in the following subsections. After discussing the various schemes of existing trust managements, an analysis summary table is provided in the follow-up.

14.4.1 Node-Centric Trust (NCT) Model

NCT schemes are trust computation methods based on the behaviour and attribution of nodes. Based on the trust value of a particular node, the system decides whether to interact with that node. Firstly, a node collects interactive records or current attribution of a neighbour, and processes them through filtering and weighting. Secondly, it computes trust values of neighbours utilising a specified trust calculation formula. Thirdly, it decides whether to interact or collaborate with the neighbor based on the trust value and a defined trust threshold. Finally, trust values of neighbours and the threshold are updated according to the effectiveness of interaction or collaboration. An overview of these mechanisms and the example mechanisms discussed in this chapter is shown in Table 14.1.

Table 14.1 NCT models

Mechanism	Category	Advantage	Disadvantage
Kerrache et al. [23]	NCT based on historical behaviour	Adaptive detection threshold, enhanced by UAVs	The potential compromise of UAVs
Alani et al. [24]	NCT based on historical behaviour	Comprehensive trust evaluation, enhanced communication range	Potential complexity and inaccuracy in sparse network
Zheng et al. [25]	NCT based on recommendation	Robust against attacks, dynamic trust update	Potential delays in trust evaluation, and vulnerable to RA
Fatemidokht et al. [26]	NCT based on recommendation	Improved connectivity, stability and ability to alleviate the RA	Vulnerable to malicious UAVs

NCT Based on Historical Behaviour

Kerrache et al. [23] propose a UAV-assisted trust management model aimed at detecting malicious nodes in VANET. This model proposes an adaptive detection threshold which is adjusted based on the observed behavior of vehicles, allowing for the dynamic identification of intelligent malicious behaviors. Furthermore, the system decays the trustworthiness of vehicles over time, rendering the trust management more flexible and effective in sparse VANET environments. Additionally, the integration of UAVs provides continuous LOS communication, which is crucial for the rapid broadcast of blacklist and selection of trusted cluster head in vehicles. The main advantage of this approach is its ability to adapt to changing behaviors and the enhanced detection capabilities provided by UAVs. However, the system may face challenges in maintaining the trustworthiness of UAVs themselves, which could be compromised.

Alani et al. [24] introduce a dual trust management model in UAV-assisted VANET to secure the system. The model evaluates trust in both V2V and U2V communications. On the one hand, the V2V trust model is based on the service satisfaction and position information of vehicles. On the other hand, the U2V model considers the coverage and average transmitted and received messages of UAVs. This model stands out due to its comprehensive trust evaluation process, which includes direct observation and mutual trust management during communication between vehicles and UAVs. Moreover, the utilization of UAVs in this scheme allows for better coverage and communication range, enhancing the overall trustworthiness of the network. The trust management model has been validated utilizing Network Simulator 2 (NS2) and Simulation of Urban Mobility (SUMO), showing promising performance in terms of higher message delivery ratio, lower end-to-end delay, and lower overhead. However, the dual trust management model may introduce additional complexity.

NCT Based on Recommendation

Zheng et al. [25] propose a trust management model for tiny federated learning in Internet of Unmanned Aerial Vehicle (IUAV) systems. This model combines direct observations and indirect recommendations of trust values to evaluate the trustworthiness of devices assisted by UAVs. Direct trust calculation incorporates the time spent in communication, computation, and block production, along with a decay function for each round of federated learning. Furthermore, the model utilizes UAVs to gather trust recommendations, enhancing the accuracy of trust calculation and solve the problem of trust propagation in sparse VANET. The main advantage of this model is its robustness against attacks and ability to prevent single points of failure in infrastructure-less VANET, which is essential for coverage and timeliness operations in IUAV systems. However, the reliance on recommendation trust through recommendations may introduce delays. The accuracy and the timeliness of trust calculation should be balanced. Moreover, the model can't resist RA, since it have no mechanism to identify false recommendations.

Fatemidokht et al. [26] propose the VANET Routing protocol with UAV-assisted (VRU). The VRU routing protocol consists of two basic parts named VRU_vu protocol and VRU_u protocol, including two distinct ways of trust routing: between vehicles with the help of UAVs and between UAVs. In VRU_vu protocol, the trust value of vehicles is evaluated and the malicious vehicles are detected with the assisted of UAVs. UAVs first collect the traffic density information through a centralized mechanism. Afterwards, the direct trust is calculated based on the density and connectivity of vehicles on the road segment, and the suitable next intersection is selected. Moreover, the VRU_vu protocol fuses the recommendations with different weights according to direct trust of recommenders. VRU_u protocol is utilized to discover the most appropriate route between UAVs by Ant Colony Optimization (ACO) algorithm, which can help with the propagation of trust recommendations. The protocol utilizes a trust-based scheme to detect malicious vehicles, and relies on UAVs to re-link disconnected road segments as well as detect malicious nodes in ITS. The VRU protocol has been shown to improve the message delivery ratio and decrease end-to-end delay. The use of UAVs in this protocol allows for better connectivity and stability in the ITS network. However, the system may be vulnerable to malicious UAVs, which could compromise the security of the network.

14.4.2 Message-Centric Trust (MCT) Model

In UAV-assisted VANET, only messages are actually received by vehicles and UAVs, therefore it is a very intuitive and effective way to detect malicious nodes based on the content of messages. MCT schemes aiming at assessing the trustworthiness of message are usually according to its quality and origin. Firstly, the meta-information (e.g., the generator, receiver, timestamps, location, size, etc.) of message is collected. Secondly, the metainformation of message is analyzed according to specific evaluation rule, and then the trustworthiness of message is calculated. Thirdly, the acceptance or rejection of message is determined according to its trust value and threshold. The calculation metric can be divided into two categories (i.e., plausibility and consistency). Here, a brief overview of mechanisms is provided in Table 14.2.

MCT Based on Plausibility

Kamel et al. [27] propose the Confidence range Tolerant misbehavior detection approach (CaTch), which is a MCT model integrating data inaccuracy in performing plausibility checks. This model stands out for its unique consideration of the confidence range in vehicular communication protocols. By attributing an uncertainty factor to misbehaving vehicles, CaTch enhances the detection quality of the system. The advantage of this model lies in its ability to differentiate between intentional and non-intentional transmission of erroneous mobility messages, which is crucial for

Table 14.2 MCT models

Mechanism	Category	Advantage	Disadvantage
Kamel et al. [27]	MCT based on plausibility	Accurately differentiates between intentional and non-intentional errors	Increased computational load in large scaled VANET due to complexity of sensor data error integration process
Le and Maple [28]	MCT based on plausibility	High precision as well as recall rates, and low confusion rate when detecting different attacks	Relies on sufficient historical trajectory data which may not be suitable for dynamic networks
Zaidi et al. [29]	MCT based on consistency	Effectively detects FMIA utilizing statistical techniques without the assistance of infrastructure and UAVs	Performance is reliant on the accuracy of statistical model, and selecting appropriate parameters in dynamic VANET is difficult
Huang et al. [30]	MCT based on consistency	Proactive and timely trust evaluation with the assistance of UAVs	Additional costs and coordination required for the deployment of UAVs

accurate misbehavior detection. However, in dense VANET, the complexity of integrating sensor data errors into the beacon messages might increase the computational load.

Le and Maple [28] propose a trust management model utilizing machine learning (K-Nearest Neighbour and Support Vector Machine) to detect malicious vehicles in VANET by message plausibility check. This model calculates the plausibility by three novel features (i.e., movement deviation, minimum distance to trajectories and minimum translation distance to trajectories) obtained from consecutive vehicle locations messages. The high precision and recall rates achieved by this method demonstrate its effectiveness. The main advantage of this trust model is its ability to reduce the overall confusion rate when classifying different types of attacks. However, the reliance on historical trajectory data for detection might pose challenges in dynamic and fast-changing network environments.

MCT Based on Consistency

Zaidi et al. [29] propose a distributed MCT model that utilizes statistic techniques for message consistency check to identify malicious vehicles. The model contrasts the traffic flow reported by a single vehicle against the flow observed by neighboring vehicles, and quantifies the consistency via hypothesis testing. This model is able to detect FMIA without the assistance of infrastructure and UAVs. Therefore, the advantage of this model is its ability to control network message congestion by reducing message transmissions, thus preventing broadcast storms. However, the performance of the trust model is affected by the accuracy of the statistical model.

Thereby, the selection of appropriate parameters in hypothesis testing is the problem in dynamic VANET environments.

Huang et al. [30] introduce a UAV-assisted Ubiquitous Trust Evaluation (UUTE) model that combines global trust with historical interaction-based local trust evaluation. In UUTE model, UAVs first collect event-related messages from highly trusted vehicles all over the area. Afterwards, UAVs fuse all the messages to form the baseline data for validating the consistency of message submitted by vehicles, in order to ensure a trusted data collection environment. The consistency is calculated through the average deviation of coming messages compare to the baseline. The advantage of UUTE model lies in its proactive and fast trust evaluation. However, the deployment of UAVs for data collection might incur additional costs and requires sophisticated coordination to ensure data accuracy.

14.4.3 Hybrid Trust (HT) Model

The HT model in UAV-assisted VANET integrates various technologies to enhance trust management, addressing specific challenges in secure and efficient communication. This section provides a detailed review of the literature on HT models, focusing on their combinations with blockchain, Fog/Edge computing, and Artificial Intelligence. A brief overview of mechanisms is provided in Table 14.3.

The Combination with Blockchain

Wang et al. [31] propose a Trusted and Efficient data sharing scheme based on Blockchain (TEBChain) to ensure secure data sharing with the assistance of UAVs in disaster scenarios. This model addresses the challenge of untrusted networks and malicious participants by recording all abnormal behavior on a decentralized ledger. The integration of blockchain with trust management ensures data integrity and accessibility, which is crucial for coordinating efficient rescue operations. The TEBChain model utilizes a lightweight consortium blockchain framework for recording and an improved Boneh-Lynn Shacham threshold signature for malicious vehicle traceability. The advantage of this model lies in its ability to maintain data integrity and provide the traceability mechanism, which traditional trust management models lack of. However, the disadvantage is the potential complexity and resource requirements for maintaining the blockchain.

Saleem et al. [32] introduce a graphical trust management model for UAV-assisted VANET utilizing a uniquely designed blockchain scheme. The blockchain scheme combines decentralized key management with lightweight authentication and key agreement. Moreover, it leverages the InterPlanetary File System (IPFS) to store traffic event-related data of messages in the blockchain through smart contracts. Additionally, the consensus algorithm is modified utilizing Practical Byzantine Fault Tolerance and steganography for better efficiency. The trust management

Table 14.3 HT models

Mechanism	Category	Advantage	Disadvantage
Wang et al. [31]	Combined with blockchain	Decentralized ledger ensures data integrity and prevents tampering	High computational and energy costs associated with blockchain operations
Saleem et al. [32]	Combined with blockchain	Enhanced privacy through steganography and secure key management via blockchain	Possible increase in computational overhead due to the integration of multiple technologies
Dewanta and Mambo [33]	Combined with Fog/Edge computing	Establishes trust without a trusted third party, reducing dependency on infrastructure	The bidding methods could be susceptible to colluded manipulations
Liu et al. [34]	Combined with Fog/Edge computing	Adapts to dynamic environments by leveraging the flexibility of UAVs as well as edge computing, and balances the cost consumption and security enhancement	The trust evaluation process could introduce latency, affecting real-time task processing
Soleymani et al. [35]	Combined with Artificial Intelligence	Fuzzy logic handles uncertainties and imprecisions in data, enhancing trust decisions	The appropriate membership of fuzzy logic algorithm must be inferred from pre-trials
Rashid et al. [36]	Combined with Artificial Intelligence	Optimizes trust propagation path utilizing hybrid optimization techniques	The optimization process may be computationally intensive and time-consuming

system benefits from the combination of blockchain, IPFS, as well as steganography, enabling distributed storage and quick access to data for trust calculation. The main advantage of this approach is its enhanced security and privacy preservation, while the potential disadvantage is the increased computational complexity due to the integration of multiple technologies.

The Combination with Fog/Edge Computing

Dewanta and Mambo [33] propose a novel distributed model to establish trust in vehicular fog computing services in rural areas where infrastructure is sparse. The model allows vehicles to utilize the computational resources of each other during travel. The trust establishment in this model is based on Bidding Price-based Transaction (BPT) method to limit potential attacks. After the service providing vehicle and requesting vehicle recognize each other, they then proceed with bidding-agreement as the compensation of transaction, in case any of them violate the transaction by launching malicious activities. The advantage of this model is its ability to establish trust without centralized infrastructure by frog computing, while the disadvantage may be the BPT mechanisms could be susceptible to colluded manipulations.

Liu et al. [34] propose a trust management model for low-cost and secure UAV-assisted edge computing in task offloading scenario. In this model, three trust metrics are considered to calculate the trustworthiness of UAVs: location privacy, effectiveness in detecting malicious hijacking, and ability to defend against black hole attacks. Moreover, the trust management is combined with a cost model, which is proposed to evaluate the overall energy consumption and delay cost of edge computing system. The cost of optimal caching and offloading will reduce some trust values, while the optimal trust value of UAVs based edge computing will increase some costs. Therefore, the joint optimization problem of security and cost can be transformed into a convex optimization problem, considering trust values as constraints. The advantage of this model is its balance of cost-effectiveness and security enhancement, while the potential disadvantage is the complexity added by the integration of trust management with task offloading decisions.

The Combination with Artificial Intelligence (AI)

Soleymani et al. [35] propose a fuzzy logic-based trust model to deal with the uncertainty and imprecision inherent in VANET. Upon receiving an event message from surrounding vehicles, the receiver checks the event message authentication by the vehicle ID. Simultaneously, it checks message lifetime by calculating the difference between the message generation time and current time. Afterwards, the receiver evaluates the trust value based on experience, plausibility, and accuracy level of location by fuzzy logic algorithm. The main advantage of this model exists in the combination of trust management and fuzzy logic, which enhances the decision-making accuracy. However, the appropriate membership function should be obtained by pre-trials, so the model can't be applied instantly if the scenario changes.

Rashid et al. [36] introduce a novel trust management model for intelligent collaborative UAV assisted VANET. Firstly, the Trust evaluation is performed by integrating direct trust and recommendation trust. Next, the model combines genetic algorithms and crow swarm optimization to find an optimal path between the vehicles to reach their respective UAVs. The combination with these AI techniques enhances trust evidence propagation, addressing the challenge of trust establishment in sparse network. The model addresses the issues of message delivering delay and overhead in the network, by optimizing the message path selection and mobility of UAVs. However, optimization process may be computationally intensive and time-consuming.

14.5 Further Discussions

The poor utilization of resources with high false positive rate: UAVs may frequently change their communication links or exhibit intermittent connectivity, which could be flagged by trust systems as potential malicious activities. High false positive rates of detection can lead to the exclusion of legitimate nodes from the network, reducing the overall network efficiency and reliability. To address this, researchers are exploring adaptive threshold techniques that can adjust to the dynamic environment of UAV-assisted VANET and reduce the false positive rate. However, the challenge lies in developing trust models that can accurately distinguish between the inherent variability of UAVs communication and actual malicious behavior.

The consensus in trust model: Achieving consensus in a trust model within a UAV-assisted VANET is challenging due to the decentralized nature of the network. Consensus refers to the agreement among network nodes on the trustworthiness of a particular node or piece of information. In a dynamic environment with mobile vehicles and UAVs, maintaining a consistent and accurate trust view across the network is difficult. Trust models must be able to handle the rapid changes in network topology and node behavior, which can affect the consensus process.

The risk of cross domain attack in UAV-assisted VANET: In UAV-assisted VANET, most existing works only consider a single subnet in the dual-layered network. The integration of UAVs with ground vehicles increases network complexity, potentially allowing attackers to exploit trust relationships or configuration vulnerabilities between different network domains. For example, UAVs often serve as relaying nodes or data aggregators, introducing risks like unauthorized access, message interception, and manipulation. Common vulnerabilities (e.g., Global Positioning System spoofing and signal jamming) in UAVs domain may compromise message relaying path and disrupt location-based services for vehicles. Simultaneously, vulnerabilities (e.g., message tampering or Sybil attacks) in the VANET can spread through UAV-assisted systems, amplifying the scope and impact of attack. The interconnected nature of these domains creates a broad attack surface, requiring robust, cross-domain security mechanisms to ensure system integrity.

The privacy problems when applying trust management: Implementing trust management in UAV-assisted VANET raises significant privacy concerns. Trust management systems often require the collection and analysis of behaviors or messages, which can potentially reveal sensitive information about the vehicle or its users. Ensuring privacy while maintaining a robust trust model is a delicate balance. Moreover, the collection of trust related data must comply with data protection regulations and user expectations.

The scale of data set in AI-based trust management: The application of AI in trust management for UAV-assisted VANET involves processing large volumes of data to make accurate trust assessments. The scale of data can be overwhelming, especially in dense traffic scenarios where numerous vehicles and UAVs are interacting. Managing and processing this data in real-time requires significant computational resources and efficient algorithms. AI-based trust management systems must be scalable and capable of handling the data generated by a growing number of network participants.

14.6 Conclusion

In conclusion, the chapter has underscored the significance of trust management in the context of UAV-assisted VANET. Firstly, this chapter introduced the architecture of UAV-assisted VANET. Secondly, it clarified the principle of various security threats (e.g., EA, JA, SA and FMIA) which can compromise the integrity and reliability of the network. The chapter also presented a comprehensive classification of trust management schemes, including NCT, MCT, and HT models, each with its advantages and disadvantages. These schemes are crucial for detecting and mitigating malicious behavior, thereby ensuring secure and efficient communication within UAV-assisted VANET. Despite the progress, there are still unresolved issues that require attention, such as the poor resource utilization with high detection false positive rate, challenge of achieving consensus in trust models, risk of cross-layer attacks, privacy concerns, and scale of dataset in AI-based trust management. These challenges necessitate further research and development to enhance the robustness of trust management models for UAV-assisted VANET.

References

1. Ameur, A. I., Lakas, A., Yagoubi, M. B., et al. (2022). Peer-to-peer overlay techniques for vehicular ad hoc networks: Survey and challenges. *Vehicular Communications, 34*, Article 100455.
2. Shi, W., Zhou, H., Li, J., et al. (2018). Drone assisted vehicular networks: Architecture, challenges and opportunities. *IEEE Network, 32*(3), 130–137.
3. Michailidis, E. T., Maliatsos, K., Skoutas, D. N., et al. (2022). Secure UAV-aided mobile edge computing for IoT: A review. *IEEE Access, 10*, 86353–86383.
4. Adil, M., Jan, M. A., Liu, Y., et al. (2022). A systematic survey: Security threats to UAV-aided IoT applications, taxonomy, current challenges and requirements with future research directions. *IEEE Transactions on Intelligent Transportation Systems, 24*(2), 1437–1455.
5. Li, W., Meng, W., & Kwok, L. F. (2021). Surveying trust-based collaborative intrusion detection: state-of-the-art, challenges and future directions. *IEEE Communications Surveys & Tutorials, 24*(1), 280–305.
6. Mazhar, S., Rakib, A., Pan, L., et al. (2024) State-of-the-art authentication and verification schemes in VANETs: A survey. *Vehicular Communications*, 100804.
7. Mishra, R., Singh, A., & Kumar, R. (2016). VANET security: Issues, challenges and solutions. In *2016 International Conference on Electrical, Electronics, and Optimization Techniques (ICEEOT)* (pp. 1050–1055).
8. Pekaric, I., Sauerwein, C., Haselwanter, S., et al. (2021). A taxonomy of attack mechanisms in the automotive domain. *Computer Standards & Interfaces, 78*, Article 103539.
9. Annabi, M., Zeroual, A., & Messai, N. (2024). Towards zero trust security in connected vehicles: A comprehensive survey. *Computers & Security*, 104018.

10. Verma, A., Saha, R., Kumar, G., et al. (2021). The security perspectives of vehicular networks: A taxonomical analysis of attacks and solutions. *Applied Sciences, 11*(10), 4682.
11. Masood, A., Lakew, D. S., & Cho, S. (2020). Security and privacy challenges in connected vehicular cloud computing. *IEEE Communications Surveys & Tutorials, 22*(4), 2725–2764.
12. Braga, D. D. S., Niemann, M., Hellingrath, B., et al. (2018). Survey on computational trust and reputation models. *ACM Computing Surveys, 51*(5), 1–40.
13. Lu, Z., Qu, G., & Liu, Z. (2018). A survey on recent advances in vehicular network security, trust, and privacy. *IEEE Transactions on Intelligent Transportation Systems, 20*(2), 760–776.
14. Guo, H., Li, J., Liu, J., et al. (2021). A survey on space-air-ground-sea integrated network security in 6G. *IEEE Communications Surveys & Tutorials, 24*(1), 53–87.
15. Li, Z., Li, S., Zhang, H., et al. (2024). *Overview of sensing attacks on autonomous vehicle technologies and impact on traffic flow*. arXiv preprint arXiv:2401.15193
16. Khan, D., & Jamil, M. (2017). Study of detecting and overcoming black hole attacks in MANET: A review. In *2017 International Symposium on Wireless Systems and Networks (ISWSN)* (pp. 1–4).
17. Hasrouny, H., Samhat, A. E., Bassil, C., et al. (2017). VANET security challenges and solutions: A survey. *Vehicular Communications, 7*, 7–20.
18. Shiu, Y. S., Chang, S. Y., Wu, H. C., et al. (2011). Physical layer security in wireless networks: A tutorial. *IEEE Wireless Communications, 18*(2), 66–74.
19. Guang, L., Assi, C., & Benslimane, A. (2008). MAC layer misbehavior in wireless networks: Challenges and solutions. *IEEE Wireless Communications, 15*(4), 6–14.
20. Nazir, R., Laghari, A.A., Kumar, K., et al. (2021) Survey on wireless network security. *Archives of Computational Methods in Engineering*, 1–20.
21. Rasool, R., Wang, H., Ashraf, U., et al. (2020). A survey of link flooding attacks in software-defined network ecosystems. *Journal of Network and Computer Applications, 172*, Article 102803.
22. Movahedi, Z., Hosseini, Z., Bayan, F., et al. (2015). Trust-distortion resistant trust management frameworks on mobile ad hoc networks: A survey. *IEEE Communications Surveys & Tutorials, 18*(2), 1287–1309.
23. Kerrache, C. A., Lakas, A., Lagraa, N., et al. (2018). UAV-assisted technique for the detection of malicious and selfish nodes in VANETs. *Vehicular Communications, 11*, 1–11.
24. Alani, S., Qader, A. A., Al-Tahai, M., et al. (2023). Trust management scheme-based intelligent communication for UAV-assisted VANETs. In *International Conference on Frontiers of Intelligent Computing* (pp. 419–432). Springer Nature.
25. Zheng, J., Xu, J., Du, H., et al. (2024). Trust management of tiny federated learning in internet of unmanned aerial vehicles. *IEEE Internet Things Journal, 11*(2), 21046–21060.
26. Fatemidokht, H., Rafsanjani, M. K., Gupta, B. B., et al. (2021). Efficient and secure routing protocol based on artificial intelligence algorithms with UAV-assisted for vehicular ad hoc networks in intelligent transportation systems. *IEEE Transactions on Intelligent Transportation Systems, 22*(7), 4757–4769.
27. Kamel, J., Kaiser, A., Ben Jemaa, I., et al. (2019). CaTch: A confidence range tolerant misbehavior detection approach. In *2019 IEEE Wireless Communications and Networking Conference (WCNC)* (pp. 1–8).
28. Le, A., & Maple, C. (2019). Shadows don't lie: n-sequence trajectory inspection for misbehaviour detection and classification in VANETs. In *2019 IEEE 90th Vehicular Technology Conference (VTC2019-Fall)* (pp 1–6).
29. Zaidi, K., Milojevic, M. B., Rakocevic, V., et al. (2015). Host-based intrusion detection for VANETs: A statistical approach to rogue node detection. *IEEE Transactions on Vehicular Technology, 65*(8), 6703–6714.
30. Huang, M., Liu, A., Xiong, N. N., et al. (2021). A UAV-assisted ubiquitous trust communication system in 5G and beyond networks. *IEEE Journal on Selected Areas in Communications, 39*(11), 3444–3458.
31. Wang, H., Wang, C., Zhou, K., et al. (2024). TEBChain: A trusted and efficient blockchain-based data sharing scheme in UAV-assisted IoV for disaster rescue. *IEEE Transactions on Network and Service Management, 21*(4), 4119–4130.

32. Saleem, Z., Firdous, U., Afzal, M. K., et al. (2023). Blockchain-based privacy preservation using steganography in drone-enabled VANETs. In *2023 IEEE Global Communications Conference (GLOBECOM)* (pp. 6862–6867).
33. Dewanta, F., & Mambo, M. (2019). Bidding price-based transaction: Trust establishment for vehicular fog computing service in rural area. In *2019 IEEE International Conference on Pervasive Computing and Communications Workshops (PerCom Workshops)* (pp. 882–887).
34. Liu, J., Xie, P., Lin, K., et al. (2024). Trust-aware task offloading for cost-effective UAV-based edge computing based on reinforcement learning. *Neural Computing & Applications*, 1–18.
35. Soleymani, S. A., Abdullah, A. H., Zareei, M., et al. (2017). A secure trust model based on fuzzy logic in vehicular ad hoc networks with fog computing. *IEEE Access, 5*, 15619–15629.
36. Rashid, S. A., Mustafa, A. S., Abbas, A. D., et al. (2023). On the design of trust and mobility based evaluation for intelligent collaborative UAVs assisted VANETs. *Bulletin of Electrical Engineering and Informatics, 12*(5), 2993–2999.

GPSR Compliance

The European Union's (EU) General Product Safety Regulation (GPSR) is a set of rules that requires consumer products to be safe and our obligations to ensure this.

If you have any concerns about our products, you can contact us on ProductSafety@springernature.com

In case Publisher is established outside the EU, the EU authorized representative is:

Springer Nature Customer Service Center GmbH
Europaplatz 3
69115 Heidelberg, Germany

Batch number: 10704307

Printed by Printforce, the Netherlands